ACT®
Math &
Science Prep

Fourth Edition

Related Titles for College-Bound Students

ACT®
Math &
Science Prep

Fourth Edition

PUBLISHING

New York

Published by Kaplan Publishing, a division of Kaplan, Inc.
750 Third Avenue
New York, NY 10017

Printed in the United States of America

10 9 8 7 6 5 4

ISBN-13: 978-1-5062-1440-5

Kaplan Publishing books are available at special quantity discounts to use for sales promotions, employee premiums, or educational purposes. Please call the Simon & Schuster special sales department at 866-506-1949.

Contents

SECTION THREE: ACT SCIENCE

SECTION FOUR: ADDITIONAL RESOURCES

SECTION FIVE: PRACTICE TESTS

Introduction to the ACT

CHAPTER 1

Understanding the ACT

Congratulations! By picking up this workbook, you're making a commitment to yourself to learn about the ACT and how you can do your very best on the Science and Math sections of the test. The information in this chapter will tell you what you need to know about the ACT. You'll know what to expect on Test Day, so you can walk into your test center feeling confident and prepared. Going into the ACT with that positive attitude is crucial. Familiarizing yourself with the test structure and working through practice problems are a huge part of creating the mind-set that will help you ace the ACT. Let's get started.

ACT STRUCTURE

The ACT is divided into five tests: English, Mathematics, Reading, Science, and Writing. You can elect to take the ACT with only the first four tests, which make up the multiple-choice portion of the ACT. The fifth test, in which you produce an essay, is an optional test.

PREDICTABILITY

No matter where or when you take the ACT, the order of the sections and the time allotted for each are always the same. This consistency works in your favor: The more you know about what to expect on Test Day, the more confident you'll feel. You may know that one section of the test, let's say, Science, usually seems more challenging for you, but at least you know that Science will always be fourth. The ACT won't ever surprise you by making the Science section the first thing you see when you open your exam booklet. Knowing the structure of the ACT will help you feel in control of your test-taking experience.

The following table summarizes the predictable structure of the ACT:

Section	Time Allotted	Number and Type of Questions
1: English	45 minutes	75 multiple choice
2: Mathematics	60 minutes	60 multiple choice
3: Reading	35 minutes	40 multiple choice
4: Science	35 minutes	40 multiple choice
5: Writing	40 minutes	1 essay prompt

WHAT IS A STANDARDIZED TEST?

Here's your first ACT practice question:

One of the most important ways to succeed on a standardized test is to:

A. do nothing but practice problems in your spare time the week before the test.

B. talk to anyone who will listen about how nervous you are.

C. choose choice (C) for any multiple-choice question you're unsure about.

D. understand what a standardized test is and why taking it doesn't have to be a demoralizing experience.

Which answer did you choose? Although some of the choices may have made you groan or grimace if you recognized they weren't true, we hope you spotted that option (D) is the best answer.

As you use this book and apply the Kaplan strategies to work through practice problems, you'll come to see that the test experience need not be demoralizing at all. Right now, however, you may be apprehensive for a variety of reasons. Your own teachers didn't write the test. You've heard the test maker includes trick answers. You feel weak in one of the content areas and don't know how you can possibly improve enough to do well on that test section. Thousands of students will be taking the test. All of these things can seem very intimidating.

Let's look carefully at that last reason. The simple fact that thousands of students from different places take the ACT is actually a good thing. It means that the test is necessarily constructed in a deliberate and predictable way. Because it's a standardized test, the ACT must include very specific content and skills that are consistent from one test date to another. The need for standardization makes it predictable, not intimidating. It's predictable not simply in the layout of the test sections in the booklet, but also in the topics that are tested and even *in the way those topics are tested*. Working the practice problems in this book will help you understand not only how each topic is tested but also how to approach the various question types.

If you feel anxious about the predominance of multiple-choice questions on the ACT, think about this fact: For multiple-choice questions, there has to be *only one right answer*, and it's *right there in front of you* in the test booklet. A question that could be interpreted differently by students from different schools, even different parts of the country, who've had different teachers and different high school courses, would never make it onto the ACT. Each question on the ACT is put there to test a specific skill. Either the question or the passage it's associated with (for English, Reading, and Science) *must* include information that allows all students to determine the correct answer.

There can be no ambiguity about which answer is best for a multiple-choice question on a standardized test. This workbook will teach you proven Kaplan strategies for finding that answer. The Kaplan strategies, along with your understanding about how the test is structured and written, will put *you* in control of your ACT Test Day experience.

ACT SCORING

SCORING FOR THE MULTIPLE-CHOICE SECTIONS: RAW SCORE, SCALED SCORE, AND PERCENTILE RANKING

Let's look at how your ACT composite score is calculated. For each multiple-choice section of the test (English, Mathematics, Reading, and Science), the number of questions you answer correctly is totaled. No points are deducted for wrong answers. The total correct for each section is called the *raw* score for that section. Thus, the highest possible raw score for a section is the total number of questions in that section.

Because each version of the ACT is different (more in the wording of the questions than in the types of questions or skills needed to answer them), a conversion from the raw score to a *scaled* score is necessary. For each version of the ACT that is written, the test maker generates a conversion chart that indicates what scaled score each raw score is equivalent to. The conversion from raw score to scaled score is what allows for accurate comparison of test scores even though there are slight variations in each version of the test. The scaled score ranges from a low of 1 to a high of 36 for each of the first four sections. Scaled scores have the same meaning for all the different versions of the ACT offered on different test dates.

The score for each of the first three sections of the ACT is broken down further into subscores. The subscores for a particular section do *not* necessarily add up to the overall score for the section. The following table lists the subscores that are reported for each section.

Test Section	Subscore Categories
English (75 questions)	Production of Writing; Knowledge of Language; Conventions of Standard English
Mathematics (60 questions)	Preparing for Higher Mathematics; Integrating Essential Skills; Modeling
Reading (40 questions)	Key Ideas and Details; Craft and Structure; Integration of Knowledge and Ideas
Science (40 questions)	Interpretation of Data; Scientific Investigation; Evaluation of Models, Inferences, and Experimental Results
Writing (1 essay prompt)	Ideas and Analysis; Development and Support; Organization; Language Use and Conventions

What most people think of as the ACT score is the composite score. Your composite score, between 1 and 36, is the average of the four scaled scores on the English, Mathematics, Reading, and Science sections of the test.

In addition to the raw score, scaled score, and section subscores, your ACT score report also includes a *percentile ranking*. This is not a score that indicates what percentage of questions you answered correctly on the test. Rather, your percentile ranking provides a comparison between your performance and that of other recent ACT test takers. Your percentile ranking indicates the percentage of ACT test takers who scored the same as or lower than you. In other words, if your percentile ranking is 80, that means that you scored the same as or higher than 80 percent of the students who took the test.

A raw-to-scaled score conversion chart is necessary to take into account slight variations in the difficulty levels of different versions of the test. In other words, it's not possible to say for every ACT test date that if you answer, for example, 55 out of 75 English questions correctly, your scaled score will always be 24. However, you should know that the variations in each test version—and therefore in the raw-to-scaled score conversion chart—are very slight and should not concern you. The following table gives some *approximate* raw score ranges for each section, the associated scaled score, and the likely percentile ranking.

Raw Score and Scale Score Approximate Equivalences					
Scaled Score	English Questions Correct (Total = 75)	Mathematics Questions Correct (Total = 60)	Reading Questions Correct (Total = 40)	Science Questions Correct (Total = 40)	Percentile Ranking
32	70	55–56	34–35	35–36	99
27	61–63	45–47	27–28	30–32	90
24	53–56	36–39	22–25	26–28	75
20	42–46	31–32	18–20	20–21	50

Don't get bogged down in the numbers in this table. We've put it here to help you relax. The big-picture message of the chart is that you can get a good ACT score even if you don't answer every question correctly. In terms of how many questions you need to get right, reaching your ACT score goal is probably not as difficult as you might think.

SCORING FOR THE WRITING SECTION

The ACT essay subscores range from 2 to 12 points, with 12 being the highest. To determine the essay subscores, two trained graders read your essay and assign it a score between 1 and 6 for each of the four domains: Ideas and Analysis, Development and Support, Organization, and Language Use and Conventions.

Not all colleges and universities require the ACT Writing test for admission, so the fifth section of the ACT is optional. Not every student takes it, and therefore the Writing score has no effect on the composite score. However, if you opt to take the Writing section of the ACT, you will receive three scores in addition to your composite score. First, you will see your essay subscores, between 2 and 12. Second, you will see a Writing section score, which is the average of the four subscores. Third, you will see an ELA score, which is a weighted composite of your Reading, English, and Writing scaled scores. This score also ranges from 1 to 36. If you choose not to do the Writing section, you will not receive essay subscores, an essay scaled score, or an ELA score.

ADDITIONAL SCORES

In addition to an ELA score, you will also receive a STEM score. This is the average of your Math and Science scaled scores and ranges from 1-36.

ACT REGISTRATION

All the information you need about ACT registration is available on the test maker's website at www.act.org. There are two ways to register for the ACT: You can do so online, or you can use a registration packet and send in your forms by mail. If you need a registration packet, you should be able to get one from your school counselor, or you can request one directly via the test maker's website.

Choosing a test day that is right for you depends on admission deadlines. Most students take the test during their junior spring and senior fall semesters, although some opt to take the ACT as early as freshman or sophomore year if they have completed most of the academic skills that are tested by the ACT. The earlier you take the test, the more opportunities you have to take the test again and increase your composite score. Check the websites of colleges to which you are applying, as well as scholarship agencies, to solidify a test date. The ACT website, www.act.org, provides a list of colleges with information about standardized testing policies. It takes about three to eight weeks to receive your scores, so be sure to check college admission deadlines before you choose a test date.

If you think that registering for the ACT on time is simply a matter of logistics and fees, think again. Individual testing centers have limited space. When you know that you're interested in a particular test date at a particular location, it is worthwhile to register as soon as possible. The earlier you register, the more likely it is that you'll be able to test at your preferred location. Many students prefer to test at their own high schools, in a familiar setting. The morning of Test Day will go much more smoothly if you don't have to worry about directions to get to an unfamiliar location. Planning ahead for ACT registration can help you avoid such unnecessary distractions.

When you register, you should read all the information the test maker provides. Learn specifically about what to bring with you, including forms of ID, pencils, acceptable calculators, and snacks for the breaks. You should also pay attention to what behaviors are and are not acceptable during the test. The more you know ahead of time about what to expect on Test Day, the more relaxed and confident you'll be going into the test. When you put that confidence together with the Kaplan strategies and practice you'll get from this book, you can look forward to higher scores on your ACT.

CHAPTER 2

ACT Strategies

Now that you've got some idea of the kind of adversary you face in the ACT, it's time to start developing strategies for dealing with this adversary. In other words, you've got to start developing your ACT mind-set.

The ACT, as you've just seen, isn't a normal test. A normal test requires that you rely almost exclusively on your memory. On a normal test, you'd see questions like this:

> The "golden spike," which joined the Union Pacific and Central Pacific Railroads, was driven in Ogden, Utah, in May 1869. Who was president of the United States at the time?

To answer this question, you have to resort to memory dredging. Either you know the answer is Ulysses S. Grant or you don't. No matter how hard you think, you'll **never** be able to answer this question if you can't remember your history.

The ACT doesn't test your long-term memory. The answer to every ACT question can be found in the test. Theoretically, if you read carefully and understand the words and concepts the test uses, you can get almost any ACT question right. Notice the difference between the previous regular test question and the following ACT-type question:

1. What is the product of n and m^2, where n is an odd number and m is an even number?

   A. An odd number
   B. A multiple of four
   C. A noninteger
   D. An irrational number
   E. A perfect square

9

Aside from the obvious difference (this question has answer choices, while the other one does not), there's another difference: The ACT question mostly tests your ability to understand a situation rather than your ability to passively remember a fact. Nobody expects you to know off the top of your head what the product of an odd number and the square of an even number is. However, the ACT test makers do expect you to be able to roll up your sleeves and figure it out.

THE ACT MIND-SET

Most students take the ACT with the same mind-set that they use for normal tests. Their brains are on "memory mode." Students often panic and give up because they can't seem to remember enough. **You don't need to remember a ton of picky little rules for the ACT, however. Don't give up on an ACT question just because your memory fails.**

On the ACT, if you understand what a question is really asking, you can almost **always** answer it. For instance, take the previous math problem. You might have been thrown by the way it was phrased. "How can I solve this problem?" you may have asked yourself. "It doesn't even have numbers in it!"

The key here, as in all ACT questions, is taking control. Take the question (by the throat, if necessary) and wrestle it into a form you can understand. Ask yourself: What's really being asked here? What does it mean when they say something like "the product of n and m^2"?

You could start by putting it into words you might use. You might say something like this: "I've got to take one number times another. One of the numbers is odd and the other is an even-number squared. Then I've got to see what kind of number I get as an answer." Once you put the question in your own terms, it becomes much less intimidating—and much easier to get right. You'll realize that you don't have to do complex algebraic computations with variables. All you have to do is substitute numbers.

So do it! Try picking some easy-to-use numbers. Say that n is 3 (an odd number) and m is 2 (an even number). Then m^2 would be 4, because 2×2 is 4. And $n \times m^2$ would be 3×4, which is 12—a multiple of 4, but not odd, not a noninteger, not an irrational number, and not a perfect square. The only answer that can be right, then, is (B).

See what we mean about creatively figuring out the answer rather than passively remembering it? True, there are some things you had to remember here—what even and odd numbers are, how variables and exponents work, and what integers and irrational numbers are. But these are very basic concepts. Most of what you're expected to know on the ACT is just that: basic. (By the way, you'll find such concepts gathered together in the very attractive 100 Key Math Concepts for the ACT section at the end of this book.)

Of course, basic doesn't always mean easy. Many ACT questions are built on basic concepts, but are tough nonetheless. The previous problem, for instance, is difficult because it requires some thought to figure out what's being asked. This isn't only true in Math. It's the same for every part of the ACT.

The creative, take-control kind of thinking we call the ACT mind-set is something you want to bring to virtually every ACT question you encounter. As you'll see, being in the ACT mind-set means reshaping the test-taking experience so that you are in the driver's seat.

It means:

- Answering questions **if you** want to (by guessing on the impossible questions rather than wasting time on them).

- Answering questions **when you** want to (by skipping tough but "doable" questions and coming back to them after you've gotten all of the easy questions done).

- Answering questions **how you** want to (by using "unofficial" ways of getting correct answers fast).

That's really what the ACT mind-set boils down to: taking control and being creative. Solving specific problems to get points as quickly and easily as you can.

What follows are the top ten strategies you need to do just that.

TEN STRATEGIES FOR MASTERING THE ACT

1. TRIAGE THE QUESTIONS

In a hospital emergency room, the triage nurse is the person who evaluates each patient and decides which ones get attention first and which ones should be treated later. You should do the same thing on the ACT.

Practicing triage is one of the most important ways of controlling your test-taking experience. It's a fact that there are some questions on the ACT that most students could **never** answer correctly, no matter how much time or effort they spent on them.

EXAMPLE

If $\sec^2 x = 4$, which of the following could be $\sin x$?

A. 1.73205

B. 3.14159

C. $\sqrt{3}$

D. $\dfrac{\sqrt{3}}{2}$

E. Cannot be determined from the given information.

Clearly, even if you could manage to come up with an answer to this question, it would take some time (if you insist on doing so, refer to the explanation in the next section). But would it be worth the time? We think not.

This question clearly illustrates our point: Triage the questions on the ACT. The first time you look at a question, make a quick decision about how hard and time-consuming it looks. Then decide whether to answer it now or skip it and do it later.

- If the question looks comprehensible and reasonably doable, do it right away.

- If the question looks tough and time-consuming, but ultimately doable, skip it, circle the question number in your test booklet, and come back to it later.

- If the question looks impossible, forget about it. Guess and move on, *never* to return.

This triage method will ensure that you spend the time needed to do all the easy questions before getting bogged down with a tough problem. Remember, every question on a subject test is worth the same number of points. You get no extra credit for test machismo.

Answering easier questions first has another benefit: It gives you confidence to answer harder ones later. Doing problems in the order you choose rather than in the order imposed by the test makers gives you control over the test. Most students don't have time to do all of the problems, so you've got to make sure you do all of the ones you can easily score on!

DO YOU KNOW YOUR TRIG?

Okay, because you're reading this, it's obvious that you want to know the answer to the trig question we just looked at. The answer is (D). Here's how we got it:

$\sec^2 x = 4$	Given
$\sec x = 2 \text{ or } -2$	Square root both sides.
$\cos x = \dfrac{1}{2} \text{ or } -\dfrac{1}{2}$	$\cos x = \dfrac{1}{\sec x}$
$\cos^2 x = \dfrac{1}{4}$	Square both sides.
$\sin^2 x = 1 - \dfrac{1}{4}$	$\sin^2 x + \cos^2 x = 1$
	$\sin^2 x = 1 - \cos^2 x$
$\sin^2 x = \dfrac{3}{4}$	
$\sin x = \sqrt{\dfrac{3}{4}} \text{ or } -\sqrt{\dfrac{3}{4}}$	Square root both sides.
$\sin x = \dfrac{\sqrt{3}}{\sqrt{4}} \text{ or } -\dfrac{\sqrt{3}}{\sqrt{4}}$	$\sqrt{4} = 2$
$\sin x = \dfrac{\sqrt{3}}{2} \text{ or } -\dfrac{\sqrt{3}}{2}$	

So (D) is correct. But if you got it right, don't congratulate yourself quite yet. How long did it take you to get it right? So long that you could have gotten the answers to two easy questions in the same amount of time?

DEVELOP A PLAN OF ATTACK

For the English, Reading, and Science sections, the best plan of attack is to do each passage as a block. Make a longish first pass through the questions (the "triage" pass), doing the easy ones, guessing on the impossible ones, and skipping any that look like they might cause trouble. Then make a second pass (the "cleanup pass"), and do those questions you think you can solve with some elbow grease.

For Math, use the same two-pass strategy, but move through the whole subject test twice. Work through the doable questions first. Most of these will probably be toward the beginning, but not all. Then come back and attack the questions that look possible but tough or time-consuming.

No matter which subject test you're working on, **make sure you grid your answers in the right place.** It's easy to misgrid when you're skipping around, so be careful. And of course: *Make sure you have an answer gridded for every question by the time the subject test is over!*

2. PUT THE MATERIAL INTO A FORM YOU CAN UNDERSTAND

ACT questions are rarely presented in the simplest, most helpful way. In fact, your main job for many questions is to figure out what the question means so you can solve it.

Because the material is presented in such an intimidating way, one of the best strategies for taking control is to recast (reword) the material into a form you can handle better. This is what we did in the math problem about "the product of n and m^2." We took the question and reworded it in a way we could understand.

MARK UP YOUR TEST BOOKLET

This strategy should be employed on all four subject tests. For example, in Reading, many students find the passages overwhelming: 85 to 90 lines of dense verbiage for each one. But the secret is to put the passages into a form you can understand and use. Circle or underline the main idea, for one thing. Make yourself a road map of the passage, labeling each paragraph so you understand how it all fits together. That way, you'll also know—later, when you're doing the questions—where in the passage to find certain types of information you need.

REWORD THE QUESTIONS

You'll also need to do some recasting of the *questions*. For instance, take this question from a Science passage.

EXAMPLE

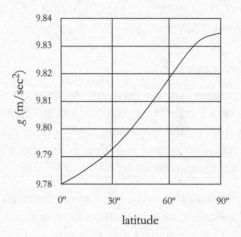

Figure 1

According to Figure 1, at approximately what latitude would calculations using an estimated value at sea level of $g = 9.80$ m/sec^2 produce the least error?

A. 0°

B. 20°

C. 40°

D. 80°

At what latitude would the calculations using a value of $g = 9.80$ m/sec^2 produce the least error? Yikes! What does that mean?

Take a deep breath. Ask yourself: Where would an estimate for g of 9.80 m/sec^2 produce the least error? At a latitude where 9.80 m/sec^2 is the actual value of g. If you find the latitude at which the value of g is 9.80 m/sec^2, then using 9.80 m/sec^2 as an estimate at that latitude would produce no error at all.

In other words, this question is asking: At what latitude does $g = 9.80$ m/sec^2? Now that's a form of the question you can understand. In that form, you can answer it easily: (C), which you can deduce just by reading the chart to see where the horizontal line at 9.80 and the curve intersect.

DRAW DIAGRAMS

Sometimes, putting the material into a usable form involves drawing with your pencil. For instance, take a look at the following math problem.

EXAMPLE

Jason bought a painting with a 1 inch wide frame. If the dimensions of the outside of the frame are 5 inches by 7 inches, which of the following could be the length of one of the sides of the painting inside the frame?

F. 3 inches

G. 4 inches

H. $4\frac{1}{2}$ inches

J. $5\frac{1}{2}$ inches

K. $6\frac{1}{2}$ inches

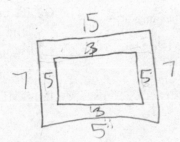

At first glance, you might be tempted to simply subtract 1 from the outside dimensions and think that the inside dimensions are 4 by 6 (and pick G). This isn't correct, though, because the frame goes all the way around—both above and below the painting, both to the right and to the left. This would have been clear if you had put the problem in a form you could understand and use.

For instance, you could make the situation graphic by sketching the painting frame (who says you don't have to be an artist to succeed at the ACT?):

When you draw the picture frame like this, you realize that if the outside dimensions are 5 by 7, the inside dimensions must be 3 by 5. Thus, the correct answer is (F).

So remember: On the ACT, put everything into a form that you can understand and use.

3. IGNORE IRRELEVANT ISSUES

It's easy to waste time on ACT questions by considering irrelevant issues. Just because an issue looks interesting, or just because you're worried about something, doesn't make it important.

EXAMPLE

. . . China was certainly one of the cradles

of civilization. <u>It's obvious that, China has a</u>

<u>long history.</u> As is the case with other ancient
                    14

cultures, the early history of China is lost in

mythology . . .

14. **F.** NO CHANGE
     G. It's obvious that China has a
           long history.
     H. Obviously; China has a long
           history.
     J. OMIT the underlined portion.

In this question, the test makers are counting on you to waste time worrying about punctuation. Does that comma belong? Can you use a semicolon here? These issues might be worrisome, but there's a bigger issue here—namely, does the sentence belong in the passage at all? No, it doesn't. If China has an ancient culture and was a cradle of civilization, it must have a long history, so the sentence really is "obvious." Redundancy is the relevant issue here, not punctuation. Choice (J) is correct.

Remember, you've got limited time, so don't get caught up in issues that won't get you a point.

4. CHECK BACK

Remember, the ACT is not a test of your memory, so don't make it one. All of the information you need is in the test itself. Don't be afraid to refer to it. Much of the information is too complex to accurately remember anyway.

In Reading and Science, always refer to the place in the passage where the answer to a question can be found (the question stem will often contain a line reference or a reference to a specific table, graph, or experiment to help you out). Your chosen answer should match the passage—not in exact vocabulary or units of measurement, perhaps, but in meaning.

EXAMPLE

Isaac Newton was born in 1642 in the hamlet of Woolsthorpe in Lincolnshire, England. But he is more famous as a man of Cambridge, where he studied and taught . . .

Which of the following does the author imply is a fact about Newton's birth?

A. It occurred in Lincoln, a small hamlet in England.

B. It took place in a part of England known for raising sheep.

C. It did not occur in a large metropolitan setting.

D. It caused Newton to seek his education at Cambridge.

You might expect the right answer to be that Newton was born in a hamlet, or in Woolsthorpe, or in Lincolnshire, but none of those is offered as a choice. Choice A is tempting, but wrong. Newton was born in Lincolnshire, not Lincoln. Choice B is actually true, but it's not the correct answer here. As its name suggests, Woolsthorpe was once known for its wool—which comes from sheep. However, the question asks for something implied in the passage.

The correct answer is (C), because a hamlet is a small village, which is not a large metropolitan setting. (It's also a famous play, but that's not among the choices.)

Checking back is especially important in Reading and Science, because the passages leave many people feeling adrift in a sea of details. Often, the wrong answers will be "misplaced details"—details taken from different parts of the passage. They are things that don't answer the question properly but that might sound good to you if you aren't careful. By checking back with the passage, you can avoid choosing such devilishly clever wrong choices.

There's another important lesson here: **Don't pick a choice just because it contains key words you remember from the passage.** Many wrong choices, such as D in the previous question, are "distortions"—they use the right words but say the wrong things about them. Look for answer choices that contain the same ideas presented in the passage.

One of the best ways to avoid choosing misplaced details and distortions is to check back with the passage.

5. ANSWER THE RIGHT QUESTION

This strategy is a natural extension of the last. As we said, **the ACT test makers often include among the wrong choices for a question the correct answer to a different question.** Under time pressure, it's easy to fall for one of these red herrings, thinking that you know what's being asked for when really you don't.

EXAMPLE

What is the value of $3x$ if $9x = 5y + 2$ and $y + 4 = 2y - 10$?

A. 5
B. 8
C. 14
D. 24
E. 72

To solve this problem, we need to find y first, even though the question asks about x (because x here is given only in terms of y). You could solve the second equation like this:

$y + 4 = 2y - 10$ Subtract y from both sides of the equation.

$4 = y - 10$ Add 10 to both sides.

$14 = y$

However, C, 14, isn't the right answer here, because the question doesn't ask for the value of *y*—it asks about *x*. We can use the value of *y* to find *x*, however, by plugging the calculated value of *y* into the first equation:

$9x = 5y + 2$ Now use the equation that has *x* in it.

$9x = 5(14) + 2$ Substitute 14 for *y*.

$9x = 70 + 2$

$9x = 72$

Nevertheless, E, 72, isn't the answer either, because the question doesn't ask for 9*x*. It doesn't ask for *x* either, so if you picked B, 8, you'd be wrong as well. Remember to refer to the question! The question asks for 3*x*, so divide 9*x* by 3:

$9x = 72$ Divide both sides of the equation by 3.

$3x = 24$

Thus, the answer is (D).

Always check the question again before choosing your answer. Doing all the right work but then getting the wrong answer can be seriously depressing, so make sure you're answering the right question.

6. LOOK FOR THE HIDDEN ANSWER

On many ACT questions, the right answer is hidden in one way or another. An answer can be hidden by being written in a way that you aren't likely to expect. For example, you might work out a problem and get 0.5 as your answer, but then find that 0.5 isn't among the answer choices. Then you notice that one choice reads "$\frac{1}{2}$." Congratulations, Sherlock. You've found the hidden answer.

There's another way the ACT can hide answers. **Many ACT questions have more than one possible right solution, though only one correct answer choice is given. The ACT will hide that answer by offering one of the less obvious possible answers to a question.**

EXAMPLE

If $3x^2 + 5 = 17$, which of the following could be the value of *x*?

A. −3
B. −2
C. 0
D. 1
E. 4

Quickly solve this very straightforward problem like this:

$3x^2 + 5 = 17$	Subtract 5 from both sides of the given equation.
$3x^2 = 12$	Divide both sides by 3.
$x^2 = 4$	Take the square root of both sides.
$x = 2$	

Having gotten an answer, you confidently look for it among the choices. But 2 isn't a choice. The explanation? This question actually has two possible solutions, not just one. The square root of 4 can be either 2 or –2. Thus, (B) is the answer.

Keep in mind that although there's only one right answer choice for each question, that right answer may not be the one that occurs to you first. A common mistake is to pick an answer that seems "sort of" like the answer you're looking for even when you know it's wrong. Don't settle for second best. If you don't find your answer, don't assume that you're wrong. Try to think of another right way to answer the question.

7. GUESS INTELLIGENTLY

An unanswered question is always wrong, but even a wild guess could be right. On the ACT, a guess can't hurt you, but it can help. In fact, smart guessing can make a big difference in your score. **Always** guess on every ACT question that you can't answer. **Never** leave a question blank.

You'll be doing two different kinds of guessing during your two sweeps through any subject test:

- Blind guessing (which you do mostly on questions you deem too hard or time-consuming even to try).

- Considered guessing (which you do mostly on questions for which you do some work on but can't make headway).

When you guess blindly, you just choose any letter you feel like choosing (many students like to choose B/G or C/H). When you guess in a considered way, on the other hand, you've usually done enough work on a question to eliminate at least one or two choices. If you can eliminate any of the choices, you'll increase the odds that you'll guess correctly.

Here are some fun facts about guessing: If you were to work on only half of the questions on the ACT and get them all right, then guess blindly on the other half of the questions, you would probably earn a composite ACT score of around 23 (assuming you had a statistically reasonable success rate on your guesses). A 23 would put you in roughly the top quarter of all students taking the ACT. It's a good score. And all you had to do was answer half the questions correctly.

On the other hand, if you were to hurry and finish all the questions, and get only half of them right, you'd probably earn only a 19, which is below average.

How? Why are you better off answering half and getting them all right instead of answering all and getting only half right?

Here's the trick. The student who answers half the questions right and skips the others can still take guesses on the unanswered questions—and odds are this student will have enough correct guesses to move up 4 points, from a 19 to a 23. However, the student who answers all the questions and gets half wrong doesn't have the luxury of taking guesses.

In short: **Guess if you can't figure out an answer for any question!**

8. BE CAREFUL WITH THE ANSWER GRID

Your ACT score is based on the answers you select on your answer grid. Even if you work out every question correctly, you'll get a low score if you misgrid your answers. So be careful! Don't disdain the process of filling in those little "bubbles" on the grid. Sure, it's mindless, but under time pressure it's easy to lose control and make mistakes.

It's important to **develop a disciplined strategy for filling in the answer grid.** We find that it's smart to grid the answers in groups rather than one question at a time. What this means is: As you figure out each question in the test booklet, circle the answer choice you come up with. Then transfer those answers to the answer grid in groups of five or more (until you get close to the end of the section, when you start gridding answers one by one).

Gridding in groups like this cuts down on errors because you can focus on this one task and do it right. It also saves time you'd otherwise spend moving papers around, finding your place, and redirecting your mind. Answering ACT questions takes deep, hard thinking. Filling out answer grids is easy, but you have to be careful, especially if you do a lot of skipping around. Shifting between "hard thinking" and "careful bookkeeping" takes time and effort.

In English, Reading, and Science, the test is divided naturally into groups of questions—the passages. For most students, it makes sense to circle your answers in your test booklet as you work them out. Then when you're finished with each passage and its questions, grid the answers as a group.

In Math, the strategy has to be different because the Math test isn't broken into natural groups. Mark your answers in the test booklet and then grid them when you reach the end of each page or two. Because there are usually about five math questions per page, you'll probably be gridding five or ten math answers at a time.

No matter what subject test you're working on, though, if you're near the end of a subject test, start gridding your answers one at a time. You don't want to be caught with ungridded answers when time is called.

During the test, the proctor should warn you when you have about five minutes left on each subject test. Don't depend on proctors, though. Yes, they're usually nice people, but they can mess up once in a while. **Rely on your own watch.** When there are five minutes left in a subject test, start gridding your answers one by one. With a minute or two left, start filling in everything you've left blank. Remember: Even one question left blank could cut your score.

9. USE THE LETTERS OF THE CHOICES TO STAY ON TRACK

One oddity about the ACT is that even-numbered questions have F, G, H, J (and, in Math, K) as answer choices, rather than A, B, C, D (and, again, E in Math). This might be confusing at first, but you can make it work for you. **A common mistake with the answer grid is to accidentally enter an answer one row up or down. On the ACT, that won't happen if you pay attention to the letter in the answer.** If you're looking for an A and you see only F, G, H, J, and K, you'll know you're in the wrong row on the answer grid.

Another advantage of having answers F through K for even-numbered questions is that it makes you less nervous about patterns in the answers. It's common to start worrying if you've picked the same letter twice or three times in a row. Because the questions have different letters, this can't happen on the ACT. Of course, you could pick the first choice (A or F) for several questions in a row. This shouldn't worry you. It's common for the answers in the same position to be correct three times in a row, and even four times in a row isn't unheard of.

10. KEEP TRACK OF TIME

During each subject test, you really have to pace yourself. On average, English, Reading, and Science questions should take about 30 seconds each. Math questions should average less than one minute each. Remember to take into account the fact that you'll probably be taking two passes through the questions.

Set your watch to 12:00 at the beginning of each subject test, so it will be easy to check your time. Again, don't rely on proctors, even if they promise that they will dutifully call out the time every 15 minutes. Proctors get distracted once in a while.

For English, Reading, and Science questions, it's useful to check your timing as you grid the answers for each passage. English and Reading passages should take about nine minutes each. Science passages should average about six minutes.

More basic questions should take less time, and harder ones will probably take more. In Math, for instance, you need to go much faster than one per minute during your first sweep. At the end, however, you may spend two or three minutes on each of the harder problems you work out.

TAKE CONTROL

You are the master of the test-taking experience. A common thread in all ten strategies above is: Take control. That's Kaplan's ACT mind-set. Do the questions in the order you want and in the way you want. Don't get bogged down or agonize. Remember, you don't earn points for suffering, but you do earn points for moving on to the next question and getting it right.

STRATEGY REFERENCE SHEET

1. Triage the questions.

2. Put the material into a form you can understand and use.

3. Ignore irrelevant issues.

4. Check back.

5. Answer the right question.

6. Look for the hidden answer.

7. Guess intelligently.

8. Be careful with the answer grid.

9. Use the letters of the choices to stay on track.

10. Keep track of time.

SECTION TWO

ACT Math

CHAPTER 3

Introduction to ACT Math

TEST BASICS

The ACT Math test is the second section of the ACT. It is a 60-question, 60-minute test. (Here's some basic math: 60 questions/60 minutes = 1 minute per question!) It is completely multiple choice in format and most questions are usually discrete (this means very few groups of questions referring to the same diagram, picture, etc.). One key difference between the Math test and the other tests on the ACT is the number of answer choices. The Math test has five answer choices, while each of the other sections has only four. This means that there is an additional choice to consider for each question.

Even though all the answers are multiple-choice, there are still some differences in the presentation of the questions that you should be prepared to deal with. For example, some questions will be basic math problems while others will be word problems. Both test the same concepts, but the word problems will require more thought. They simply add another step, because you will have to turn the word problems into a basic math format. An example of this follows:

If a sweater that originally costs $40 is on sale for 15% off, what is the total discount given?

A. $2.00

B. $3.00

C. $4.00

D. $5.00

E. $6.00

You need to turn these words into an equation. Ask yourself, what is this problem really asking? It wants you to find 15% of $40. So the equation is $40 × 0.15 = $6.00. This means that you are given a discount of $6.00. The answer is (E).

CONTENT

The ACT Math test covers material that you are typically expected to have learned up through the end of your junior year, which includes pre-algebra, elementary algebra, intermediate algebra, plane geometry, coordinate geometry, and trigonometry. The relative number of questions on these topics does not change very much from year to year. There are usually about 14 pre-algebra, 10 elementary algebra, 9 intermediate algebra, 9 coordinate geometry, 14 plane geometry, and 4 trigonometry questions. This may vary by a question or two from year to year, but overall it is relatively unchanged. You are expected to know the basic formulas that are central to each of these content areas as well as when to use specific formulas.

SCORING

ACT Math test scaled scores range from 1 to 36. Along with your overall Math score, you will receive three subscores: Preparing for Higher Mathematics, which is broken down further into separate scores for Number & Quantity, Algebra, Functions, Geometry, and Statistics & Probability; Integrating Essential Skills; and Modeling. One thing that is important to note about the ACT scoring system is that **there is no penalty for wrong answers.** This is very important because it means that wrong answers cannot hurt your score. This also means that you should always **answer every question!**

TIPS FOR THE ACT MATH TEST

Sometimes the textbook approach will be the best way to find the answer. Other times, however, applying common sense or a strategy will get you to the correct answer more quickly and easily. The key is to be open to creative approaches to problem solving. This usually involves taking advantage of the multiple-choice format of the question.

Two methods in particular are extremely useful when you don't see—or would rather not use—the textbook approach to solving a question. These strategies aren't always quicker than more traditional methods, but they're a great way to make confusing problems more concrete. If you can apply these strategies, you're guaranteed to arrive at the correct answer every time you use them. Let's examine these strategies now.

PICKING NUMBERS

Picking numbers is an extremely handy strategy for making sense of "abstract" problems—ones that deal with variables rather than numbers. Picking Numbers makes abstract problems concrete, and comes in particularly handy when there are variables in the question and in the answer choices. Here's how to apply the method to this type of question:

Step 1: Pick simple numbers to stand in for the variables.
Step 2: Answer the question using the numbers you picked.
Step 3: Try out all the answer choices using the numbers you picked, eliminating those that give you a different result.
Step 4: If more than one choice remains, pick a different set of numbers and repeat steps 1–3.

Let's try this strategy on the following problem.

> Money collected by c charities is to be divided equally. According to donation records, p people gave d dollars each. How much money will each charity receive?
>
> **A.** $\dfrac{c}{pd}$
>
> **B.** $\dfrac{pd}{c}$
>
> **C.** $pd + c$
>
> **D.** $\dfrac{dc}{p}$
>
> **E.** $(p - c)d$

If the mere thought of this problem gives you a headache, Picking Numbers can provide you with a strategic way to quickly get to the answer. The key is to pick numbers that make the math easy for you. Because the money will be divided evenly among the charities, pick a number that is a factor of the total amount of money donated ($p \times d$) to make the math go smoothly. Try 2 for p, 8 for d, and 4 for c. Now the question asks: If 2 people each donated 8 dollars and all the money is divided equally among 4 charities, how much money, in dollars, will each charity receive? The answer to this question is 4 dollars, so replace p with 2, d with 8, and c with 4 in each of the answer choices, and see which one equals 4.

A. $\dfrac{c}{pd} = \dfrac{4}{2 \times 8} = \dfrac{4}{16} = \dfrac{1}{4}$, which is way too small.

B. $\dfrac{pd}{c} = \dfrac{2 \times 8}{4} = \dfrac{16}{4} = 4$. That's perfect!

C. $pd + c = 2 \times 8 + 4 = 16 + 4 = 20$. That's too big.

D. $\dfrac{dc}{p} = \dfrac{8 \times 4}{2} = \dfrac{32}{2} = 16$. That's too big.

E. $(p - c)d = (2 - 4) \times 8 = (-2) \times 8 = -16$. That won't work.

Only (B) works, so it must be correct.

BACKSOLVING

Sometimes, a problem will be extremely confusing, but the situation doesn't lend itself to Picking Numbers. When a question has numbers in the choices, refers to an unknown value in the question, or is confusing but asks a straightforward question, you are probably best off trying to *Backsolve* from the choices. When you Backsolve, you simply plug the choices back into the question until you find the one that works. If you do this systematically, it shouldn't take much time. Here's the system:

Step 1: Start with B or D.

By starting with B or D first, you have a 40 percent chance of getting the correct answer in one try. (You'll understand why after reading Step 2.)

Step 2: Eliminate choices you know are too big or too small.

The first choice you test is either too small, too large, or correct. Because choices are ordered by size, if you started with B and it's too large, A would be correct. Likewise, if you started with D and it's too small, E would be correct.

Step 3: Test the choice (B or D) that you did *not* start with.

Step 3 is only necessary if you haven't found the answer yet.

If B is too small or D is too large, you'll have three choices left. In either case, testing the middle remaining choice will immediately reveal the correct answer. For example, if you started with B and it was too small, you would be left with C, D, and E. If D turns out to be too small, E would be correct. If it's too large, C would be correct.

By predicting if the answer might run small or large (to help you decide whether to start with B or D) and following this system, you won't have to test more than two of the five choices. In fact, you'll often only have to test one. Let's try out this strategy on the following problem:

> Thirty people paid a total of $330 for admission to a concert. If each adult paid $15 and each child paid $5, how many adults were admitted to the concert?
>
> A. 15
>
> B. 18
>
> C. 20
>
> D. 22
>
> E. 24

If algebra isn't your strong suit, this problem can be a bit difficult to set up. The question, however, is clear: "How many adults were admitted to the concert?" There's no way to Pick Numbers here, but the numbers in the answer choices suggest that this is an excellent opportunity to Backsolve. Adult tickets seem expensive compared to children's tickets, so let's start with B.

If 18 adults each paid $15 for admission, the adults would've paid a total of $15 × 18 = $270. This leaves $330 − $270 = $60 for 30 − 18 = 12 children. You can calculate that $5 × 12 *does* equal $60. Bingo! That works, so (B) is correct.

Had you started with D instead, 22 adults would pay $15 × 22 = $330. That leaves nothing for 30 − 22 = 8 children, so D is too large.

There are three exceptions to following the above order for testing choices:

1. Choice B or D results in a complex calculation.

2. Some of the other answer choices are extremely easy to test.

3. The question asks for the smallest or largest possible value.

If plugging in B or D results in an ugly calculation (i.e., if it results in a number that doesn't divide evenly and there are other answer choices that *would* divide evenly), test an easier answer choice instead.

If you're lucky enough to get answer choices with extremely easy numbers to test, such as 0 or 1, test them first because you can evaluate them much faster than most other numbers.

And if the question asks for the smallest or largest possible value of something, you'll want to test the smallest or largest value in the answer choices first.

Remember that on the ACT, all questions are worth the same amount of points regardless of difficulty. If you are stuck on a question, DO NOT spend five minutes trying to get the answer. Instead, move on and answer all the questions you can and come back to that question during a second pass through the test. This will ensure that you answer the most questions correctly. If you are reaching the end of the test and are running out of time, make an educated guess for questions you still have not answered. No matter what, you always want to make sure you have an answer for every single question.

Another strategy that will help you on the ACT is to use your calculator only when necessary. Calculators do not have brains and they will not tell you when you accidentally press 8 even though you meant to press 7. This is a mistake that you can catch yourself if you are writing your scratchwork in your test booklet. Just about all of the questions on the Math test can be solved without the use of a calculator, and using one too much can actually slow you down on the test. If you are going to bring a calculator to the test, be sure it is one you are familiar with and make sure you have extra batteries.

Special note about calculators: Not all calculators are allowed for the ACT Math test. Before test day, check to make sure your calculator is allowed.

With the 60-minute time limit, many students feel the need to rush through the test because they fear they will run out of time at the end. Time is a factor, but you must still be sure to read each question thoroughly and make sure you understand what it is asking. For example, a question might ask you to find the circumference of a circle, but gives you the diameter. If you are moving too quickly you might use the diameter in your calculations, even though the formula calls for the use of the radius. The answer you get might even be one of the answer choices because the test makers are aware of the common mistakes that students make. However, by reading thoroughly and working carefully, you can avoid mistakes like this. Finally, if you have time at the end of the test, go back and check your work. If there are any questions you are not sure about, double-check your calculations.

Now that you are familiar with the structure, scoring, and a few strategies for taking the ACT Math test, let's move on to review the major content of the test.

CHAPTER 4

ACT Math Basics

The ACT Math test will cover six main content areas:

Content Area	Number of Questions
Pre-Algebra	14
Elementary Algebra	10
Intermediate Algebra	9
Coordinate Geometry	9
Plane Geometry	14
Trigonometry	4

As you can see, a little more than half of the test covers algebra concepts. As you review the content below, keep in mind the relative number of questions that are asked on each topic and plan your studying accordingly.

PRE-ALGEBRA

Pre-algebra questions will test your ability to work with numbers. The following are the basic concepts that you will need to know for the ACT.

ORDER OF OPERATIONS

PEMDAS = **P**lease **E**xcuse **M**y **D**ear **A**unt **S**ally: This mnemonic will help you remember the order of operations.

P = Parentheses

E = Exponents

M = Multiplication

D = Division

} in order from left to right

A = Addition

S = Subtraction

} in order from left to right

EXAMPLE

$30 - 5 \times 4 + (7 - 3)^2 \div 8$

First, perform any operations within **parentheses**. (If the expression has parentheses within parentheses, work from the innermost out.)	$30 - 5 \times 4 + 4^2 \div 8$
Next, evaluate terms that contain **exponents**.	$30 - 5 \times 4 + 16 \div 8$
Then, perform all **multiplication** and **division** from left to right.	$30 - 20 + 2$
Finally, perform all **addition** and **subtraction** from left to right.	$10 + 2 = 12$

LAWS OF OPERATIONS

Commutative law: Addition and multiplication are both **commutative**; it doesn't matter **in what order** the operation is performed. For example, $5 + 8 = 8 + 5$ and $2 \times 6 = 6 \times 2$.

Division and subtraction are **not** commutative. For example, $3 - 2 \neq 2 - 3$ and $6 \div 2 \neq 2 \div 6$.

Associative law: Addition and multiplication are also **associative**; the terms can be **regrouped** without changing the result.

EXAMPLE

$$(a + b) + c = a + (b + c) \qquad (a \times b) \times c = a \times (b \times c)$$
$$(3 + 5) + 8 = 3 + (5 + 8) \qquad (4 \times 5) \times 6 = 4 \times (5 \times 6)$$
$$8 + 8 = 3 + 13 \qquad\qquad 20 \times 6 = 4 \times 30$$
$$16 = 16 \qquad\qquad\qquad 120 = 120$$

Distributive law: The **distributive law** of multiplication allows you to "distribute" a factor among the terms being added or subtracted.

EXAMPLE

$$a(b + c) = ab + ac$$
$$4(3 + 7) = 4 \times 3 + 4 \times 7$$
$$4 \times 10 = 12 + 28$$
$$40 = 40$$

Division can be distributed in a similar way.

EXAMPLE

$$\frac{3 + 5}{2} = \frac{3}{2} + \frac{5}{2}$$
$$\frac{8}{2} = 1\frac{1}{2} + 2\frac{1}{2}$$
$$4 = 4$$

Don't get carried away, though. When there is a sum or difference in the **denominator**, no distribution is possible.

EXAMPLE

$\dfrac{9}{4 + 5}$ is NOT equal to $\dfrac{9}{4} + \dfrac{9}{5}$.

FRACTIONS

$4 \leftarrow$ numerator
$—\ \leftarrow$ fraction bar (means "divided by")
$5 \leftarrow$ denominator

Equivalent fractions: The value of a number is unchanged if you multiply the number by 1. In a fraction, multiplying the numerator and denominator by the same nonzero number is the same as multiplying the fraction by 1, so the value of the fraction is unchanged. Similarly, dividing the top and bottom by the same nonzero number leaves the value of the fraction unchanged.

EXAMPLE

$$\frac{1}{2} = \frac{1 \times 2}{2 \times 2} = \frac{2}{4} \quad \text{and} \quad \frac{5}{10} = \frac{5 \div 5}{10 \div 5} = \frac{1}{2}$$

Canceling and reducing: Generally speaking, when you work with fractions on the ACT, you'll need to put them in **lowest terms**. This means the numerator and the denominator are not divisible by any common integer greater than 1. The fraction $\frac{1}{2}$ is in lowest terms, but the fraction $\frac{3}{6}$ is not, because 3 and 6 are both divisible by 3. The process used to write a fraction in lowest terms is called **reducing**, which simply means dividing out any common multiples from both the numerator and denominator. This process is also commonly called **canceling.**

EXAMPLE

Reduce $\frac{15}{35}$ to lowest terms.

First, determine the largest common factor of the numerator and denominator. Then, divide the top and bottom by that number to reduce the fraction (or cancel the common factor).

$$\frac{15}{35} = \frac{3 \times 5}{7 \times 5} = \frac{3 \times 5 \div 5}{7 \times 5 \div 5} = \frac{3}{7} \quad \text{or} \quad \frac{15}{35} = \frac{3 \times 5}{7 \times 5} = \frac{3 \times \cancel{5}}{7 \times \cancel{5}} = \frac{3}{7}$$

Addition and subtraction: You can't add or subtract two fractions directly unless they have the same denominator. Therefore, before adding (or subtracting), you must find a common denominator. A common denominator is a **common multiple** of all the denominators of the fractions. The **least common denominator** is the **least common multiple** (the smallest positive number that is a multiple of all the terms).

EXAMPLE

$$\frac{3}{5} + \frac{2}{3} - \frac{1}{2} \text{ Denominators are 5, 3, 2.} \quad \text{LCM} = 5 \times 3 \times 2 = 30 = \text{LCD}$$

Multiply the numerator and denominator of each fraction by the value that raises its respective denominator to the LCD.

$$\left(\frac{3}{5} \times \frac{6}{6}\right) + \left(\frac{2}{3} \times \frac{10}{10}\right) - \left(\frac{1}{2} \times \frac{15}{15}\right)$$
$$= \frac{18}{30} + \frac{20}{30} - \frac{15}{30}$$

Combine the numerators by adding or subtracting and keep the LCD as the denominator.

$$= \frac{18 + 20 - 15}{30} = \frac{23}{30}$$

Multiplication: To multiply two or more fractions, multiply the numerators together and multiply the denominators together. To make the numbers easier to work with, you can cross-cancel (diagonally) or cancel top-to-bottom as needed.

EXAMPLE

$$\frac{10}{9} \times \frac{3}{4} \times \frac{8}{15}$$

First, reduce (cancel) diagonally and vertically: $\dfrac{^2\cancel{10}}{_3\cancel{9}} \times \dfrac{^1\cancel{3}}{_1\cancel{4}} \times \dfrac{\cancel{8}^2}{\cancel{15}_3}$

Then, multiply numerators and denominators: $\dfrac{2 \times 1 \times 2}{3 \times 1 \times 3} = \dfrac{4}{9}$

Division: Dividing is the same as multiplying by the **reciprocal** of the divisor. To get the reciprocal of a fraction, invert it by interchanging the numerator and the denominator. The reciprocal of the fraction $\dfrac{3}{7}$ is $\dfrac{7}{3}$.

EXAMPLE

$$\frac{4}{3} \div \frac{4}{9} = \frac{4}{3} \times \frac{9}{4} = \frac{^1\cancel{4}}{_1\cancel{3}} \times \frac{\cancel{9}^3}{\cancel{4}_1} = \frac{1 \times 3}{1 \times 1} = 3$$

PERCENTS

Percents are one of the most commonly used mathematical relationships and are quite popular on the ACT. *Percent* is just another word for *hundredth*. For example, 27% (27 percent) means:

27 hundredths

$\dfrac{27}{100}$

0.27

27 out of every 100 things

27 parts out of a whole of 100 parts

Here are some useful formulas to learn before Test Day:

- Percent × whole = part
- Percent = $\dfrac{\text{part}}{\text{whole}} \times 100\%$
- Percent change = $\dfrac{\text{amount of change}}{\text{initial amount}} \times 100\%$

Keep in mind that when using percentages in calculations, you must write the percent as a decimal number.

RATIOS

A **ratio** is a comparison of two quantities by division.

Ratios may be written as a fraction $\left(\dfrac{x}{y}\right)$, with a colon (*x*:*y*), or in words (the ratio of *x* to *y*). We recommend the first way, since ratios can be treated as fractions for the purposes of computation.

Ratios can (and in most cases, should) be reduced to lowest terms just as fractions are reduced.

In a ratio of two numbers, the numerator is often associated with the word *of*, and the denominator with the word *to*. For example, the ratio **of** 3 **to** 4 is $\dfrac{\text{of } 3}{\text{to } 4} = \dfrac{3}{4}$.

Ratios can compare parts to parts or parts to wholes. For example, suppose you make a fruit salad using 6 oranges, 3 apples, and 2 pears. The ratio of apples to pears is 3:2. The ratio of apples to all the fruit is 3:11.

Ratios are frequently encountered as parts of a **proportion**. A proportion is simply an equation in which two ratios are set equal to each other. They are an efficient way to solve certain problems, but you must exercise caution when setting them up. Watching the units of each piece of the proportion is critical.

RATES

A **rate** is a ratio that relates two different kinds of quantities. Speed (the ratio of distance traveled to time elapsed) is one such example.

Questions that involve rates usually use the word *per,* as in "miles per hour," "cost per item," and so forth. Because *per* means "for one" or "for each," rates are expressed as ratios reduced to a denominator of 1. For example, if John travels 50 miles in two hours, his average rate is:

$$\frac{50 \text{ miles}}{2 \text{ hours}} = 25 \text{ miles per hour}$$

Note: Questions are often written in terms of "average rate," because it may be improbable (as in the case of speed) that the rate has been constant over the given period of time.

MEAN, MEDIAN, MODE

MEAN

The arithmetic **mean**, or **average**, of a group of numbers is defined as the sum of the terms divided by the number of terms.

$$\text{Average} = \frac{\text{Sum of Terms}}{\text{Number of Terms}}$$

EXAMPLE

Henry buys three items costing $2.00, $0.75, and $0.25. What is the average price?

$$\begin{aligned}
\text{Average price} &= \frac{\text{Sum of prices}}{\text{Number of prices}} \\
&= \frac{\$2.00 + \$0.75 + \$0.25}{3} \\
&= \frac{\$3.00}{3} \\
&= \$1.00
\end{aligned}$$

If you know the average of a group of values and the number of values in the group, you can find the **sum** of the values using the following formula:

Sum of Values = Average Value × Number of Values

EXAMPLE

The average daily temperature for the first full week in January was 31 degrees. If the average temperature for the first six days was 30 degrees, what was the temperature on the seventh day?

The sum for all seven days is $31 \times 7 = 217$ degrees.

The sum of the first six days is $30 \times 6 = 180$ degrees.

The temperature on the seventh day is $217 - 180 = 37$ degrees.

For evenly spaced numbers, the average is the middle value. The average of consecutive integers 6, 7, and 8 is 7. The average of 5, 10, 15, and 20 is 12.5 (midway between the middle values 10 and 15).

It might be useful to try to think of the average as the "balanced" value. In other words, the total deficit of all the values below the average will balance out the total surplus of all the values that exceed the average. The average of 3, 5, and 10 is 6. Three is 3 less than 6 and 5 is 1 less than 6, for a total deficit of $3 + 1 = 4$. This is the same amount by which 10 is greater than 6.

EXAMPLE

The average of 3, 4, 5, and x is 5. What is the value of x?

Think of each value in terms of its position relative to the average, 5.

   3 is 2 less than the average

   4 is 1 less than the average

   5 is at the average

Together, the three numerical terms have a total deficit of $1 + 2 + 0 = 3$. Therefore, x must be 3 **more** than the average to restore the balance at 5. So x is $3 + 5 = 8$.

MEDIAN

On the ACT, you might see a reference to the **median**. If a group of numbers is arranged in numerical order, the median is the middle value if there is an odd number of terms in the set or the average of the two middle terms if there is an even number of terms in the set. The median of the numbers 1, 4, 5, 6, and 100 is 5, while the median of the numbers 2, 3, 7, 9, 22, and 34 is $\dfrac{7 + 9}{2} = \dfrac{16}{2} = 8$.

EXAMPLE

What is the median of 5, 101, 53, 2, 8, 4, and 11?

Rearranged in numerical order, the list is 2, 4, 5, 8, 11, 53, and 101. The list contains an odd number of terms, 7, so the median is the middle number, which is 8.

In an evenly spaced list of numbers, such as a set of consecutive integers, the median is equal to the mean. However, the median can also be quite different from the mean as you've seen from the examples above.

MODE

The **mode** is rarely tested on the ACT. It refers to the term that appears most frequently in a set. If Set $A = \{1, 5, 7, 1, 3, 4, 1, 3, 0\}$, the mode would be 1 as it shows up three times— more than any other term.

If more than one term is tied for most frequent, every one of them is a mode.

EXAMPLE

What is the mode of 3, 23, 12, 23, 3, 7, 0, 5?

Both 3 and 23 each appear twice, which is more often than any other term, so both 3 and 23 are modes of this list.

If no term shows up more than once, there is no mode.

PROBABILITY

Probability measures the likelihood of an event taking place. It can be expressed as a fraction ("The probability of snow tomorrow is $\frac{1}{2}$"), a decimal ("There is a 0.5 chance of snow tomorrow"), or a percent ("The probability of snow tomorrow is 50%").

When expressed as a fraction, a probability can be read as "x chances in y," where x is the numerator and y is the denominator. So a $\frac{2}{3}$ probability of winning a car could be read as "2 chances in 3 to win a car."

To compute a probability, divide the number of desired outcomes by the number of possible outcomes.

$$\text{Probability} = \frac{\text{Number of Desired Outcomes}}{\text{Number of Possible Outcomes}}$$

EXAMPLE

If you have 12 shirts in a drawer and 9 of them are white, what is the probability of randomly picking a white shirt?

When randomly picking a shirt in this situation, there are 12 possible outcomes, 1 for each shirt. Of these 12, 9 are white, so there are 9 desired outcomes.

Therefore, the probability of picking a white shirt at random is $\frac{9}{12} = \frac{3}{4}$. The probability can also be expressed as 0.75 or 75%.

A probability of 0 means that an event has no chance of happening. A probability of 1 means that an event will always happen.

NUMBER LINE AND ABSOLUTE VALUE

A **number line** is a straight line that extends infinitely in either direction, on which real numbers are represented as points.

As you move to the right on a number line, the values increase.

Conversely, as you move to the left, the values decrease.

Zero separates the positive numbers (to the right of zero) and the negative numbers (to the left of zero) along the number line. Zero is neither positive nor negative.

The **absolute value** of a number is just the number without its sign. It is written as two vertical lines. For example, $|-3| = |+3| = 3$.

The absolute value can be thought of as the number's distance from zero on the number line. In the example, -3 and 3 are both 3 units from zero, so their absolute values are both 3.

ELEMENTARY ALGEBRA

Elementary algebra will build on some of the concepts introduced in pre-algebra while introducing some new ones as well. The following concepts are important to know for test day.

RULES OF OPERATIONS WITH POWERS

In the term $3x^2$, 3 is the **coefficient**, x is the **base**, and 2 is the **exponent**. The exponent refers to the number of times the base is a factor of the expression. For example, 4^3 has 3 factors of 4, or $4^3 = 4 \times 4 \times 4$.

An integer times itself is the **square** of that integer. ($y \times y$ is the square of y, or y^2)

An integer times itself twice is the **cube** of that integer. ($4 \times 4 \times 4$ is the cube of 4, or 4^3)

To multiply two terms with the same base, keep the base and add the exponents.

$$m^4 \times m^7 = m^{4+7} = m^{11}$$

EXAMPLE

$$
\begin{aligned}
2^2 \times 2^3 &= (2 \times 2)(2 \times 2 \times 2) \\
&= (2 \times 2 \times 2 \times 2 \times 2) \\
&= 2^5
\end{aligned}
\qquad \text{or} \qquad
\begin{aligned}
2^2 \times 2^3 &= 2^{2+3} \\
&= 2^5
\end{aligned}
$$

To divide two terms with the same base, keep the base and subtract the exponent of the denominator from the exponent of the numerator.

$$d^{10} \div d^7 = d^{10-7} = d^3$$

EXAMPLE

$$
\begin{aligned}
4^4 \div 4^2 &= \frac{4 \times 4 \times 4 \times 4}{4 \times 4} \\
&= \frac{4 \times 4}{1} \\
&= 4^2
\end{aligned}
\qquad \text{or} \qquad
\begin{aligned}
4^4 \div 4^2 &= 4^{4-2} \\
&= 4^2
\end{aligned}
$$

To raise a power to another power, multiply the exponents.

$$(p^5)^3 = p^{5 \times 3} = p^{15}$$

EXAMPLE

$$
\begin{aligned}
\left(3^2\right)^4 &= (3 \times 3)^4 \\
&= (3 \times 3)(3 \times 3)(3 \times 3)(3 \times 3) \\
&= 3^8
\end{aligned}
\qquad \text{or} \qquad
\begin{aligned}
\left(3^2\right)^4 &= 3^{2 \times 4} \\
&= 3^8
\end{aligned}
$$

Any nonzero number raised to the zero power is equal to 1. So, $a^0 = 1$ as long as $a \neq 0$, while 0^0 is undefined.

To evaluate a negative exponent, take the reciprocal of the base and change the sign of the exponent.

$$a^{-n} = \frac{1}{a^n} \text{ or } \left(\frac{1}{a}\right)^n$$

EXAMPLE

$$2^{-3} = \left(\frac{1}{2}\right)^3 = \frac{1}{2^3} = \frac{1}{8}$$

A fractional exponent indicates a **root**.

$(a)^{\frac{1}{n}} = \sqrt[n]{a}$ (Read "the nth root of a." If no "n" is present, the radical sign means a square root.)

EXAMPLE

$$8^{\frac{1}{3}} = \sqrt[3]{8} = 2$$

RULES OF OPERATIONS WITH ROOTS

When it comes to the four basic arithmetic operations, radicals are treated in much the same way as variables.

Addition and subtraction: Only like radicals can be added to or subtracted from one another.

EXAMPLE

$$\begin{aligned}
\sqrt{3} + 4\sqrt{2} - \sqrt{2} - 3\sqrt{3} &= \left(4\sqrt{2} - \sqrt{2}\right) + \left(2\sqrt{3} - 3\sqrt{3}\right) \left[\text{Note}: \sqrt{2} = 1\sqrt{2}\right] \\
&= 3\sqrt{2} + \left(-\sqrt{3}\right) \\
&= 3\sqrt{2} - \sqrt{3}
\end{aligned}$$

Multiplication and division: To multiply or divide one radical by another, multiply or divide the numbers outside the radical signs separately from the numbers inside the radical signs, much like you would with coefficients and variables.

EXAMPLE

$$\left(6\sqrt{3}\right) \times \left(2\sqrt{5}\right) = (6 \times 2)\left(\sqrt{3} \times \sqrt{5}\right) = 12\sqrt{3 \times 5} = 12\sqrt{15}$$

EXAMPLE

$$12\sqrt{15} \div 2\sqrt{5} = (12 \div 2)\left(\sqrt{15} \div \sqrt{5}\right) = 6\left(\frac{\sqrt{15}}{\sqrt{5}}\right) = 6\sqrt{3}$$

EXAMPLE

$$\frac{4\sqrt{18}}{2\sqrt{6}} = \left(\frac{4}{2}\right)\left(\frac{\sqrt{18}}{\sqrt{6}}\right) = 2\left(\sqrt{\frac{18}{6}}\right) = 2\sqrt{3}$$

If the number inside the radical is a multiple of a perfect square, the expression can be simplified by factoring out the perfect square.

EXAMPLE

$$\sqrt{72} = \sqrt{36 \times 2} = \sqrt{36} \times \sqrt{2} = 6\sqrt{2}$$

Elementary algebra on the ACT will also move beyond the realm of simple numbers by introducing variables. The following is important to know when dealing with variables.

ALGEBRAIC TERMINOLOGY

Terms: A **term** is a numerical constant or the product (or quotient) of a numerical constant and one or more variables. Examples of terms are $3x$, $4x^2yz$, and $\frac{2a}{c}$.

Expressions: An **algebraic expression** is a combination of one or more terms. Terms in an expression are separated by either $+$ or $-$ signs. Examples of expressions are $3xy$, $4ab + 5cd$, and $x^2 - 1$.

A number without any variables is called a **constant term**.

In the term $3xy$, the number **3** is called a **coefficient**.

In a simple term where no coefficient is listed, such as z, the coefficient is **1**.

An expression with one term, such as $3xy$, is called a **monomial.**

An expression with two terms, such as $4a + 2d$, is a **binomial.**

An expression with three terms, such as $xy + z - a$, is a **trinomial.**

The general name for these types of expressions is **polynomial.**

OPERATIONS WITH POLYNOMIALS

All of the laws of arithmetic operations, such as the commutative, associative, and distributive laws, are applicable to polynomials as well.

Commutative law:

$$2x + 5y = 5y + 2x$$
$$5a \times 3b = 3b \times 5a = 15ab$$

Associative law:

$$(2x + 3x) + 5x = 2x + (3x + 5x) = 10x$$
$$4s \times (7j \times 2p) = (4s \times 7j) \times 2p = 56\,sjp$$

Distributive law:

$$3a(2b - 5c) = (3a \times 2b) - (3a \times 5c) = 6ab - 15ac$$

Note: The product of two binomials can be calculated by applying the distributive law twice.

EXAMPLE

$$(x + 5)(x - 2) = x(x - 2) + 5(x - 2)$$
$$= x \times x + x \times (-2) + 5 \times x + 5 \times (-2)$$
$$= x^2 - 2x + 5x - 10$$
$$= x^2 + 3x - 10$$

A simple mnemonic for this is **F**irst **O**uter **I**nner **L**ast, or **FOIL.**

FACTORING ALGEBRAIC EXPRESSIONS

Factoring a polynomial means expressing it as a product of two or more simpler expressions.

Common monomial factor: When there is a monomial factor common to every term in the polynomial, it can be factored out by using the distributive law.

EXAMPLE

$2a + 6ac = 2a(1 + 3c)$ ($2a$ is the greatest common factor of $2a$ and $6ac$.)

Difference of two perfect squares: The difference of two squares can be factored into a product: $a^2 - b^2 = (a - b)(a + b)$.

EXAMPLE

$$9x^2 - 1 = (3x)^2 - (1)^2 = (3x + 1)(3x - 1)$$

Polynomials of the form $a^2 + 2ab + b^2$: Any polynomial of this form is equivalent to the square of a binomial. Notice that $(a + b)^2 = a^2 + 2ab + b^2$ (try FOIL).

Factoring such a polynomial is simply reversing this procedure.

EXAMPLE

$x^2 + 6x + 9 = (x)^2 + 2(x)(3) + (3)^2 = (x + 3)^2$

Polynomials of the form $a^2 - 2ab + b^2$: Any polynomial of this form is equivalent to the square of a binomial as well. Here, though, the binomial is the difference of two terms: $(a - b)^2 = a^2 - 2ab + b^2$.

EXAMPLE

$x^2 - 4x + 4 = (x)^2 - 2(x)(2) + (2)^2 = (x - 2)^2$

Polynomials of the form $ax^2 + bx + c$: Polynomials of this form can often be factored into a product of two binomials. The product of the first term in each binomial must equal the first term of the polynomial. The product of the last terms of the binomials must equal the third term of the polynomial. The sum of the remaining products must equal the second term of the polynomial. Factoring can be thought of as the FOIL method backwards.

EXAMPLE

$x^2 - 3x + 2$

To factor this into two binomials, each containing an x term, start by writing down what you know.

$x^2 - 3x + 2 = (x + \ldots)(x + \ldots)$

Next, you need to fill in the missing term to the right of each binomial. The **product** of the two missing terms will be the last term in the polynomial: 2. The **sum** of the two missing terms will be the coefficient of the second term of the polynomial: -3. Try the possible factors of 2 until we get a pair that adds up to -3. There are two possibilities: 1 and 2, or -1 and -2. Since $(-1) + (-2) = -3$, fill -1 and -2 into the empty spaces.

Thus, $x^2 - 3x + 2 = (x - 1)(x - 2)$.

Note: Using FOIL on a factored polynomial is a great way to check your work.

INEQUALITIES

Inequality symbols:

> $>$ greater than

> $<$ less than

> $\geq$ greater than or equal to

> $\leq$ less than or equal to

- $x > 4$ means all numbers greater than 4.

- $x < 0$ means all numbers less than zero (the negative numbers).

- $x \geq -2$ means x can be -2 or any number greater than -2.

- $x \leq \dfrac{1}{2}$ means x can be $\dfrac{1}{2}$ or any number less than $\dfrac{1}{2}$.

Solving inequalities: Inequalities behave the same way as normal equations with one exception: **Multiplying or dividing by a negative number reverses the inequality's direction.**

EXAMPLE

Find the values of x that represent the solution set for the inequality $3 - \dfrac{x}{4} \geq 2$.

1. Multiply both sides by 4 to get $12 - x \geq 8$.

2. Subtract 12 from both sides to get $-x \geq -4$.

3. Divide both sides by -1 and change the direction of the inequality to get $x \leq 4$.

Note: The solution set to an inequality is not a single value but a range of possible values. Here, the values include 4 and all numbers below 4.

INTERMEDIATE ALGEBRA

Intermediate algebra continues to build upon the concepts introduced thus far while adding several more advanced ones. The following will be important to know for the ACT.

SIMULTANEOUS EQUATIONS

Sometimes a problem on the ACT will involve an equation with more than one variable. In general, if you want to find numerical values for all the variables, you will need as many **distinct** equations as you have variables. Two equations are distinct if neither simplifies into the other. $x^2 + 2xy + 5y^2 = 5$ and $3x^2 + 6xy + 15y^2 = 15$ are *not* distinct because multiplying the first equation by 3 results in the second.

If you had only one equation with two variables, such as $x - y = 7$, there would be an infinite number of solution sets because each unique value of x has a different corresponding value for y. If $x = 8$, $y = 1$ (because $8 - 1 = 7$). If $x = 12$, $y = 5$ (because $12 - 5 = 7$). And so forth.

If you are given two distinct equations with two variables, you can combine them to get a unique solution set. This is known as a **system of equations**. There are two algebraic ways to do this:

METHOD I: COMBINATION

Combination (also known as elimination) involves adding or subtracting a multiple, positive or negative, of one equation from the other. The idea is to choose a multiple such that one of the variables is eliminated, solve for the remaining variable, then plug the resulting value back into the *other* equation to solve for the other variable if necessary.

EXAMPLE

If $3a + 2b = 12$ and $5a + 4b = 23$, what is the value of a?

1. Start by lining up the equations, one under the other.

$$3a + 2b = 12$$
$$5a + 4b = 23$$

2. Multiply the top equation by -2 to get $-4b$ (which will cancel with $4b$ in the bottom equation).

$$-6a + (-4b) = -24$$
$$5a + 4b = 23$$

3. Add to eliminate the b term.

$$-6a + \cancel{(-4b)} = -24$$
$$\underline{5a + \cancel{4b} = 23}$$
$$-a \qquad = -1$$
$$a = 1$$

METHOD II: SUBSTITUTION

Substitution involves solving for one of the variables in terms of the other in one equation, then substituting the resulting expression back into the other equation.

EXAMPLE

Solve for m and n when $m = 4n + 2$ and $3m + 2n = 16$.

1. The first equation gives m in terms of n, so substitute $4n + 2$ for m in the second equation.

$$3(4n + 2) + 2n = 16$$
$$12n + 6 + 2n = 16$$

2. Solve for n.

$$14n = 10$$
$$n = \frac{10}{14} = \frac{5}{7}$$

3. Substitute $\frac{5}{7}$ for n in the first equation to solve for m.

$$m = 4n + 2$$
$$m = 4\left(\frac{5}{7}\right) + 2$$
$$= \frac{20}{7} + \frac{14}{7}$$
$$= \frac{34}{7}$$

QUADRATIC EQUATIONS

When the polynomial $ax^2 + bx + c$ equals 0, it is given a special name—a **quadratic equation**. As an equation, you can find the value(s) of x that make it true.

EXAMPLE

What values of x satisfy the equation $x^2 - 3x + 2 = 0$?

To find the solutions, or roots, start by factoring it. You can factor $x^2 - 3x + 2$ into $(x - 2)(x - 1)$, making the quadratic equation $(x - 2)(x - 1) = 0$.

You now have an equation where the product of two binomials equals 0. This is true when either one of the factors is 0. Therefore, to find the roots, set the two binomials equal to 0 and solve for x. In other words, either $x - 2 = 0$ or $x - 1 = 0$ (or both). Solving for x gives $x = 2$ or $x = 1$. To check the math, plug 1 and 2 back into the original equation and make sure that both variables satisfy the equation.

$$2^2 - 3(2) + 2 = 0 \qquad 1^2 - 3(1) + 2 = 0$$
$$4 - 6 + 2 = 0 \qquad 1 - 3 + 2 = 0$$
$$0 = 0 \qquad 0 = 0$$

FUNCTIONS

Classic function notation problems may also appear on the ACT. An algebraic expression of only one variable may be defined as a function, such as f or g, of that variable. You could be asked to find a specific domain or range value of a function, to describe the graph of the function, or a variety of other questions about the function in general.

EXAMPLE

What is the minimum value of the function $f(x) = x^2 - 1$?

For the function $f(x) = x^2 - 1$, when x is 1, $f(1) = 1^2 - 1 = 0$. In other words, by inputting 1 into the function, the output $f(1) = 0$. Every input value has only one output (though the reverse is not necessarily true). The set of all the input values is called the **domain** of the function, and the set of all the outputs is called the **range**.

Here, you are asked to find the minimum value, that is, the smallest possible range value of the function. Minimums are usually found using math that is much too complex for the ACT, so any minimum (or maximum) value problem on the test can be solved in one of two simple ways: either plug the answer choices into the function and find which gives you the lowest value, or apply some critical thinking before testing any values. In the case of $f(x) = x^2 - 1$, the function will be at a minimum when x^2 is as small as possible. Because x^2 gets larger the farther x is from 0, x^2 is as small as possible when $x = 0$. Consequently, the smallest value of $x^2 - 1$ occurs when $x = 0$. So the minimum value of the function is $f(0) = 0^2 - 1 = -1$.

MATRICES

Matrices do not appear on the ACT test very often and those that do will be very straight-forward and simple. Take matrices A and B below:

$$A = \begin{bmatrix} 3 & 4 \\ 6 & 2 \end{bmatrix} \quad B = \begin{bmatrix} 1 & 2 \\ 5 & 7 \end{bmatrix}$$

To add or subtract matrices, simply add or subtract the corresponding entries; $A + B$ would be:

$$\begin{bmatrix} 3+1 & 4+2 \\ 6+5 & 2+7 \end{bmatrix} = \begin{bmatrix} 4 & 6 \\ 11 & 9 \end{bmatrix}$$

PLANE GEOMETRY

Plane geometry deals with a broad range of topics. The following is a list of what you can expect to see on the ACT.

LINES AND ANGLES

A **line** is a one-dimensional geometric abstraction—infinitely long with no width. It is not physically possible to **draw** a line, as any physical line would have a finite length and some width, no matter how long and thin we tried to make it. Two points determine a straight line or, in other words, given any two points, there is exactly one straight line that passes through them.

Segments: A **line segment** is a section of a straight line of finite length with two endpoints. A line segment is named by its endpoints, as in segment AB. The **midpoint** is the point that divides a line segment into two equal parts.

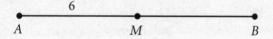

EXAMPLE

In the figure above, A and B are the endpoints of $\overline{AB}$ and M is its midpoint $\left(\overline{AM} = \overline{MB}\right)$. What is the length of AB? $\overline{AM}$ is 6, meaning $\overline{MB}$ is also 6, so $\overline{AB} = 6 + 6 = 12$.

Two lines are **parallel** if they lie on the same plane and never intersect each other regardless of how far they are extended. If line ℓ_1 is parallel to line ℓ_2, you write $\ell_1 \| \ell_2$.

Angles: An **angle** is formed whenever two lines or line segments intersect at a point. The point of intersection is called the **vertex** of the angle. Angles are measured in degrees (°).

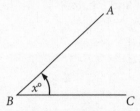

Angle x, $\angle ABC$, and $\angle B$ all denote the same angle in the diagram above.

An **acute angle** is an angle whose degree measure is between 0° and 90°. A **right angle** is an angle whose degree measure is exactly 90°. An **obtuse angle** is an angle whose degree measure is between 90° and 180°. A **straight angle** is an angle whose degree measure is exactly 180°.

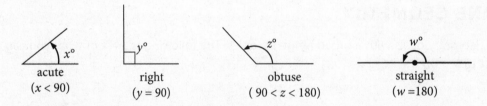

The sum of the measures of the angles on one side of a straight line is 180°.

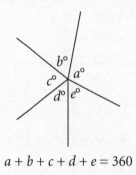

The sum of the measures of the angles around a point (i.e., a full circle) is 360°.

$$a + b + c + d + e = 360$$

Two lines are **perpendicular** if they intersect at a 90° angle. The shortest distance from a point to a line is the line segment drawn from the point to the line such that it is perpendicular to the line. If line ℓ_1 is perpendicular to line ℓ_2, you write $\ell_1 \perp \ell_2$. If $\ell_1 \perp \ell_2$ and $\ell_2 \perp \ell_3$, then $\ell_1 \| \ell_3$:

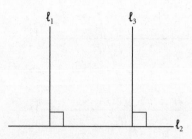

Two angles are **supplementary** if together they make up a straight angle, i.e., if the sum of their measures is 180°. Two angles are **complementary** if together they make up a right angle, i.e., if the sum of their measures is 90°.

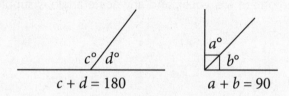

A line or line segment **bisects** an angle if it splits the angle into two equal halves. $\overline{BD}$ below bisects $\angle ABC$, and $\angle ABD$ has the same measure as $\angle DBC$. The two smaller angles are each half the size of $\angle ABC$.

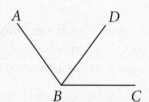

Vertical angles are a pair of opposite angles formed by two intersecting line segments. At the point of intersection, two pairs of vertical angles are formed. Angles a and c below are vertical angles, as are b and d.

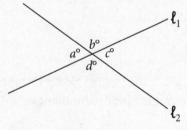

Vertical angles have the same degree measure. In the diagram above, $a = c$ and $b = d$. In addition, since ℓ_1 and ℓ_2 are straight lines, $a + b = c + d = a + d = b + c = 180$. In other words, each angle is supplementary to each of its two adjacent angles.

If two parallel lines intersect with a third line (called a *transversal*), each of the parallel lines will intersect the third line at the same angle. In the following figure, $a = e$ (corresponding angles relative to the transversal), $a = c$ (vertical angles), and $e = g$ (vertical angles). Therefore, $a = c = e = g$ and $b = d = f = h$.

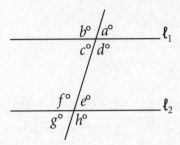

If $\ell_1 \parallel \ell_2$, then $a = c = e = g$ and $b = d = f = h$.

In other words, when two parallel lines intersect with a third line, all acute angles formed are equal, all obtuse angles formed are equal, and any acute angle is supplementary to any obtuse angle.

TRIANGLES

GENERAL TRIANGLES

A **triangle** is a closed figure with three angles and three straight sides. The sum of the measures of the angles in a triangle is **180°**.

Each **interior angle** is supplementary to an adjacent **exterior angle**. The degree measure of an exterior angle is equal to the sum of the measures of the two nonadjacent (remote) interior angles, or 180° minus the measure of the adjacent interior angle.

In the figure below, a, b, and c are interior angles, so $a + b + c = 180$. Angle d is supplementary to c as well, so $d + c = 180$, $d + c = a + b + c$, and $d = a + b$. Thus, the exterior angle d is equal to the sum of the two remote interior angles—a and b.

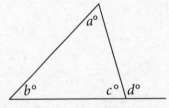

The **altitude** (or height) of a triangle is the perpendicular distance from a vertex to the side opposite the vertex. The altitude can fall inside the triangle, outside the triangle, or along one of the sides.

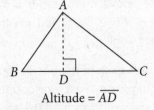

Altitude = $\overline{AD}$

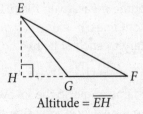

Altitude = $\overline{EH}$

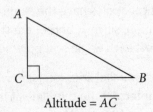

Altitude = $\overline{AC}$

Sides and angles: The length of any side of a triangle is less than the sum of the lengths of the other two sides and greater than their positive difference.

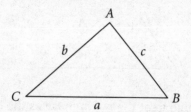

$$b + c > a > |b - c|$$

$$a + b > c > |a - b|$$

$$a + c > b > |a - c|$$

If the lengths of two sides of a triangle are unequal, the **greater angle** lies **opposite the longer side** and vice versa. In the figure above, if $m\angle A > m\angle B > m\angle C$ then $a > b > c$.

Area of a triangle: The **area** of a triangle refers to the space it takes up.

The area of a triangle is $\dfrac{1}{2}$ base × height.

EXAMPLE

If a triangle has a base of 4 and an altitude of 3, what is the area of the triangle?

$$A = \frac{1}{2}bh$$

$$= \frac{1}{2} \times 4 \times 3 = 6$$

Remember that the height (or altitude) is perpendicular to the base. Therefore, when two sides of a triangle are perpendicular to each other, the area is easy to find. In a right triangle, the two sides that form the 90° angle are called the **legs**, and the area is one-half the product of the legs:

$$A = \frac{1}{2}bh$$

$$= \frac{1}{2}l_1 \times l_2$$

EXAMPLE

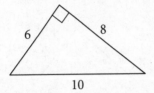

In the triangle above, you could treat the hypotenuse as the base, because that's how the figure is drawn. Then you would need to find the distance from the hypotenuse to the opposite vertex in order to determine the area of the triangle. A more straightforward method is to notice that this is a **right** triangle with legs of lengths 6 and 8, which allows you to use the alternative formula for the area:

$$A = \frac{1}{2}l_1 \times l_2 = \frac{1}{2} \times 6 \times 8 = 24$$

Perimeter of a triangle: The **perimeter** of a triangle is the distance around the triangle. In other words, the perimeter is equal to the sum of the lengths of the sides.

EXAMPLE

In the triangle below, the sides are of lengths 5, 6, and 8. Therefore, the perimeter is 5 + 6 + 8, or 19.

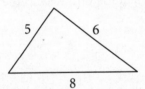

Isosceles triangles: An **isosceles triangle** is a triangle that has two sides of equal length. The two equal sides are called the **legs,** and the third side is called the **base.**

Because the two legs have the same length, the two angles opposite the legs must have the same measure. In the figure below, $\overline{PQ} = \overline{PR}$ and $\angle R = \angle Q$.

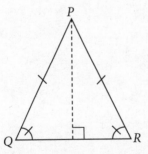

Equilateral triangles: An **equilateral triangle** has three sides of equal length and three 60° angles.

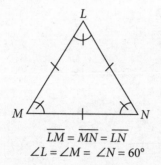

$$\overline{LM} = \overline{MN} = \overline{LN}$$
$$\angle L = \angle M = \angle N = 60°$$

Similar triangles: Triangles are **similar** if they have the same shape (that is, corresponding angles have the same measure). For instance, any two triangles whose angles measure 30°, 60°, and 90° are similar. In similar triangles, corresponding sides are in the same ratio. Triangles are **congruent** if corresponding angles have the same measure and corresponding sides have the same length.

EXAMPLE

What is the perimeter of $\triangle DEF$ below?

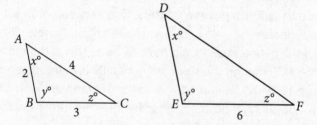

Each triangle has an $x°$ angle, a $y°$ angle, and a $z°$ angle, so they are similar and corresponding sides are in the same ratio. $\overline{BC}$ and $\overline{EF}$ are corresponding sides—each is opposite the $x°$ angle. Because $\overline{EF}$ is twice the length of $\overline{BC}$, *each* side of $\triangle DEF$ will be twice the length of its corresponding side in $\triangle ABC$.

Therefore, $\overline{DE} = 2\big(\overline{AB}\big) = 4$ and $\overline{DF} = 2\big(\overline{AC}\big) = 8$. The perimeter of $\triangle DEF$ is $4 + 6 + 8 = 18$.

The ratio of the **areas** of two similar triangles is the **square** of the ratio of their corresponding lengths. In the example above, because each side of $\triangle DEF$ is twice the length of its corresponding side in $\triangle ABC$, the area of $\triangle DEF$ must be $2^2 = 4$ times the area of $\triangle ABC$.

$$\frac{\text{Area } \triangle DEF}{\text{Area } \triangle ABC} = \left(\frac{DE}{AB}\right)^2 = \left(\frac{2}{1}\right)^2 = 4$$

Right triangles: A right triangle has one interior angle of 90°. The longest side, which lies opposite the right angle, is called the **hypotenuse**. The other two sides are called the **legs**.

PYTHAGOREAN THEOREM

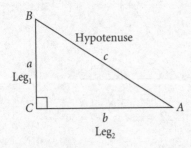

$$\left(\text{Leg}_1\right)^2 + \left(\text{Leg}_2\right)^2 = \left(\text{Hypotenuse}\right)^2$$

or

$$a^2 + b^2 = c^2$$

The **Pythagorean theorem** holds for all right triangles, and it states that the square of the hypotenuse is equal to the sum of the squares of the legs.

Some sets of integers happen to satisfy the Pythagorean theorem. These sets of integers are commonly referred to as **Pythagorean triplets**. One very common set that you might remember is 3, 4, and 5. Because $3^2 + 4^2 = 5^2$, if you have a right triangle with legs of lengths 3 and 4, the length of the hypotenuse would have to be 5. This is the most common kind of right triangle on the ACT, though the sides are generally presented as a multiple of 3, 4, and 5, such as 6, 8, and 10, or 12, 16, and 20. Memorize this ratio and you'll be well on your way to speeding through triangle problems that feature this triplet on the ACT. Another triplet that occasionally appears on the ACT is 5, 12, and 13.

Whenever you're given the lengths of two sides of a right triangle, the Pythagorean theorem allows you to find the third side.

EXAMPLE

What is the length of the hypotenuse of a right triangle with legs of lengths 9 and 10?

The Pythagorean theorem states that the square of the length of the hypotenuse equals the sum of the squares of the lengths of the legs. Here, the legs are 9 and 10, so:

$$
\begin{aligned}
\text{Hypotenuse}^2 &= 9^2 + 10^2 \\
&= 81 + 100 \\
&= 181 \\
\text{Hypotenuse} &= \sqrt{181}
\end{aligned}
$$

EXAMPLE

What is the length of the hypotenuse of an isosceles right triangle with legs of length 4?

Because the triangle is isosceles, you know two of the sides have the same length. The hypotenuse can't be the same length as one of the legs (the hypotenuse must be the longest side), so it must be that the two legs are equal. Therefore, the two legs each have length 4, and you can use the Pythagorean theorem to find the hypotenuse.

$$\text{Hypotenuse}^2 = 4^2 + 4^2$$
$$= 16 + 16$$
$$= 32$$
$$\text{Hypotenuse} = \sqrt{32}$$
$$= \sqrt{16} \times \sqrt{2}$$
$$= 4\sqrt{2}$$

You can always use the Pythagorean theorem to find the lengths of the sides in a right triangle. There are, however, two special right triangles that always have leg lengths in the same ratios.

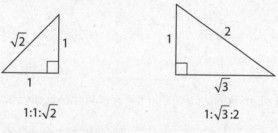

1:1:$\sqrt{2}$ 1:$\sqrt{3}$:2

(for isosceles right triangles)   (for 30-60-90 triangles)

These two types of special right triangles are tested often enough that it is to your benefit to memorize these proportions, as that would allow you to blow past such problems without having to do any math. If you forget them on test day, rest assured that you can still solve the problem by using the Pythagorean theorem (it would just take longer).

POLYGONS

A **polygon** is a closed figure whose sides are straight line segments.

The **perimeter** of a polygon is the sum of the lengths of its sides.

A **vertex** of a polygon is the point where two adjacent sides meet.

A **diagonal** of a polygon is a line segment connecting two nonadjacent vertices.

A **regular polygon** has sides of equal length and interior angles of equal measure.

The number of sides determines the specific name of the polygon. A **triangle** has three sides, a **quadrilateral** has four sides, a **pentagon** has five sides, and a **hexagon** has six sides. Triangles and quadrilaterals are by far the most important polygons on the ACT.

Interior and exterior angles: A polygon can be divided into triangles by drawing diagonals from a given vertex to all other nonadjacent vertices. For instance, the pentagon below can be divided into three triangles. Since the sum of the interior angles of each triangle is 180°, the sum of the interior angles of a pentagon is 3 × 180° = 540°.

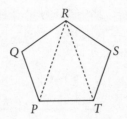

EXAMPLE

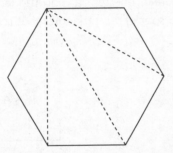

What is the measure of an interior angle in the regular hexagon above?

All angles are equal, so each is equal to one-sixth the sum of the angles. Since we can draw four triangles in a six-sided figure, the sum of the interior angles is 4 × 180° = 720°. Therefore, each angle measures $\frac{720}{6} = 120°$.

The formula to represent the total measure of an n-sided polygon's interior angles is $180(n - 2)$.

QUADRILATERALS

While there are many quadrilaterals in math, the only two that you need to worry about are **rectangles** and **squares**.

A **rectangle** is a quadrilateral with two sets of parallel sides and four right angles. A **square** is an equilateral **rectangle**.

Note the following formulas for test day:

The area of a rectangle is Length × Width.   The area of a square is Side².

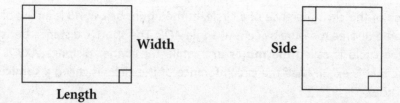

CIRCLES

Circle: The set of all points in a plane at the same distance from a certain point. This point is called the **focus** and lies at the **center** of the circle.

A circle is labeled by its center point: Circle O means the circle with center point O. Two circles of different size with the same center are **concentric**.

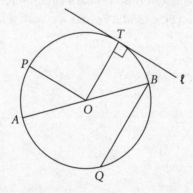

Diameter: A line segment that connects two points on the circle and passes through the center of the circle. In circle O, $\overline{AB}$ is a diameter.

Radius: A line segment from the center of the circle to any point on the circle. The radius of a circle is one-half the length of its diameter. In circle $O, \overline{OA}, \overline{OB}, \overline{OP}$, and $\overline{OT}$ are all radii.

Chord: A line segment joining two points on the circle. In circle O, $\overline{QB}$ and $\overline{AB}$ are chords. The longest chord in a circle is a diameter.

Central angle: An angle formed by two radii. $\angle AOP, \angle POB$, and $\angle BOA$ are three of circle O's central angles.

Tangent: A line that touches only one point on the circumference of the circle. A line drawn tangent to a circle is perpendicular to the radius at the point of tangency. Line ℓ is tangent to circle O at point T.

Circumference and arc length: The distance around a circle is its **circumference**. The number π (pi) is the ratio of a circle's circumference to its diameter. The value of π is usually approximated to 3.14. For the ACT, it is generally sufficient to remember that π is a little more than 3.

Since π equals the ratio of the circumference to the diameter, a formula for the circumference is:

Circumference $= \pi d = 2\pi r$

An **arc** is a portion of the circumference of a circle. In the figure below, AB is an arc of the circle, with the same degree measure as central angle AOB. The shorter distance between A and B along the circle is called the **minor arc**, while the longer distance AXB is the **major arc**. An arc that is exactly half the circumference of the circle is called a **semicircle** (meaning half a circle).

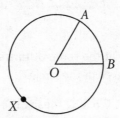

The length of an arc is the same fraction of a circle's circumference as its degree measure is of the degree measure of the circle (360°). For an arc with a central angle measuring n degrees, the following applies:

$$\text{Arc length} = \left(\frac{n}{360}\right)(\text{circumference})$$

$$= \frac{n}{360} \times 2\pi r$$

EXAMPLE

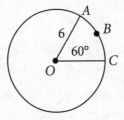

What is the length of arc ABC of the circle with center O above?

The radius is 6, so the circumference is $2\pi r = 2 \times \pi \times 6 = 12\pi$.

Since ∠AOC measures 60°, the arc is $\frac{60}{360} = \frac{1}{6}$ the circumference.

Therefore, the length of the arc is $\frac{12\pi}{6} = 2\pi$.

Area of a circle: The area of a circle is given by the formula Area $= \pi r^2$.

A **sector** is a portion of a circle bounded by two radii and an arc. In the following circle with center *O*, *OAB* is a sector. To determine the area of a sector of a circle, use the same method used to find the length of an arc. Determine what fraction of 360° is in the degree measure of the central angle of the sector, then multiply that fraction by the area of the circle. In a sector whose central angle measures *n* degrees, the following applies:

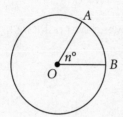

$$\text{Area of sector} = \left(\frac{n}{360}\right) \times (\text{Area of circle})$$

$$= \frac{n}{360} \times \pi r^2$$

EXAMPLE

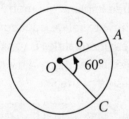

What is the area of sector *AOC* in the circle above?

A 60° "slice" is $\frac{60}{360} = \frac{1}{6}$ of the circle, so sector *AOC* has an area of

$$\frac{1}{6} \times \pi r^2 = \frac{1}{6} \times \pi \times 6^2 = \frac{1}{6} \times 36\pi = 6\pi.$$

COORDINATE GEOMETRY

As indicated by its name, **coordinate geometry** is geometry on the coordinate plane. This mostly includes equations and graphs of lines, circles, and triangles.

The important formulas to remember for lines are the following:

Slope-intercept form: $y = mx + b$, where m is the slope and b is the y-intercept of the line.

Midpoint formula: The midpoint of a line segment bounded by the points (x_1, y_1), (x_2, y_2) is $\left(\dfrac{x_1 + x_2}{2}, \dfrac{y_1 + y_2}{2}\right)$.

Distance formula: The distance between two points, (x_1, y_1) and (x_2, y_2), is

$$d = \sqrt{(x_2 - x_1)^2 + (y_2 - y_1)^2}.$$

When dealing with parallel and perpendicular lines, remember that parallel lines have the same slope and perpendicular lines have opposite reciprocal slopes. For example, if the slope of a given line is 5, the slope of a parallel line is also 5. However, the slope of a line perpendicular to these lines is $-\dfrac{1}{5}$. In more general terms, if the slope of a given line is $\dfrac{1}{a}$, the slope of a line perpendicular to it is $-a$. Also, remember that horizontal lines have a slope of zero and vertical lines have a slope that is undefined.

The same equations for circles described in plane geometry also apply to coordinate geometry. Additionally, you should know how to relate equations to graphs and how to graph inequalities. A good way to graph linear inequalities is to simply replace the inequality sign with an equal sign and graph the line. Then, depending on which inequality symbol the problem uses, shade the appropriate side of the line.

If the inequality sign is:	Line included:	Shaded Area:
$<$	No	Below line
$>$	No	Above line
$\leq$	Yes	Below line
$\geq$	Yes	Above line

Commit these relationships to memory, and you will be able to solve inequality problems quickly. Use a dashed line to represent a line that is not included in a graph ($<$ or $>$).

You should also know some basic graphs that may appear on the ACT. These include the common conic graphs, parabolas, and circles.

A **parabola** is the graph of a quadratic function, such as $y = x^2$. The graph of a parabola looks like a U:

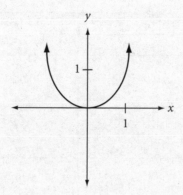

The equation of a circle is $(x - h)^2 + (y - k)^2 = r^2$, where r is the radius of the circle and (h,k) is its center.

The equation of an ellipse is $\dfrac{(x - h)^2}{a^2} + \dfrac{(y - k)^2}{b^2} = 1$, where a represents half the length of the horizontal axis and b represents half the length of the vertical axis. Just as in a cirlce, (h,k) is the center of the ellipse.

TRIGONOMETRY

Trigonometry can seem intimidating, as it is a fairly advanced math subject, but the ACT only tests the very basics of it.

The basic trigonometry functions are the sine, cosine, and tangent functions. These can be remembered using the SOHCAHTOA mnemonic device.

The **sine** function: $\sin \theta = \dfrac{\text{opposite}}{\text{hypotenuse}}$

The **cosine** function: $\cos \theta = \dfrac{\text{adjacent}}{\text{hypotenuse}}$

The **tangent** function: $\tan \theta = \dfrac{\text{opposite}}{\text{adjacent}} = \dfrac{\sin \theta}{\cos \theta}$

Values of the trig functions for certain angles (called benchmark angles) should be committed to memory, as they appear often.

Angle (°)	Sine	Cosine	Tangent
0	0	1	0
30	$\dfrac{1}{2}$	$\dfrac{\sqrt{3}}{2}$	$\dfrac{\sqrt{3}}{3}$
45	$\dfrac{\sqrt{2}}{2}$	$\dfrac{\sqrt{2}}{2}$	1
60	$\dfrac{\sqrt{3}}{2}$	$\dfrac{1}{2}$	$\sqrt{3}$
90	1	0	undefined

In addition to these three basic formulas, there are three inverse formulas: secant, cosecant, and cotangent.

The **secant** formula: $\sec \theta = \dfrac{1}{\cos \theta} = \dfrac{\text{hypotenuse}}{\text{adjacent}}$

The **cosecant** formula: $\csc \theta = \dfrac{1}{\sin \theta} = \dfrac{\text{hypotenuse}}{\text{opposite}}$

The **cotangent** formula: $\cot \theta = \dfrac{1}{\tan \theta} = \dfrac{\cos \theta}{\sin \theta}$

There are also some trigonometric identities that you should be familiar with. The most common are the **Pythagorean identities**:

$\sin^2 \theta + \cos^2 \theta = 1$

$1 + \tan^2 \theta = \sec^2 \theta$

$1 + \cot^2 \theta = \csc^2 \theta$

CHAPTER 5

Pre-Algebra Practice

1. What is the average of 230, 155, 320, 400, and 325?

   A. 205
   B. 286
   C. 288
   D. 300
   E. 430

2. Sarah has a wooden board that is 12 feet long. If she cuts three 28-inch pieces from the board, how many inches of board will she have left?

   F. 14
   G. 28
   H. 36
   J. 60
   K. 84

3. If $4x + 18 = 38$, then $x =$

   A. 3
   B. 4.5
   C. 5
   D. 12
   E. 20

4. John weighs 1.5 times as much as Ellen. If John weighs 165 pounds, how much does Ellen weigh, in pounds?

   F. 100
   G. 110
   H. 150
   J. 165
   K. 247.5

5. What is the average of 237, 482, 375, and 210?

   A. 150
   B. 185
   C. 210
   D. 326
   E. 351

6. If $\sqrt[3]{x} = 4$, then $x =$

   F. 4
   G. 12
   H. 16
   J. 36
   K. 64

7. If $x^2 + 14 = 63$, then x could $=$

   A. 4.5
   B. 7
   C. 11
   D. 14
   E. 24.5

8. Which of the following is equivalent to $\sqrt{54}$?

   F. $2\sqrt{3}$
   G. $3\sqrt{6}$
   H. 15
   J. $9\sqrt{3}$
   K. $9\sqrt{6}$

9. What whole number is closest to the value of $\sqrt{90} \times \sqrt{32}$?

   A. 7
   B. 11
   C. 36
   D. 39
   E. 54

10. $5.2^3 + 6.8^2 =$

    F. 46.24
    G. 73.28
    H. 94.872
    J. 140.608
    K. 186.848

11. If x is a real number such that $x^3 = 512$, what is the value of x^2?

    A. 8
    B. 16
    C. 48
    D. 64
    E. 135

12. $3^3 \div 9 + (6^2 - 12) \div 4 =$

    F. 3
    G. 6.75
    H. 9
    J. 11
    K. 12

13. If bananas cost $0.24 and oranges cost $0.38, what is the total cost of x bananas and y oranges?

    A. $(x + y)(\$0.24 + \$0.38)$
    B. $\$0.24x + \$0.38y$
    C. $\$0.62(x + y)$
    D. $\$0.38x + \$0.24y$
    E. $\dfrac{\$0.24}{x} + \dfrac{\$0.38}{y}$

14. If $4x + 13 = 16$, what is the value of x?

    F. 0.25
    G. 0.55
    H. 0.70
    J. 0.75
    K. 7.25

15. The chart below shows the total pizza slice sales for a pizza restaurant over a weekend. If pepperoni slices cost $2.45 and cheese slices cost $1.95, how much did the restaurant earn on Friday from pepperoni and cheese slices?

	Friday	Saturday	Sunday
Cheese	38	46	36
Sausage	43	52	47
Pepperoni	41	49	44

    A. $174.55

    B. $178.00

    C. $179.45

    D. $199.45

    E. $207.80

16. If $0.75x - 13 = 2$, what is the value of x?

    F. 11.25

    G. 14.67

    H. 15

    J. 16

    K. 20

17. In lowest terms, $\dfrac{3}{7} \times \dfrac{4}{3} =$

    A. $\dfrac{9}{28}$

    B. $\dfrac{7}{21}$

    C. $\dfrac{3}{7}$

    D. $\dfrac{4}{7}$

    E. $\dfrac{12}{21}$

18. If $2\sqrt{4n} + 7 = 15$, what is the value of n?

    F. $\sqrt{2}$

    G. 2

    H. $2\sqrt{2}$

    J. 4

    K. 8

19. If t is 5 more than s, and s is 3 less than r, what is t when $r = 3$?

    A. −5

    B. −2

    C. 1

    D. 5

    E. 8

20. A car rental agency charges $40 per day for the first 7 days, and $35 a day for each day after that. How much would Joe be charged if he rented a car for 10 days?

    F. $375

    G. $385

    H. $395

    J. $405

    K. $415

21. While away at school, Eileen receives an allowance of $400 each month, 35 percent of which she uses to pay her bills. If she budgets 30 percent of the remainder for shopping, allots $130 for entertainment, and saves the rest of the money, what percentage of her allowance is she able to save each month?

    A. 2.5%

    B. 13%

    C. 20%

    D. 35%

    E. 52%

22. If pencils cost \$0.50 each and notebooks cost \$3 each, which of the following represents the cost, in dollars, of p pencils and n notebooks?

    F. $2pn$

    G. $3.5pn$

    H. $3.5(n + p)$

    J. $3n + 0.50p$

    K. $2(n + 0.50p)$

23. If $3x + 2 = 14$, what is the value of $5x - 6$?

    A. -1

    B. 14

    C. 19

    D. 26

    E. 32

24. If $4^{4x+6} = 64^{2x}$, what is the value of x?

    F. 1

    G. 2

    H. 3

    J. 4

    K. 5

ANSWERS AND EXPLANATIONS

1.	B	13.	B
2.	J	14.	J
3.	C	15.	A
4.	G	16.	K
5.	D	17.	D
6.	K	18.	J
7.	B	19.	D
8.	G	20.	G
9.	E	21.	B
10.	K	22.	J
11.	D	23.	B
12.	H	24.	H

1. B

Plug the terms into the average formula and solve:

$$\text{Average} = \frac{\text{sum of terms}}{\text{number of terms}}$$
$$= \frac{230 + 155 + 320 + 400 + 325}{5}$$
$$= \frac{1430}{5}$$
$$= 286$$

That's (B).

2. J

Begin by converting Sarah's 12 feet of wood into inches. There are 12 inches in a foot, so Sarah has 12×12 inches $= 144$ inches. Cutting off three 28-inch pieces removes $3 \times 28 = 84$ inches, which leaves her with $144 - 84 = 60$ inches. That's (J).

3. C

To find the value of x, isolate it on one side of the equation, then solve.

$$4x + 18 = 38$$
$$4x = 20$$
$$x = 5$$

Choice (C) is correct.

4. G

Because John weighs *more* than Ellen, begin by eliminating J and K, as doing so will reduce the chance of a miscalculation error. According to the problem, John's 165 pounds represents 1.5 times Ellen's weight. Therefore, Ellen's weight must be $\frac{165}{1.5} = 110$ lb. Choice (G) is correct.

5. D

To find the average of the numbers, plug them into the average formula and solve:

$$\text{Average} = \frac{\text{sum of terms}}{\text{number of terms}}$$
$$= \frac{237 + 482 + 375 + 210}{4}$$
$$= \frac{1304}{4}$$
$$= 326$$

Choice (D) is correct.

6. K

Cube both sides to solve for x:

$$\sqrt[3]{x} = 4$$
$$x = 64$$

7. B

Isolate the variable, then solve for x:

$$x^2 + 14 = 63$$
$$x^2 = 49$$
$$x = \pm 7$$

Choice (B) is 7.

8. G

You *could* use your calculator to solve this problem, but there's a much easier way. Begin by eliminating H, as 15 is not a perfect square. Choices J and K can also be eliminated, as $9^2 = 81$, which is way too large. To simplify the radical, factor out a perfect square from 54. The largest factor of 54 that's also a perfect square is 9, so $\sqrt{54} = \sqrt{9 \times 6} = 3\sqrt{6}$, which is (G).

9. E

You *could* punch the expression into your calculator, but it may actually be quicker to estimate. $\sqrt{90} \approx \sqrt{81} = 9$ and $\sqrt{32} \approx \sqrt{36} = 6$, so $\sqrt{90} \times \sqrt{32} \approx 9 \times 6 = 54$.

With the calculator, the actual value is 53.6656. That's closest to (E).

10. K

When the choices are spaced far apart, estimation is generally the quickest way to the correct answer. To estimate, round 5.2 to 5 and 6.8 to 7. Because $5^3 + 7^2 = 125 + 49 = 174$, the correct answer will be fairly close to 174. That would be (K).

11. D

$x^3 = 512$, so $x = \sqrt[3]{512} = 8$. Be careful not to stop too soon. The problem asks for x^2, not x. $8^2 = 64$, which is (D).

12. H

To solve this problem, follow the order of operations (PEMDAS).

First, evaluate the parentheses: $3^3 \div 9 + (\mathbf{6^2 - 12}) \div 4 = 3^3 \div 9 + (\mathbf{36 - 12}) \div 4 = 3^3 \div 9 + \mathbf{24} \div 4$

Next, simplify the exponent: $\mathbf{3^3} \div 9 + 24 \div 4 = \mathbf{27} \div 9 + 24 \div 4$

Then, take care of any multiplication and/or division, from left to right: $27 \div 9 + 24 \div 4 = 3 + 6$

Finally, take care of any addition and/or subtraction, from left to right: $3 + 6 = 9$

So (H) is correct.

13. B

Each banana costs $0.24, so the price of x bananas is $0.24x$. Similarly, each orange costs $0.38, so the price of y oranges is $0.38y$. Therefore, the total price of x bananas and y oranges is $0.24x + 0.38y$. That's (B).

14. J

Isolate the variable term, then solve for x:

$$\begin{aligned} 4x + 13 &= 16 \\ 4x &= 3 \\ x &= \frac{3}{4} \end{aligned}$$

Choice (J) is correct.

15. A

Take a moment to familiarize yourself with the table before diving into the problem. The question asks for Friday's pepperoni and cheese sales, so be sure that you are reading data from the correct column. On Friday, 41 pepperoni slices were sold at $2.45 per slice and 38 cheese slices were sold at $1.95 per slice, for sales totaling ($2.45 × 41) + ($1.95 × 38) = $100.45 + $74.10 = $174.55. That's (A).

16. K

If the decimal makes this problem seem difficult, convert it into a fraction before solving for x:

$$\begin{aligned} 0.75x - 13 &= 2 \\ \frac{3x}{4} &= 15 \\ 3x &= 60 \\ x &= 20 \end{aligned}$$

That's (K).

17. D

Remember to check for opportunities to simplify fractions before multiplying or dividing with them. Doing so gives you smaller numbers to calculate, which can save you a lot of extra work. In $\frac{3}{7} \times \frac{4}{3}$, the 3 in the numerator of the first fraction and the 3 in the denominator of the second fraction can be cancelled out, reducing the problem to $\frac{1}{7} \times \frac{4}{1} = \frac{4}{7}$. That's (D).

18. J

The algebra here isn't too difficult; you just need to be careful about dealing with values outside the radical. Isolate the radical expression on one side of the equation, then square both sides to eliminate it. Backsolving, or plugging the answers into the question, would also work well.

$$3\sqrt{4n} + 7 = 15$$
$$2\sqrt{4n} = 8$$
$$\sqrt{4n} = 4$$
$$4n = 14$$
$$n = 4$$

Choice (J) is correct.

19. D

Read the stem carefully to translate the sentence into the correct equations.

$$t = s + 5$$
$$s = r - 3$$

When $r = 3$, $s = 0$, so:

$$t = 0 + 5$$
$$t = 5$$

That's (D).

20. G

When a question involves different rates, figure out how much is charged at each rate and add the totals together. The rental would include 7 days charged at $40, and 3 days charged at $35. Therefore, the first 7 days would cost 7 × $40 = $280. The next 3 days would cost 3 × $35 = $105. The total is $280 + $105 = $385, making (G) correct.

21. B

Take this carefully, step-by-step, as you translate from English to math. She spends 35% of $400 to pay bills; that is, $\frac{35}{100} \times \$400 = \$140$. That leaves her with $400 − $140 = $260. She budgets 30% of this for shopping; that is, $\frac{30}{100} \times \$260 = \$78$. That leaves $260 − $78 = $182. Of this, she allots $130 for entertainment, leaving $182 − $130 = $52 to save. Finally, $52 as a percent of the original $400 is $\frac{\$52}{\$400} \times 100\% = 13\%$, which is (B).

22. J

Early questions are worth just as many points as late questions. Take your time to translate from English to math correctly.

Total cost of pencils: $0.50p$

Total cost of notebooks: $3n$

Sum of both: $3n + 0.50p$

Choice (J) is correct.

23. B

Use the equation to solve for x, and then plug that value into the expression $5x - 6$.

$$3x + 2 = 14$$
$$3x = 12$$
$$x = 4$$
$$5x - 6 = 5(4) - 6 = 14$$

So, (B) is correct.

24. H

Don't reach for that calculator just yet. All that typing might waste valuable time. Rewrite the equation so that both sides have the same base number, 4. Once there, you can set the exponents equal to each other.

$$64 = 4^3$$
$$4^{4x + 6} = 64^{2x} = \left(4^3\right)^{2x} = 4^{6x}$$
$$4x + 6 = 6x$$
$$6 = 2x$$
$$x = 3$$

That's (H).

Alternatively, you could try plugging in the answers, because your calculator can handle these powers. For example, in F, $4^{4x + 6} = 4^{4(1) + 6} = 4^{10} = 1{,}048{,}576$, while $64^{2(1)} = 64^2 = 4{,}096$, so F is incorrect.

CHAPTER 6

Elementary Algebra Practice

1. What is 6% of 1,250?

   A. 75

   B. 150

   C. 208

   D. 300

   E. 750

2. On her first three geometry tests, Sarah scored 89, 93, and 84. If there are four tests total and Sarah needs at least a 90 average for the four tests, what is the lowest score she can receive on the final test?

   F. 86

   G. 90

   H. 92

   J. 94

   K. 95

3. The relationship between Fahrenheit and Celsius is $F = \dfrac{9}{5}C + 32$. If the temperature is 68° Fahrenheit, what is the temperature in degrees Celsius?

   A. 14°

   B. 20°

   C. 32°

   D. 64.8°

   E. 68°

4. The eighth grade girls' basketball team played a total of 13 games this season. If they scored a total of 364 points, what was their average score per game?

   F. 13

   G. 16

   H. 20

   J. 28

   K. 30

5. If $6x + 4 = 11x - 21$, what is the value of x?

    A. 2
    B. 3
    C. 4
    D. 5
    E. 6

6. A jacket with an original price of $160 is on sale for 15% off. What is the sale price?

    F. $120
    G. $136
    H. $140
    J. $144
    K. $155

7. If the average of 292, 305, 415, and x is 343, what is the value of x?

    A. 315
    B. 339
    C. 360
    D. 364
    E. 382

8. A jar contains 8 red marbles, 14 blue marbles, 11 yellow marbles, and 6 green marbles. If a marble is selected at random, what is the probability that it will be green?

    F. $\dfrac{2}{39}$

    G. $\dfrac{2}{13}$

    H. $\dfrac{3}{13}$

    J. $\dfrac{8}{39}$

    K. $\dfrac{11}{39}$

9. Each day, Laura bikes to school in the morning and bikes home in the afternoon. If she bikes at a speed of 12 miles per hour and the school is 3 miles from her house, how long does it take her to bike to school and back?

    A. 12 minutes
    B. 15 minutes
    C. 24 minutes
    D. 28 minutes
    E. 30 minutes

10. A local high school is raffling off a college scholarship to the students in its junior class. If a girl has a 0.55 chance of winning the scholarship and 154 of the juniors are girls, how many of the juniors are boys? (Assume that every junior has an equal chance to win.)

    F.   85
    G. 126
    H. 154
    J. 161
    K. 280

11. At a summer camp, students may choose between a sports elective—basketball, baseball, or soccer—and an exercise elective—yoga or Pilates. The table below shows the number of students enrolled in each elective. If each student enrolled in exactly one elective, what percentage of students enrolled in an exercise elective?

	Elective	Number of Students Enrolled
Sports	Basketball	21
	Baseball	18
	Soccer	13
Exercise	Yoga	13
	Pilates	15

   A. 28%

   B. 35%

   C. 51%

   D. 60%

   E. 65%

12. During spring break, Robert drove 240 miles to his vacation home. If he drove 60 miles per hour for the first half of the trip and 40 miles per hour for the remaining half, what was his average speed, in miles per hour, for the duration of his trip?

   F. 40

   G. 44

   H. 46

   J. 48

   K. 50

13. In a science class, the midterm is worth 30%, the final exam is worth 50%, and a class project is worth 20%. If Jason scored 86% on the midterm, 95% on the final, and 89% on the project, what was his final grade in the class, rounded to the nearest integer?

   A. 90

   B. 91

   C. 92

   D. 93

   E. 94

14. The expression $3x^2y(xy^2 + 4x^3y)$ is equivalent to which of the following?

   F. $3xy + 12x$

   G. $xy^2 + 4x^3y$

   H. $15x^8y^5$

   J. $3x^3y^3 + 12x^5y^2$

   K. $3xy^2 + 12x^3y$

15. What is the sum of the prime factors of 60?

   A. 12

   B. 15

   C. 16

   D. 19

   E. 24

16. What is the ones digit when 2^{326} is written without an exponent?

   F. 0

   G. 2

   H. 4

   J. 6

   K. 8

17. The normal price for a pair of skis is $399. If the skis are 10% off and the shop charges 8.75% sales tax, what is the total sale price of the skis?

    A. $359.10
    B. $390.52
    C. $394.01
    D. $400.50
    E. $433.91

18. If $2x + 4 = b$, then $6x + 12 =$

    F. $b + 3$
    G. $b + 12$
    H. $3b$
    J. $3b + 3$
    K. $3b + 12$

19. If $\dfrac{y}{y - 3} = \dfrac{42}{39}$, then what does y equal?

    A. 39
    B. 41
    C. 42
    D. 45
    E. 81

20. When $2x$ is subtracted from 48 and the difference is divided by $x + 3$, the result is 4. What is the value of x?

    F. 2
    G. 5
    H. 6
    J. 8
    K. 12

21. If $5 = m^x$, then $5m =$

    A. $m^{x + 1}$
    B. $m^{x + 2}$
    C. $m^{x + 5}$
    D. m^{5x}
    E. m^{2x}

22. If $3^5 = x$, which of the following expressions is equal to 3^{11}?

    F. $243x$
    G. $3x^2$
    H. $9x^4$
    J. $27x^3$
    K. x^6

23. The average (arithmetic mean) of x and y is 7, and the average of x, y, and z is 10. What is the value of z?

    A. 23
    B. 17
    C. 16
    D. 11
    E. 3

24. The average (arithmetic mean) of 7, 20, and x is 20. What is the value of x?

    F. 20
    G. 27
    H. 32
    J. 33
    K. 40

ANSWERS AND EXPLANATIONS

1.	A	13.	B
2.	J	14.	J
3.	B	15.	A
4.	J	16.	H
5.	D	17.	B
6.	G	18.	H
7.	C	19.	C
8.	G	20.	H
9.	E	21.	A
10.	G	22.	G
11.	B	23.	C
12.	J	24.	J

1. A

The quickest way to solve this problem is to estimate. While you may or may not know 6% of 1,250 off the top of your head, 10% of 1,250 is 125. Because 6% < 10%, the correct answer must be less than 125. Only (A) works.

To solve this the more traditional way, multiply 1,250 by the decimal form of 6%: 1,250 × 0.06 = 75.

2. J

When a question about averages involves variables, it often helps to think in terms of a sum instead. For Sarah's exam scores to average at least a 90, they must sum to at least 90 × 4 = 360. She already has an 89, a 93, and an 84, so she needs at least 360 − (89 + 93 + 84) = 360 − 266 = 94 points on her last test. (J) is correct.

3. B

You are given the equation to convert Fahrenheit to Celsius, so plug 68 in for F and solve for C:

$$F = \frac{9}{5}C + 32$$

$$68 = \frac{9}{5}C + 32$$

$$36 = \frac{9}{5}C$$

$$20 = C$$

This matches (B).

4. J

The basketball team scored 364 points in 13 games, so they scored an average of $\frac{364}{13} = 28$ points per game. Choice (J) is correct.

5. D

Isolate the variable term, then solve for x:

$$6x + 4 = 11x - 21$$
$$4 = 5x - 21$$
$$25 = 5x$$
$$5 = x$$

That's (D).

6. G

The original price of the jacket is $160, so a 15% sale is a discount of $160 × 0.15 = $24. Therefore, the sale price of the jacket is $160 − $24 = $136. This matches (G).

7. C

If the variable makes the average seem difficult to calculate, consider the sum instead. The average of the four numbers is 343, so their sum must be 343 × 4 = 1,372. Three of the numbers are 292, 305, and 415, so the final number is 1,372 − (292 + 305 + 415) = 1,372 − 1,012 = 360, or (C).

8. G

The probability that an event will occur is given by the formula:

$$\text{Probability} = \frac{\text{number of desired outcomes}}{\text{number of possible outcomes}}$$

In this problem, a desired outcome is getting a green marble while a possible outcome is simply getting any marble. There are 8 + 14 + 11 + 6 = 39 total marbles in the jar. Of these, 6 are green, so the probability of getting a green marble is $\frac{6}{39}$, which simplifies to $\frac{2}{13}$. That's (G).

9. E

Laura lives 3 miles from school, so biking to school and back means a total distance of $3 \times 2 = 6$ miles. Because Laura bikes at a speed of 12 miles per hour and 6 is half of 12, it must take her half an hour, or 30 minutes, to bike both legs of the journey. Choice (E) is correct.

10. G

There's a lot going on in this problem, so find the most concrete point and begin there. You are told that the chance of a girl winning the scholarship is 0.55. Because this is more than half, there must be more girls than boys in the junior class, so eliminate H, J, and K. 0.55 isn't much greater than 0.5, so if you were low on time on Test Day, (G) would be a great guess.

To solve this problem the more traditional way, use the probability formula:

$$\text{Probability} = \frac{\text{number of desired outcomes}}{\text{number of possible outcomes}}$$

Because 0.55 is the probability that a girl would win, the 154 girls are the desired outcomes, and the total number of students (boys and girls) is the number of possible outcomes. Call this total x and use this formula to solve for x:

$$0.55 = \frac{154}{x}$$
$$x = \frac{154}{0.55}$$
$$x = 280$$

With 280 total students and 154 girls, there must be $280 - 154 = 126$ boys. That's (G).

11. B

There are a total of $21 + 18 + 13 + 13 + 15 = 80$ students at the camp. Of these, 13 are taking yoga and 15 are taking Pilates, for a total of $13 + 15 = 28$ students taking an exercise elective. Therefore, $\frac{28}{80}$ of the students are taking an exercise elective, which corresponds to:

$$\frac{28}{80} \times 100\% = \frac{7}{20} \times 100\% = 35\%$$

Choice (B) is correct.

12. J

On Test Day, the correct answer to a problem about average speed will **never** be the average of the two speeds. To find Robert's average speed, use the following formula:

$$\text{Average} = \frac{\text{total distance}}{\text{total time}}$$

You are given the total distance of 240 miles, but you will need to determine the total time. Robert drove the first 120 miles at 60 miles per hour, so it took him $\frac{120}{60} = 2 \, \text{hours}$ to do so. He then drove the other 120 miles at 40 miles per hour, so that portion took him $\frac{120}{40} = 3 \, \text{hours}$. Hence, it took him a total of $2 + 3 = 5$ hours to drive all 240 miles, for an average speed of $\frac{240}{5} = 48 \, \text{miles per hour}$. Choice (J) is correct.

13. B

This problem involves a **weighted** average, so you cannot simply use the regular formula for finding an average to solve it. There are many ways to tackle this kind of problem but, because all the weights are multiples of 10%, the easiest way is to simply add an instance of each grade for every 10% of weight it carries.

The midterm is worth 30% of Jason's grade, so the 86% counts three times.

The final is worth 50% of Jason's grade, so the 95% counts five times.

The class project is worth 20% of his grade, so the 89% counts twice.

This gives Jason a sum of $86 + 86 + 86 + 95 + 95 + 95 + 95 + 95 + 89 + 89 = 911$. There are ten terms, so the average is $\frac{911}{10} = 91.1$, which is closest to (B).

14. J

To multiply exponential terms with the same base, add the exponents. Simplify the expression by distributing the $3x^2y$:

$$3x^2y(xy^2 + 4x^3y) = (3x^2y \times xy^2) + (3x^2y \times 4x^3y)$$
$$= 3x^3y^3 + 12x^5y^2$$

That looks like (J).

15. A

To find the prime factors of 60, use a prime factorization tree:

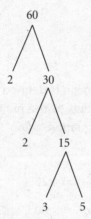

The prime factorization of 60 is $2 \times 2 \times 3 \times 5$, and the sum of these factors is $2 + 2 + 3 + 5 = 12$.

Choice (A) is correct.

16. H

You cannot evaluate 2^{326} on your calculator, but fortunately, you only need to find the ones digit. To find it, examine the powers of 2 for a pattern:

$2^1 = 2$	$2^5 = 32$
$2^2 = 4$	$2^6 = 64$
$2^3 = 8$	$2^7 = 128$
$2^4 = 16$	$2^8 = 256$

As you can see, the ones digit repeats every four terms. According to this pattern, every exponent with a multiple of 4 will have a ones digit of 6. The largest multiple of 4 that's less than 326 is 324, so 2^{324} will have a ones digit of 6. Therefore, 2^{325} will have a ones digit of 2 and 2^{326} will have a ones digit of 4. This makes (H) the correct choice.

17. B

This problem requires a couple of steps, so take them one at a time. The skis are discounted 10%, so the pre-tax price is $\$399 \times 0.9 = \$359.10$. This sale price receives an 8.75% tax, so the total sale price is $\$359.10 \times 1.0875 = \$390.52$. Choice (B) is correct.

18. H

When a problem asks for the value of an expression, look for a shortcut. You could solve the given expression for x in terms of b, then plug that into the second equation, but you'll solve the problem faster if you can find a way to relate $2x + 4$ and $6x + 12$ directly.

$$2x + 4 = b$$
$$3(2x + 4) = 3(b)$$
$$6x + 12 = 3b$$

Choice (H) is correct.

19. C

Questions near the beginning of the test are not trying to trick you or test your ability to think critically, but rather whether you can follow directions carefully. Remember this and you can rack up easy points on Test Day. If you happen to notice that the difference between the numerator and denominator is the same for both fractions, then you can tell that $y = 42$ and $y - 3 = 39$. If you didn't notice this, then you could either Backsolve or cross-multiply and solve for y. Either way, you should get (C).

20. **H**

Translate carefully from English to math.

$$\frac{48 - 2x}{x + 3} = 4$$

$$48 - 2x = 4x + 12$$

$$36 = 6x$$

$$6 = x$$

Choice (H) is correct.

21. **A**

You're given 5 and asked for $5m$—which means you want to multiply 5 by m. But remember that if you multiply the left side of the equation by m, you must do the same to the right side. You could also pick a number for m.

$$5 = m^x$$

$$5m = m(m^x)$$

$$5m = m^1 m^x$$

$$5m = m^{1 + x}$$

That's (A).

22. **G**

Don't waste time trying to write fancy expressions for 3^5 in terms of x—just convert the answer choices into powers of 3, and plug in 3^5 for each x.

F: $243x = 3^5(3^5) = 3^{10}$

(G): $3x^2 = 3(3^5)^2 = 3(3^{10}) = 3^{11}$

H: $9x^4 = 3^2(3^5)^4 = 3^2(3^{20}) = 3^{22}$

J: $27x^3 = 3^3(3^5)^3 = 3^3(3^{15}) = 3^{18}$

K: $x^6 = (3^5)^6 = 3^{30}$

Choice (G) is correct.

23. **C**

Remember that you can't always solve for all the variables in an equation. Concentrate on setting up an equation for finding each average, and then work your way toward solving for z. Write two separate equations, then plug the expression for $x + y$ from the first one into the second one:

$$\frac{x + y}{2} = 7$$

$$x + y = 14$$

$$\frac{x + y + z}{3} = 10$$

$$x + y + z = 30$$

$$(x + y) + z = 30$$

$$(14) + z = 30$$

$$z = 16$$

This matches (C).

24. **J**

The arithmetic mean is defined as the sum of a set of numbers divided by the number of terms in the set. Write an equation using this definition, and then solve for x:

$$\frac{7 + 20 + x}{3} = 20$$

$$7 + 20 + x = 60$$

$$x = 33$$

That's (J).

CHAPTER 7

Intermediate Algebra Practice

1. For all x, $(x+4)(x-4) + (2x+2)(x-2) = ?$

   A. $x^2 - 2x - 20$
   B. $3x^2 - 12$
   C. $3x^2 - 20$
   D. $3x^2 - 2x - 20$
   E. $3x^2 + 2x - 20$

2. If $s^2 - 4s - 6 = 6$, what are the possible values of s?

   F. $-2, -6$
   G. $-2, 6$
   H. $2, -6$
   J. $-2, 3$
   K. $2, -3$

3. In a high school senior class, the ratio of girls to boys is 5:3. If there are a total of 168 students in the senior class, how many girls are there?

   A. 63
   B. 84
   C. 100
   D. 105
   E. 147

4. If a car travels at 80 miles per hour for x hours and 60 miles per hour for y hours, what is the car's average speed, in miles per hour, for the total distance traveled?

   F. $\dfrac{480}{xy}$

   G. $\dfrac{80}{x} + \dfrac{60}{y}$

   H. $\dfrac{80}{x} \times \dfrac{60}{y}$

   J. $\dfrac{80x + 60y}{x + y}$

   K. $\dfrac{80 + 60}{x + y}$

5. If the first and second terms of a geometric sequence are 3 and 12, which expression gives the value of the 24th term of the sequence?

   A. $a_{24} = 3^4 \times 12$

   B. $a_{24} = 3^4 \times 23$

   C. $a_{24} = 4^3 \times 12$

   D. $a_{24} = 4^{23} \times 3$

   E. $a_{24} = 4^{24} \times 3$

6. If $3^{3x + 3} = 27^{\left(\frac{2}{3}x - \frac{1}{3}\right)}$, then $x = ?$

   F. -4

   G. $-\dfrac{7}{4}$

   H. $-\dfrac{10}{7}$

   J. -1

   K. 2

7. The complex number i is defined as $i^2 = -1$. Based on this definition, $(i + 1)^2(i - 1) = ?$

   A. $i - 1$

   B. $i - 2$

   C. $-i + 2$

   D. $-2i + 2$

   E. $-2i - 2$

8. A playground is $(x + 7)$ units long and $(x + 3)$ units wide. If a square of side length x is sectioned off from the playground to make a sandpit, which of the following could be the remaining area of the playground?

   F. $x^2 + 10x + 21$

   G. $10x + 21$

   H. $x^2 + 21$

   J. $2x + 10$

   K. 21

9. If u is an integer, then $(u - 3)^2 + 5$ must be:

   A. an even integer.

   B. an odd integer.

   C. a positive integer.

   D. a negative integer.

   E. a perfect square.

10. An international phone call costs x cents for the first 5 minutes and t cents for each minute after that. What is the cost, in cents, of a call lasting exactly v minutes where $v > 5$?

    F. $\dfrac{5x + t}{v}$

    G. $(5x + t)v$

    H. $5x + tv$

    J. $x + tv$

    K. $x + t(v - 5)$

11. Which of the following gives the solution to the inequality $|x - 3| + 6 < 15$?

    A. $-6 < x < 12$

    B. $-6 < x < 9$

    C. $-9 < x < 9$

    D. $x < -9$

    E. $x < -6$

12. In one afternoon, Brian sold 25% of his chocolate bars. If he had 72 bars left, how many chocolate bars did he have to begin with?

    F. 24

    G. 47

    H. 90

    J. 96

    K. 288

13. What is the median of the following list of numbers: 3, 8, 5, 13, 9, 15, 3?

    A. 3

    B. 8

    C. 9

    D. 13

    E. 15

14. What value of x satisfies the system of equations below?

    $$3y + 4x = 10$$
    $$2y - 4x = 0$$

    F. -2

    G. -1

    H. 0

    J. 1

    K. 2

15. A jar contains 15 red marbles, 10 green marbles, and 11 blue marbles. What is the probability that a marble chosen at random from the jar will NOT be green?

    A. $\dfrac{5}{18}$

    B. $\dfrac{13}{36}$

    C. $\dfrac{5}{12}$

    D. $\dfrac{7}{12}$

    E. $\dfrac{13}{18}$

16. As museum souvenirs, each of 50 students had the option of buying either mugs or dinosaurs or both. If 37 bought mugs, 28 bought dinosaurs, and every student bought at least one souvenir, how many students bought both mugs and dinosaurs?

    F. 11

    G. 12

    H. 13

    J. 15

    K. 17

17. If $x^2 - 11x + 24 = 0$, which of the following is a possible solution for x?

    A. -8

    B. -6

    C. -3

    D. 3

    E. 4

18. John has y dollars to spend on some new CDs from Music Plus, an online record store. He can buy any CDs at the members' price of x dollars each. To be a member, John has to first pay a one-time fee of $19. Which of the following expressions represents the number of CDs John can purchase from Music Plus?

    F. $\dfrac{xy}{19}$

    G. $\dfrac{y + 19}{x}$

    H. $\dfrac{2y - x}{19}$

    J. $\dfrac{y - 19}{x}$

    K. $\dfrac{19 - x}{y}$

19. If $\dfrac{a}{b} = 4$, $a = 8c$, and $c = 9$, what is the value of b?

    A. 2
    B. 8
    C. 18
    D. 36
    E. 72

20. A class of 30 students had an average (arithmetic mean) of 92 points on a geography test out of a possible 100. If 10 of the students had a perfect score, what was the average score for the remaining students?

    F. 58
    G. 87
    H. 88
    J. 90
    K. 92

21. If $x^{-\frac{2}{3}} = \dfrac{1}{36}$, then what does x equal?

    A. −6

    B. $\dfrac{1}{6}$

    C. 6

    D. 18

    E. 216

22. For any positive odd integer n, how many positive even integers are less than n?

    F. $\dfrac{n - 1}{2}$

    G. $\dfrac{n}{2}$

    H. $\dfrac{n + 1}{2}$

    J. $n - 1$

    K. $2n - 1$

23. If $\dfrac{p + q}{s} = 9$, $\dfrac{q}{p} = 4$, and $\sqrt{q} = 6$, what is the value of s?

    A. $\dfrac{5}{6}$

    B. 5

    C. 9

    D. 13

    E. 36

24. If $f(x) = \dfrac{x^3 - 6}{x^2 - 2x + 6}$, then what is $f(6)$?

   F.   0
   G.   3
   H.   6
   J.   7
   K.   35

25. The value of $3x + 9$ is how much more than the value of $3x - 2$?

   A.   7
   B.   11
   C. $3x + 7$
   D. $3x + 11$
   E. $6x + 7$

26. The spare change on a dresser is composed of pennies, nickels, and dimes. If the ratio of pennies to nickels is 2:3 and the ratio of pennies to dimes is 3:4, what is the ratio of nickels to dimes?

   F.   9:8
   G.   5:7
   H.   4:5
   J.   3:4
   K.   2:3

27. What are all the values of x for which $(x - 2)(x + 5) = 0$?

   A.   −5
   B.   −2
   C.   2 and −5
   D.   −2 and 5
   E.   2 and 5

28. If Jorge earns \$2,000 a month and spends \$600 a month on rent, what percent of Jorge's monthly earnings does he spend on rent?

   F.   25%
   G.   30%
   H.   35%
   J.   40%
   K.   45%

29. If a is an odd negative number and b is a positive even number, which of the following must be even and positive?

   A. $a + b$

   B. $-ab$

   C. ab

   D. $\dfrac{b}{a}$

   E. $b - a$

30. If the sum of four numbers is between 53 and 57, then the average (arithmetic mean) of the four numbers could be which of the following?

   F. $11\dfrac{1}{2}$

   G.   12

   H. $12\dfrac{1}{2}$

   J.   13

   K.   14

31. If $3a + 3(b + 1) = c$, what is $b + 1$ in terms of a and c ?

A. $\dfrac{c}{9a}$

B. $\dfrac{c}{3} - a$

C. $\dfrac{c}{3} + a$

D. $\dfrac{c}{3} - 3a$

E. $\dfrac{c}{3} + 3a$

32.

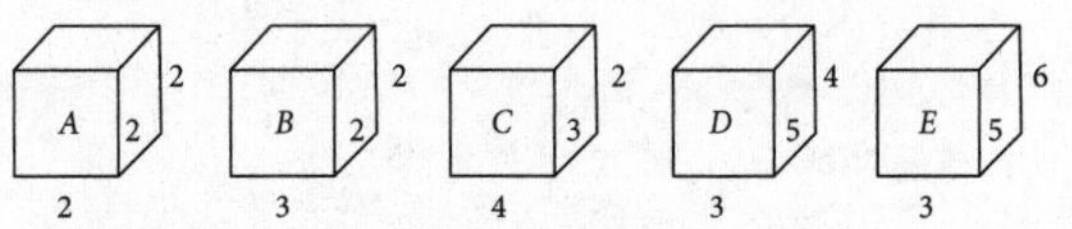

Note: Figures not drawn to scale.

Which of the rectangular solids shown above has a volume closest to the volume of a right circular cylinder with radius 4 and height 2?

F. A

G. B

H. C

J. D

K. E

33. Which of the following is divisible by 5 and 7, but is NOT divisible by 10?

A. 28

B. 50

C. 90

D. 105

E. 135

34. If 0.05 percent of n is 5, what is 5 percent of n?

F. 900

G. 700

H. 500

J. 0.007

K. 0.005

ANSWERS AND EXPLANATIONS

1.	D	18.	J
2.	G	19.	C
3.	D	20.	H
4.	J	21.	E
5.	D	22.	F
6.	F	23.	B
7.	E	24.	J
8.	G	25.	B
9.	C	26.	F
10.	K	27.	C
11.	A	28.	G
12.	J	29.	B
13.	B	30.	K
14.	J	31.	B
15.	E	32.	K
16.	J	33.	D
17.	D	34.	H

1. D

This problem is long, but it actually isn't that complicated. The order of operations says that all of the multiplication should be taken care of first. Let's begin with the first two factors:

$$(x + 4)(x - 4)$$

(If you noticed the difference of squares here, that will save you some time. If not, use FOIL.)

First: $x \times x = x^2$ Outer: $x \times (-4) = -4x$

Inner: $4 \times x = 4x$ Last: $4 \times (-4) = -16$

Combine like terms: $x^2 + (-4x) + 4x + (-16) = x^2 - 16$

Now for the other two factors:

$$(2x + 2)(x - 2)$$

First: $2x \times x = 2x^2$ Outer: $2x \times (-2) = -4x$

Inner: $2 \times x = 2x$ Last: $2 \times (-2) = -4$

Combine like terms: $2x^2 + (-4x) + 2x + (-4) = 2x^2 - 2x - 4$

Finally, add the two polynomials:

$$(x^2 - 16) + (2x^2 - 2x - 4) = 3x^2 - 2x - 20$$

That looks like (D).

2. G

To solve a quadratic equation, first set it equal to 0, and then factor if possible. Begin by subtracting 6 from both sides of the equation which results in $s^2 - 4s - 12 = 0$. To factor this, you need two factors of -12 that add up to -4. The only pair of factors that meets this criterion is -6 and 2, so the equation factors as $(s + 2)(s - 6) = 0$. For this equation to be true, either $s + 2$ or $s - 6$ (or both) must be 0. Therefore, the two possible values of s are -2 and 6. Choice (G) is correct.

3. D

The ratio of girls to boys is 5:3, so the ratio of girls to the total number of seniors is 5:(3 + 5) or 5:8. Call the number of girls in the senior class x. Set up a proportion and cross-multiply to solve for x:

$$\frac{5}{8} = \frac{x}{168}$$
$$8x = 840$$
$$x = 105$$

There are 105 girls in the senior class, which is (D).

4. J

With variables in both the question stem and the answer choices, this problem is perfect for Picking Numbers. Pick 2 for x and 3 for y. Now the problem reads: "If a car travels at 80 miles per hour for 2 hours and 60 miles per hour for 3 hours, what is the car's average speed, in miles per hour, for the total distance traveled?"

In this case, the car would have traveled $80 \times 2 = 160$ miles and $60 \times 3 = 180$ miles, for a total of $160 + 180 = 340$ miles in five hours. The average speed is therefore $\dfrac{340}{5} = 68$ miles per hour. Plug 2 in for x and 3 in for y for each of the choices and see which equals 68:

F: $\quad \dfrac{480}{2 \times 3} = \dfrac{480}{6} = 80$. Eliminate.

G: $\quad \dfrac{80}{2} + \dfrac{60}{3} = 40 + 20 = 60$. Eliminate.

H: $\quad \dfrac{80}{2} \times \dfrac{60}{3} = 40 \times 20 = 800$. Eliminate.

(J): $\quad \dfrac{80(2) + 60(3)}{2 + 3} = \dfrac{160 + 180}{5} = \dfrac{340}{5} = 68.$

$\qquad$ This works.

K: $\quad \dfrac{80 + 60}{2 + 3} = \dfrac{140}{5} = 28$. Eliminate.

Only (J) works, so it must be correct.

5. D

In a geometric sequence, use the formula: $a_n = a_1(r^{n-1})$, where a_n is the nth term in the sequence, a_1 is the first term in the sequence, and r is the amount by which each preceding term is multiplied to get the next term.

The first two terms in this sequence are 3 and 12, so r is $\dfrac{12}{3} = 4$. Now that you have r, you can plug each known value in the equation and solve for a_{24}:

$$a_{24} = 3(4^{24 - 1}) = 3 \times 4^{23}$$

That's (D).

6. F

When an equation has complicated looking exponents, try to rewrite the equation so that either the bases or the exponents themselves are the same. In this problem, the two bases seem different at first glance but, because 27 is actually 3^3, you can rewrite the equation as:

$$3^{3x + 3} = 3^{3\left(\frac{2}{3}x - \frac{1}{3}\right)}$$

This simplifies to $3^{3x + 3} = 3^{2x - 1}$. Now that the bases are equal, set the exponents equal to each other and solve for x:

$$
\begin{aligned}
3x + 3 &= 2x - 1 \\
x + 3 &= -1 \\
x &= -4
\end{aligned}
$$

That's (F).

7. E

A complex number can seem scary on the ACT, but this problem defines it for you, so treat it like you would any other variable that you plug numbers into. In this problem, the key is swapping every i^2 with a -1. Begin by simplifying the first factor in the expression:

$$
\begin{aligned}
(i + 1)^2 &= (i + 1)(i + 1) = i^2 + 2i + 1 \\
&= -1 + 2i + 1 = 2i
\end{aligned}
$$

Multiplying this by the second factor gives:

$$2i(i - 1) = 2i^2 - 2i = 2(-1) - 2i = -2 - 2i$$

That's the same as (E).

8. G

This is an area problem with a twist—you're cutting a piece out of the rectangle. To find the area of the remaining space, subtract the area of the sandpit from the area of the original playground. Recall that the area of a rectangle is length $\times$ width. The dimensions of the original playground are $x + 7$ and $x + 3$, so its area is $(x + 7)(x + 3) = x^2 + 10x + 21$. The sandpit is a square with side x, so its area is x^2. Remove the pit from the playground and the remaining area is $x^2 + 10x + 21 - x^2 = 10x + 21$. Choice (G) is correct.

9. C

When a problem tests a number property, the easiest way to solve it is usually to Pick Numbers. Because u

is an integer, pick some integers for u. If $u = 2$, then $(u - 3)^2 + 5 = (2 - 3)^2 + 5 = (-1)^2 + 5 = 1 + 5 = 6$. This eliminates B, D, and E. If $u = 3$, then $(u - 3)^2 + 5 = (3 - 3)^2 + 5 = 5$. This eliminates A, leaving (C) as the correct answer.

10. **K**

With so many variables in both the question stem and the answer choices, try Picking Numbers. Pick 10 for x, 3 for t, and 9 for v. Now the problem reads, "An international phone call costs 10 cents for the first 5 minutes and 3 cents for each minute after. What is the cost, in cents, of a call lasting exactly 9 minutes?"

The first 5 minutes costs 10 cents and there are $9 - 5 = 4$ additional minutes, each of which costs 3 cents, for a total cost of $10 + (4 \times 3) = 10 + 12 = 22$ cents. Substitute 10 for x, 3 for t, and 9 for v into each of the choices to see which equals 22:

F: $\dfrac{5(10) + 3}{9} = \dfrac{53}{9} = 5\dfrac{8}{9}$. Eliminate.

G: $(5(10) + 3)(9) = (50 + 3)(9) = (53)(9) = 477$. Eliminate.

H: $(5 \times 10) + (3 \times 9) = 50 + 27 = 77$. Eliminate.

J: $10 + (3 \times 9) = 10 + 27 = 37$. Eliminate.

(K): $10 + 3(9 - 5) = 10 + 3(4) = 10 + 12 = 22$. This works.

Only (K) works, so it must be correct.

11. **A**

The absolute value of a number is its positive distance from 0 on the number line. In other words, the absolute value of a number, much like counting numbers, can never be negative. To solve this problem, begin by isolating the absolute value term on one side:

$$|x - 3| + 6 < 15$$

$$|x - 3| < 9$$

This means that $|x - 3|$ lies somewhere between -9 and 9 on the number line, which can be written as $-9 < x - 3 < 9$. Adding 3 to all parts of the inequality yields $-6 < x < 12$. That's (A).

12. **J**

With numbers in the answer choices, this problem is an excellent candidate to Backsolve. Brian is left with 72 bars *after* selling 25% of his original stock, so he must have started with *more* than 72 bars. Eliminate F and G. Backsolve the remaining three choices. Begin in the middle with (J):

Starting with 96 bars and selling 25% leaves him with $96 - 0.25(96) = 96 - 24 = 72$. That works, so (J) is correct.

13. **B**

The median of a set of numbers with an odd number of terms is the middle number. However, this only applies to a set of numbers written in numerical order. To find the median of this set, first rewrite the numbers in numerical order: 3, 3, 5, 8, 9, 13, 15.

The middle term is 8, so the answer is (B).

14. **J**

To solve for a variable in a system of equations, you need as many *distinct* equations as you have variables. In this problem, you have two variables, x and y, and two distinct equations. To solve for x, use either combination or substitution. Using combination:

$$
\begin{array}{r}
3y + 4x = 10 \\
+(2y - 4x = 0) \\
\hline
5y = 10 \\
y = 2
\end{array}
$$

Plugging 2 in for y into the second equation yields:

$$2y - 4x = 0$$
$$2(2) - 4x = 0$$
$$4 - 4x = 0$$
$$-4x = -4$$
$$x = 1$$

To use substitution, solve for y in terms of x in one of the equations:

$$2y - 4x = 0$$
$$2y = 4x$$
$$y = 2x$$

Then plug $2x$ in for y into the other equation:

$$3y + 4x = 10$$
$$3(2x) + 4x = 10$$
$$6x + 4x = 10$$
$$10x = 10$$
$$x = 1$$

In either case, $x = 1$, which is (J).

15. **E**

The probability that an event will occur is given by the formula:

$$\text{Probability} = \frac{\text{number of desired outcomes}}{\text{number of possible outcomes}}$$

In this problem, a desired outcome is getting a marble that isn't green while a possible outcome is simply getting any marble. There are $15 + 10 + 11 = 36$ total marbles in the jar. Of these, $15 + 11 = 26$ aren't green, so the probability of getting a nongreen marble is $\frac{26}{36}$, which simplifies to $\frac{13}{18}$. That's (E).

16. **J**

In this problem, some of the students bought mugs, some bought dinosaurs, and some bought both. The problem is asking for the number of students who bought both. Thirty-seven bought mugs and 28 bought dinosaurs, which accounts for a total of $37 + 28 = 65$ students. Because there are only 50 students in total, $65 - 50 = 15$ students were counted twice. These 15 are the ones who bought both mugs and dinosaurs. That's (J).

17. **D**

If you aren't comfortable with factoring quadratics, a bit of critical thinking may be all you need to solve this problem. Because "$x^2 + 24$" must be a positive value, the only way for the equation to equal 0 is for the middle term, $-11x$, to be negative—that is, x needs to be positive. Eliminate A, B, and C. Back-solve the remaining two choices to find the answer. Try (D) first.

(D): $x^2 - 11x + 24 = (3)^2 - 11(3) + 24 = 9 - 33 + 24 = 0$.

This works, so it must be correct.

18. **J**

Do not be intimidated by all the variables. If you don't see how to turn the word problem into an equation, try picking your own values for x and y. Remember, John has to pay the $19 first before spending any leftover money on CDs, so he has $y - 19$ dollars to buy CDs with. Since each CD costs x dollars, he can buy $\frac{y - 19}{x}$ CDs, which makes (J) correct.

19. **C**

Systematically solve for the variables until you have what you're looking for. Start with c, because its value is given in the question stem. Use c to find a, then a to find b.

$$a = 8c = 8 \times 9 = 72$$
$$\frac{72}{b} = 4$$
$$72 = 4b$$
$$18 = b$$

Choice (C) is correct.

20. **H**

The average is the sum of the terms divided by the number of terms. If 10 students scored a 100, then 20 students each got a score of y (assume each gets the same score just to make things simpler), so:

$$\frac{100(10) + 20y}{30} = 92$$
$$1{,}000 + 20y = 2{,}760$$
$$20y = 1{,}760$$
$$y = 88$$

That's (H).

21. **E**

Don't feel intimidated by negative or fractional exponents. Convert numbers that are raised to negative exponents into fractions. Express any fractional exponents as the product of a unit fraction and a whole number so that it is easier to reduce the equation. Backsolving, or plugging the answers into the question stem, also works well for this type of question.

$$x^{-\frac{2}{3}} = \frac{1}{36}$$
$$\frac{1}{x^{\frac{2}{3}}} = \frac{1}{36} \quad \text{(Flip both fractions.)}$$
$$x^{\frac{2}{3}} = 36$$
$$\left(x^{\frac{1}{3}}\right)^2 = 36$$
$$x^{\frac{1}{3}} = 6$$
$$\left(x^{\frac{1}{3}}\right)^3 = 6^3$$
$$x = 216$$

Choice (E) is correct.

22. **F**

Picking Numbers works great here. Select numbers that follow all the rules in the question stem and are easy to work with. Pick an odd integer: 11. How many positive even integers are less than 11? There's 2, 4, 6, 8, and 10 for a total of 5. Plug 11 into the answer choices until you find the one that produces a value

of 5. Choice (F) is the only one that does, so it must be correct.

23. **B**

A good idea for a problem like this is to create a mental road map. Figure out what you're looking for, see what you need to know to find that, then work out how to get from the information in the question to the information you need. In this case, you need to start with q to get to s. Substitute carefully.

$$\sqrt{q} = 6, \text{ so } q = 36$$
$$\frac{q}{p} = 4, \frac{36}{p} = 4, \text{ so } p = 9$$
$$\frac{p + q}{s} = 9$$
$$\frac{9 + 36}{s} = 9$$
$$\frac{45}{s} = 9, \text{ so } s = 5$$

This matches (B).

24. **J**

Substitute $x = 6$ into the function and simplify. Be sure to follow the correct order of operations (PEMDAS).

$$f(6) = \frac{6^3 - 6}{6^2 - 2(6) + 6} = \frac{216 - 6}{36 - 12 + 6} = \frac{210}{30} = 7$$

Choice (J) is correct.

25. **B**

When you find the language confusing, try to put it in concrete terms. If you wanted to know how much more 9 was than 7, what would you do? You would subtract: $9 - 7 = 2$ more. So you need to subtract these two algebraic expressions.

$$(3x + 9) - (3x - 2) = 3x + 9 - 3x + 2 = 11$$

Choose (B).

26. F

Don't fall into the "obvious answer" trap of just taking the number of nickels and the number of dimes from the two ratios to come up with 3:4. To find the ratio of nickels to dimes, you need to get the two ratios in proportion to one another by getting the same number of pennies in each. Multiplying the first ratio by 3 gives you 6:9; multiplying the second by 2 gives you 6:8. Because the number corresponding to pennies is now the same in each ratio, the ratio of nickels to dimes can now be found. The ratio of nickels to dimes is 9:8, which is (F).

27. C

For the given equation to equal 0, either $(x - 2)$ or $(x + 5)$ equals 0.

$$x - 2 = 0$$
$$x = 2$$

or

$$x + 5 = 0$$
$$x = -5$$

Choice (C) is correct.

28. G

A percent is the ratio of $\frac{\text{part}}{\text{whole}}$ with 100 in the denominator.

$$\frac{\text{part}}{\text{whole}} = \frac{\$600}{\$2,000}$$
$$= \frac{3}{10}$$
$$= 30\%$$

This means (G) is correct.

29. B

An easy way to solve this problem is to pick numbers for a and b and then test the answer choices.

$$a = -1$$
$$b = 2$$

A: $a + b = -1 + 2 = 1$; Positive but not even

(B): $-ab = -(-1)(2) = 2$; Positive and even

C: $ab = (-1)(2) = -2$; Even but not positive

D: $\dfrac{b}{a} = \dfrac{2}{-1} = -2$; Even but not positive

E: $b - a = 2 - (-1) = 3$; Positive but not even

Choice (B) is correct.

30. K

The average is the sum of the terms divided by the number of terms. The question stem gives you a range for the sum of the terms, so divide the smallest and the greatest by the number of terms (4) to determine the range for the average.

$$\frac{53}{4} = 13\frac{1}{4}$$
$$\frac{57}{4} = 14\frac{1}{4}$$

The average is between $13\frac{1}{4}$ and $14\frac{1}{4}$. Only (K) works.

31. B

Don't solve for b. On Test Day, you will see questions that ask you to solve for expressions with variables, such as $b + 1$. This usually means there has to be an easy way to factor or rearrange the variables in the equation in order to solve for $b + 1$. Remember, when you're solving for a variable, you have to remove quantities that are added and subtracted first, then get rid of what's multiplied/divided. In other words, you need to move the $3a$ before you can divide both sides by the 3 in front of $b + 1$.

$$3(b + 1) = c - 3a$$
$$(b + 1) = \frac{c - 3a}{3} = \frac{c}{3} - \frac{3a}{3} = \frac{c}{3} - a$$

Choice (B) is correct.

32. K

The formula for the volume of a right circular cylinder is $\pi r^2 h$. Because the question asks for "closest volume," you can approximate the value of π.

First, find the volume of the cylinder.

$$V = \pi r^2 h$$

$$= \pi (4)^2 (2) = 32\pi \approx 96$$

Find the rectangular solid with the volume closest to 96.

A: $2 \times 2 \times 2 = 8$

B: $3 \times 2 \times 2 = 12$

C: $4 \times 3 \times 2 = 24$

D: $3 \times 4 \times 5 = 60$

E: $3 \times 5 \times 6 = 90$

The volume of rectangular solid E is closest to 96, which makes (K) the correct choice.

33. D

Use the process of elimination to find the solution.

A: Not divisible by 5

B: Not divisible by 7

C: Not divisible by 7

(D): Divisible by 5 and 7, but not divisible by 10. Correct.

E: Not divisible by 7

Choice (D) is the only choice that fits the criteria.

34. H

When dealing with percent problems, remember that *of* means you will use multiplication.

0.05 percent of *n* is 5

$$\frac{0.05}{100} \times n = 5$$

$$0.0005 \times n = 5$$

$$n = \frac{5}{0.0005} = 10,000$$

$$5 \text{ percent of } n = 0.05 \times 10,000 = 500$$

Choice (H) is correct.

CHAPTER 8

Coordinate Geometry Practice

1. Which of the following represents the solution to the inequality $-1 \geq -\dfrac{3}{5}x + 2$?

A.
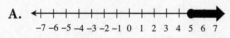
   $-7\ -6\ -5\ -4\ -3\ -2\ -1\ \ 0\ \ 1\ \ 2\ \ 3\ \ 4\ \ 5\ \ 6\ \ 7$

B.
   $-7\ -6\ -5\ -4\ -3\ -2\ -1\ \ 0\ \ 1\ \ 2\ \ 3\ \ 4\ \ 5\ \ 6\ \ 7$

C.
   $-7\ -6\ -5\ -4\ -3\ -2\ -1\ \ 0\ \ 1\ \ 2\ \ 3\ \ 4\ \ 5\ \ 6\ \ 7$

D.
   $-7\ -6\ -5\ -4\ -3\ -2\ -1\ \ 0\ \ 1\ \ 2\ \ 3\ \ 4\ \ 5\ \ 6\ \ 7$

E.
   $-7\ -6\ -5\ -4\ -3\ -2\ -1\ \ 0\ \ 1\ \ 2\ \ 3\ \ 4\ \ 5\ \ 6\ \ 7$

2. Given that $f(x) = \dfrac{1}{3}x + 13$ and $g(x) = 3x^2 + 6x + 12$, what is the value of $f(g(x))$?

F. $x^2 + 12x + 4$

G. $\dfrac{x^2}{3} + 2x + 194$

H. $x^2 + 2x + 17$

J. $x^2 + 2x + 25$

K. $x^2 + 6x + 25$

3. What is the length of side AC in triangle ABC graphed on the coordinate plane below?

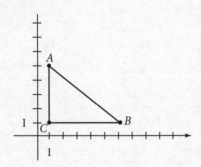

A. 3
B. 4
C. 5
D. 6
E. 7

4. If $f(x) = 3\sqrt{x^2 + 3x + 4}$, what is the value of $f(4)$?

F. 4
G. 12
H. $3\sqrt{2}$
J. $4\sqrt{2}$
K. $12\sqrt{2}$

5. What is the equation of a line that is perpendicular to the line $y = \frac{2}{3}x + 5$ and contains the point $(4, -3)$?

A. $y = \frac{2}{3}x + 4$

B. $y = -\frac{2}{3}x - \frac{1}{3}$

C. $y = -\frac{2}{3}x + 3$

D. $y = -\frac{3}{2}x + 3$

E. $y = -\frac{3}{2}x - 9$

6. Which of the following is the equation of the graph below?

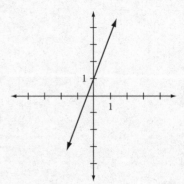

F. $y = 2x + 1$
G. $y = -2x + 1$
H. $y = -2x - 1$
J. $y = \frac{1}{2}x + 1$
K. $y = -\frac{1}{2}x + 1$

7. In the figure below, at which point does $\overline{XZ}$ intersect with its perpendicular bisector?

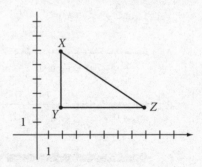

A. (4,5)
B. (2,4)
C. (5,2)
D. (5,4)
E. (3,2)

8. What is the length of a line segment with endpoints (3,−6) and (−2,6)?

    F. 1
    G. 5
    H. 10
    J. 12
    K. 13

9. What is the midpoint of the line segment in the graph below?

    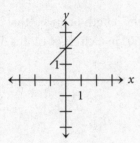

    A. (0,1)
    B. (0,2)
    C. (1,2)
    D. (1,1)
    E. (2,0)

10. What is the *x*-intercept of the line given by $3x + y = 9$?

    F. −3
    G. 1
    H. 2
    J. 3
    K. 6

11. If each tick mark on each axis of each graph represents one unit, which of the following could be a graph of the equation $y = ax^2 − 4$, given that $a > 1$?

    A.

    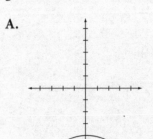

    B.

    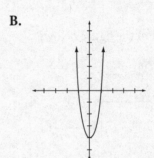

    C.

    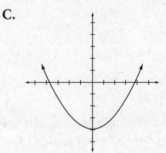

    D.

    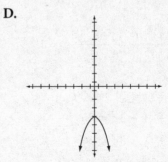

    E.

    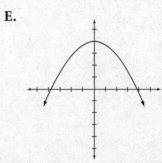

12. What is the equation of a line that has a y-intercept of -3 and is parallel to the line $3x = 4 + 5y$?

F. $y = -\dfrac{3}{5}x + 3$

G. $y = -\dfrac{5}{3}x + 3$

H. $y = -\dfrac{5}{3}x - 3$

J. $y = \dfrac{3}{5}x + 3$

K. $y = \dfrac{3}{5}x - 3$

13. What is the area of the figure below?

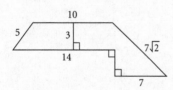

A. $39 + 7\sqrt{2}$

B. 60.5

C. 64.5

D. 91

E. 108.5

14. Which of the following represents the equation of the biggest circle that could fit inside an ellipse given by the following equation:

$$\frac{(x - 4)^2}{16} + \frac{(y + 3)^2}{25} = 1$$

F. $(x - 4)^2 + (y + 3)^2 = 400$

G. $(x - 4)^2 + (y + 3)^2 = 25$

H. $(x - 4)^2 + (y + 3)^2 = 16$

J. $x^2 + (y + 3)^2 = 25$

K. $(x - 4)^2 + y^2 = 16$

15. What is the equation of a line that passes through the origin and is perpendicular to the line $2y = 4x - 6y + 4$?

A. $y = \dfrac{1}{2}x + \dfrac{1}{2}$

B. $y = \dfrac{1}{2}x$

C. $y = 2x$

D. $y = -2x$

E. $y = -4x$

16. What is the length of an arc that has a central angle of $60°$ in a circle of radius 6?

F. 2π

G. 4π

H. 6π

J. 12π

K. 15π

17. What is the value of $f(g(3))$ if $f(x) = 2x - 4$ and $g(x) = 3x^2 - 2$?

A. 10

B. 25

C. 26

D. 40

E. 46

18. Line l has an undefined slope and contains the point $(-2,3)$. Which of the following points is also on line l?

F. $(0,3)$

G. $(5,5)$

H. $(0,0)$

J. $(3,-2)$

K. $(-2,5)$

19. If $g(t) = 2t - 6$, then at what value of t does the graph of $g(t)$ cross the x-axis?

   A. −6

   B. −3

   C. 0

   D. 2

   E. 3

20.

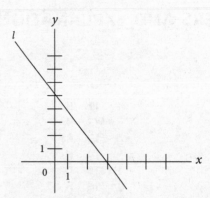

What is the equation of line l in the figure above?

   F. $y = -\dfrac{5}{3}x + 3$

   G. $y = -\dfrac{5}{3}x + 5$

   H. $y = -\dfrac{3}{5}x + 5$

   J. $y = \dfrac{3}{5}x + 3$

   K. $y = \dfrac{3}{5}x + 5$

ANSWERS AND EXPLANATIONS

1.	A	11.	B
2.	H	12.	K
3.	B	13.	B
4.	K	14.	H
5.	D	15.	D
6.	F	16.	F
7.	D	17.	E
8.	K	18.	K
9.	B	19.	E
10.	J	20.	G

1. A

Ignore the number lines initially and focus on the inequality. Inequalities work like normal equations in all aspects except one—when multiplying or dividing by a negative number, remember to flip the inequality sign. To solve this inequality, isolate x:

$$-1 \geq -\frac{3}{5}x + 2$$

$$-3 \geq -\frac{3}{5}x$$

$$5 \leq x$$

"Greater than or equal to 5" is represented by a closed circle at 5 with the arrow pointing toward the right. That's (A).

2. H

With nested functions, work from the inside out. To solve this problem, substitute the entire $g(x)$ function for x in the function $f(x)$, then simplify:

$$f(g(x)) = \frac{1}{3}\left(3x^2 + 6x + 12\right) + 13$$

$$f(g(x)) = x^2 + 2x + 4 + 13$$

$$f(g(x)) = x^2 + 2x + 17$$

Choice (H) is correct.

3. B

Normally, to find the length of a line segment on the coordinate plane, you need to use the Distance formula. However, because A (1,5) and C (1,1) have the same x-coordinate, a much faster way is to simply subtract the y-coordinate of C from the y-coordinate of A. The length of segment AC is $5 - 1 = 4$. That's (B).

4. K

To find $f(4)$, plug in 4 for x into the function $f(x)$ and simplify:

$$
\begin{aligned}
f(x) &= 3\sqrt{x^2 + 3x + 4} \\
&= 3\sqrt{4^2 + 3(4) + 4} \\
&= 3\sqrt{16 + 12 + 4} \\
&= 3\sqrt{32} \\
&= 3 \times 4\sqrt{2} \\
&= 12\sqrt{2}
\end{aligned}
$$

So the answer is (K).

5. D

Perpendicular lines have negative reciprocal slopes. Because the line in the problem has a slope of $\frac{2}{3}$, the line you are looking for must have a slope of $-\frac{3}{2}$. The problem also says that this line contains the point (4,−3). Plugging all of this information into the equation of a line, $y = mx + b$, will allow you to find the final missing piece—the y-intercept:

$$
\begin{aligned}
y &= mx + b \\
-3 &= -\frac{3}{2}(4) + b \\
-3 &= -6 + b \\
3 &= b
\end{aligned}
$$

With a slope of $-\frac{3}{2}$ and a y-intercept of 3, the line is $y = -\frac{3}{2}x + 3$. That matches (D).

6. F

The line crosses the y-axis at (0,1), so its y-intercept is 1. Eliminate H. A quick look at the four remaining choices reveals that each has a different slope, so finding the slope of the line will be enough to answer the question. You already have one point, (0,1), so pick another point, such as (1,3), and plug both into the formula for slope:

$$\text{Slope} = \frac{y_2 - y_1}{x_2 - x_1}$$
$$= \frac{3 - 1}{1 - 0}$$
$$= \frac{2}{1}$$
$$= 2$$

The slope is 2 and the y-intercept is 1, so the line is $y = 2x + 1$. That's (F).

7. D

Don't let the triangle fool you—its presence is wholly irrelevant, as you only need $\overline{XZ}$ itself to answer the question. A bisector of a line segment cuts the segment in half, so it must intersect that segment at its midpoint. Therefore, we are actually looking for the midpoint of $\overline{XZ}$. According to the figure, the coordinates of $\overline{XZ}$ are (2,6) for X and (8,2) for Z. Plug these points into the Midpoint formula and simplify:

$$\text{Midpoint} = \left(\frac{x_1 + x_2}{2}, \frac{y_1 + y_2}{2}\right)$$
$$= \left(\frac{2 + 8}{2}, \frac{6 + 2}{2}\right)$$
$$= \left(\frac{10}{2}, \frac{8}{2}\right)$$
$$= (5, 4)$$

Choice (D) is correct.

8. K

To find the distance between two points, plug the coordinates into the Distance formula and simplify:

$$\text{Distance} = \sqrt{(x_2 - x_1)^2 + (y_2 - y_1)^2}$$
$$= \sqrt{(-2 - 3)^2 + (6 - (-6))^2}$$
$$= \sqrt{(-5)^2 + 12^2}$$
$$= \sqrt{25 + 144}$$
$$= \sqrt{169}$$
$$= 13$$

Choice (K) is correct.

9. B

Plug the given points into the Midpoint formula and simplify:

$$\text{Midpoint} = \left(\frac{x_1 + x_2}{2}, \frac{y_1 + y_2}{2}\right)$$
$$= \left(\frac{1 + (-1)}{2}, \frac{3 + 1}{2}\right)$$
$$= \left(\frac{0}{2}, \frac{4}{2}\right)$$
$$= (0, 2)$$

Choice (B) is correct.

10. J

The x-intercept of a line is the point at which it crosses the x-axis—that is, the value of x when $y = 0$. To find the x-intercept of the given line, plug 0 in for y and solve for x:

$$3x + y = 9$$
$$3x + 0 = 9$$
$$3x = 9$$
$$x = 3$$

Therefore, (J) is correct.

11. B

This problem only *looks* tough. Because *a* is a positive number, the parabola should look like a right-side-up "U." Therefore, A, D, and E can be eliminated, because they all look like an *upside-down* "U." To choose between the remaining choices, set $x = 0$ to find the *y*-intercept:

$$y = ax^2 - 4$$

$$y = a(0)^2 - 4$$

$$y = 0 - 4$$

$$y = -4$$

The *y*-intercept is −4. Unfortunately, this occurs in both B and C, so compare the two graphs to see what else you need to know. The graph in B is narrow, while the graph in C is wide. Rules of transformations state that when the coefficient of the squared variable (here, *a*) is greater than 1, the graph is stretched vertically, creating a graph that is relatively narrow, so (B) is the correct choice. (Note: For C to be correct, the value of *a* would be a fraction or decimal between 0 and 1.)

12. K

Begin by converting $3x = 4 + 5y$ into $y = mx + b$ form:

$$3x = 4 + 5y$$

$$3x - 4 = 5y$$

$$\frac{3}{5}x - \frac{4}{5} = y$$

The slope of this line is $\frac{3}{5}$. Parallel lines have the same slope, so the line you are looking for has a slope of $\frac{3}{5}$ and a *y*-intercept of −3. That's (K).

13. B

To find the total area of this composite figure, you need to break it down into simpler figures. The figure has been broken down below.

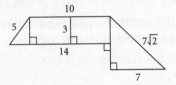

Start with the small right triangle on the left. It has a height of 3 and a hypotenuse of 5, so it must be a 3-4-5 right triangle with a base of 4. Plug these values into the formula to find its area:

$$\text{Area} = \frac{1}{2}bh$$

$$= \frac{1}{2}(4)(3)$$

$$= \frac{1}{2}(12)$$

$$= 6$$

Next, find the area of the large rectangle in the middle. The rectangle has a base of 10 and a height of 3, so plug these values into the formula to find its area:

$$\text{Area} = bh$$

$$= (10)(3)$$

$$= 30$$

Finally, find the area of the larger right triangle on the right. It has a base of 7 and a hypotenuse of $7\sqrt{2}$, so it must be a 45-45-90 right triangle with a height of 7. Plug these values into the formula to find its area:

$$\text{Area} = \frac{1}{2}bh$$

$$= \frac{1}{2}(7)(7)$$

$$= \frac{1}{2}(49)$$

$$= 24.5$$

To find the total area, add the areas of all three pieces: Area = 6 + 30 + 24.5 = 60.5. That's (B).

14. **H**

To solve this problem, you need to know both the equation for an ellipse, $\dfrac{(x - h)^2}{a^2} + \dfrac{(y - k)^2}{b^2} = 1$, and the equation for a circle, $(x - h)^2 + (y - k)^2 = r^2$. To inscribe the largest possible circle, the circle and ellipse must both be centered at the same point. This eliminates J and K. Because a circle has a constant radius, the diameter of the largest possible circle inscribed within an ellipse will be equal to the ellipse's shorter axis. In $\dfrac{(x - 4)^2}{16} + \dfrac{(y + 3)^2}{25}$, $a^2 = 16$ and $b^2 = 25$, so the horizontal ($2a$) and vertical ($2b$) axes have lengths of $2 \times 4 = 8$ and $2 \times 5 = 10$, respectively. Because $8 < 10$, the circle has a diameter of 8 and a radius of $\dfrac{8}{2} = 4$. Therefore, $r^2 = 4^2 = 16$, and the equation of the circle is $(x - 4)^2 + (y + 3)^2 = 16$. Choice (H) is correct.

15. **D**

Begin by converting $2y = 4x - 6y + 4$ into $y = mx + b$ form:

$$2y = 4x - 6y + 4$$
$$8y = 4x + 4$$
$$y = \frac{1}{2}x + \frac{1}{2}$$

The slope of this line is $\dfrac{1}{2}$, so a line perpendicular to it that passes through the origin will have a slope of –2 and a y-intercept of 0. That's (D).

16. **F**

Plug the given values into the arc length formula and simplify:

$$
\begin{aligned}
\text{Arc length} &= \frac{\theta}{360} \times 2\pi r \\
&= \frac{60}{360} \times 2\pi(6) \\
&= \frac{1}{6} \times 12\pi \\
&= 2\pi
\end{aligned}
$$

So the arc length is 2π, which is (F).

17. **E**

With nested functions, work from the inside out. Begin by evaluating $g(3)$:

$$
\begin{aligned}
g(3) &= 3(3)^2 - 2 \\
&= 3(9) - 2 \\
&= 27 - 2 \\
&= 25
\end{aligned}
$$

Now use this value to find $f(g(3))$:

$$
\begin{aligned}
f(25) &= 2(25) - 4 \\
&= 50 - 4 \\
&= 46
\end{aligned}
$$

Choice (E) is correct.

18. **K**

Remember, slope is $\dfrac{\text{rise}}{\text{run}}$ or $\dfrac{y_2 - y_1}{x_2 - x_1}$, so a slope of zero means the line is horizontal (rise = 0), and an undefined slope means the line is vertical (run = 0). This line is vertical, so the x-coordinate stays the same. The only answer choice that has the same x-coordinate as the point (–2,3) is (K).

19. **E**

The graph of a function crosses the x-axis when the value of the function equals zero. Find the value of t at which $g(t) = 0$ by solving the equation $2t - 6 = 0$. You could also use Backsolving, plugging each possible value of t into the function to see which one gives a value of 0.

$$2t - 6 = 0$$

$$2t = 6$$

$$t = 3$$

This means (E) is correct.

20. **G**

Slope-intercept form of a line is defined as $y = mx + b$, where m is the slope of the line and b is the y-intercept. Use the graph to determine these values.

Slope: $-\dfrac{5}{3}$

y-intercept: 5

Equation of the line: $y = -\dfrac{5}{3}x + 5$

Choice (G) is correct.

CHAPTER 9

Plane Geometry Practice

1. What is the area of a circle with a diameter of 8?

    A. 4π

    B. 8π

    C. 16π

    D. 32π

    E. 64π

2. A rectangle has a side length of 8 and a perimeter of 24. What is the area of the rectangle?

    F. 16

    G. 24

    H. 32

    J. 64

    K. 96

3. Isosceles triangle ABC has an area of 48. If $\overline{AB} = 12$, what is the perimeter of ABC?

    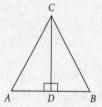

    A. 30

    B. 32

    C. 36

    D. 48

    E. 64

4. The rectangular backyard of a house measures 130 feet by 70 feet. If the backyard is completely fenced in, what is the length, in feet, of the fence?

    F. 130

    G. 140

    H. 200

    J. 260

    K. 400

5. In the figure below, lines *m* and *l* are parallel and $m\angle a = 68°$. What is the measure of $\angle f$?

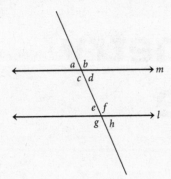

   A. 22°

   B. 68°

   C. 80°

   D. 112°

   E. 136°

6. A boy who is 4 feet tall stands in front of a tree that is 24 feet tall. If the tree casts a shadow that is 18 feet long on the ground and the two shadows end at the same point, what is the length of the boy's shadow, in feet?

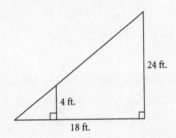

   F. 3

   G. 4

   H. 5

   J. 6

   K. 8

7. The length of the hypotenuse of right triangle *RST* is 16. If the measure of $\angle R = 30°$, what is the length of *RS*?

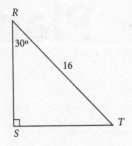

   A. 4

   B. 8

   C. $8\sqrt{3}$

   D. $\dfrac{16}{\sqrt{2}}$

   E. 12

8. Circle *O* has a radius of 5 and $m\angle AOB = 45°$. What is the length of arc *AB*?

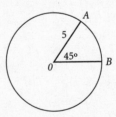

   F. 5

   G. $\dfrac{4\pi}{5}$

   H. 3

   J. $\dfrac{5\pi}{4}$

   K. $\dfrac{25\pi}{8}$

9. What is the length of the diagonal of a square with sides of length 7?

A. 7

B. $7\sqrt{2}$

C. 14

D. $14\sqrt{2}$

E. 21

10. What is the perimeter of a regular hexagon with a side of length 11?

F. 33

G. 44

H. 66

J. 72

K. 96

11. A rectangle has a perimeter of 28 and its longer side is 2.5 times the length of its shorter side. What is the length of the diagonal of the rectangle, rounded to the nearest tenth?

A. 4.0

B. 10.0

C. 10.8

D. 12.4

E. 20.4

12. In the figure below, $\overline{MN}$ and $\overline{PQ}$ are parallel. Point A lies on $\overline{MN}$ and points B and C lie on $\overline{PQ}$. If $AB = AC$ and $m\angle MAB = 55°$, what is the measure of $\angle ACB$?

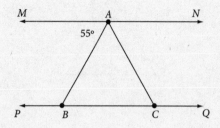

F. 35°

G. 55°

H. 65°

J. 70°

K. 80°

13. The chord in the circle below is 8 units long. If the chord is 3 units from the center of the circle, what is the area of the circle?

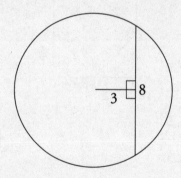

A. 9π

B. 16π

C. 18π

D. 25π

E. 36π

14. If isosceles triangle *QRS* below has a base of length 16 and sides of length 17, what is the area of the triangle?

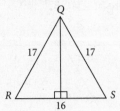

   F.   50
   G.   80
   H.   110
   J.   120
   K.   240

15. Square *QRST* is inscribed inside square *ABCD*. If *SR* = 5, what is the area of triangle *QAT*?

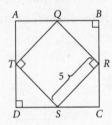

   A. 3
   B. 3.125
   C. 4.50
   D. 5
   E. 6.25

16. A circle with radius 5 is inscribed in a square. What is the difference between the area of the square and the area of the circle?

   F.   25π
   G.   50π
   H.   50 − 25π
   J.   100 − 25π
   K.   100

17. In the figure below, chord *FG* has a length of 12 and triangle *GOH* has an area of 24. What is the area of the circle centered at *O*?

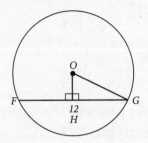

   A.   30π
   B.   36π
   C.   50π
   D.   64π
   E.   100π

18.

In the figure above, lines *m* and *n* are parallel. What is the value of *z* ?

   F.   65
   G.   115
   H.   125
   J.   145
   K.   170

19.

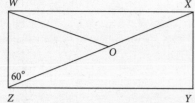

In the figure above, *WXYZ* is a rectangle. Point *O* is the midpoint of the diagonal $\overline{XZ}$. What is the measure of ∠*WOX* ?

A.   30°

B.   60°

C.   90°

D. 120°

E. 150°

20.

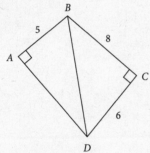

In quadrilateral *ABCD* shown, if $\overline{CD} = 6$, $\overline{BC} = 8$, and $\overline{AB} = 5$, what is the length of $\overline{AD}$?

F.   4

G. $3\sqrt{5}$

H. $5\sqrt{3}$

J.   10

K.   15

21.

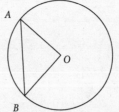

Point *O* is the center of the circle in the figure shown. If the measure of angle *AOB* = 70°, what is the measure of angle *ABO* ?

A.   40°

B.   50°

C.   55°

D.   70°

E.   110°

22.

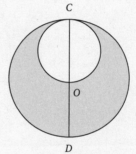

The larger circle has center *O* and diameter $\overline{CD}$. The smaller circle has diameter $\overline{OC}$. If the area of the larger circle is 25π, what is the perimeter of the shaded region?

F.   5π

G. 10π

H. 15π

J. 20π

K. 25π

23.

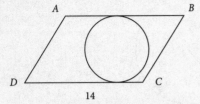

In the figure shown, the circle is tangent to sides
$\overline{AB}$ and $\overline{DC}$ of parallelogram $ABCD$, which has
an area of 168. What is the area of the circle?

A. 12π

B. 26π

C. 36π

D. 49π

E. 144π

ANSWERS AND EXPLANATIONS

1. C	**13. D**
2. H	**14. J**
3. B	**15. E**
4. K	**16. J**
5. D	**17. E**
6. F	**18. G**
7. C	**19. D**
8. J	**20. H**
9. B	**21. C**
10. H	**22. H**
11. C	**23. C**
12. G	

1. C

A circle with diameter 8 has a radius of $\frac{8}{2} = 4$. To find the area of this circle, plug this radius into the area formula and simplify:

$$\begin{aligned} \text{Area} &= \pi r^2 \\ &= \pi (4)^2 \\ &= 16\pi \end{aligned}$$

Choice (C) is correct.

2. H

The perimeter of a rectangle is twice its length plus twice its width, or perimeter = $2l + 2w$. To find the area, you must first determine the value of w, so plug in the values for perimeter and length to solve:

$$\begin{aligned} \text{Perimeter} &= 2l + 2w \\ 24 &= 2(8) + 2w \\ 24 &= 16 + 2w \\ 8 &= 2w \\ 4 &= w \end{aligned}$$

So the width is 4. The area of the rectangle is length × width, or 8 × 4 = 32. That's (H).

3. B

With only one known side, you cannot find the perimeter directly; you need to figure out more sides first. Given the area of triangle ABC and its base, the first step is to find height $\overline{CD}$:

$$\begin{aligned} \text{Area} &= \frac{1}{2}bh \\ 48 &= \frac{1}{2}(12)h \\ 48 &= 6h \\ 8 &= h \end{aligned}$$

So $\overline{CD} = 8$. Triangle ABC is an isosceles triangle, so $\overline{CD}$ also happens to be the perpendicular bisector of $\overline{AB}$, meaning $\overline{AD} = \overline{DB} = 6$. With legs of 6 and 8, each of the smaller right triangles must be 3-4-5 right triangles, making the hypotenuse of each— $\overline{AC}$ and $\overline{CB}$ —10. Therefore, the perimeter of triangle ABC is 10 + 10 + 12 = 32. Choice (B) is correct.

4. K

This question sounds more complex than it really is; it is only asking for the perimeter of a rectangle with the given dimensions, so plug them into the perimeter formula and simplify:

$$\begin{aligned} \text{Perimeter} &= 2l + 2w \\ &= 2(130) + 2(70) \\ &= 260 + 140 \\ &= 400 \end{aligned}$$

Choice (K) is correct.

5. D

When two parallel lines are cut by a transversal, half of the angles will be acute and half will be obtuse. All the acute angles will have the same measure. The same is true of all the obtuse angles. Furthermore, the acute angles will be supplementary to the obtuse angles. $\angle a$ is an acute angle measuring 68° while $\angle f$ is an obtuse angle, so $\angle a$ must be supplementary to $\angle f$. Therefore, $m\angle f = 180° - 68° = 112°$.

That's (D).

6. F

While this problem may *look* like a geometry problem at first glance, a closer look reveals that each of the three angles of one triangle is congruent to its corresponding angle in the other. The triangles are thus similar and similar triangles have proportional sides, so this is actually a proportion problem. The boy's height is proportional to the tree's height in the same way that the boy's shadow is proportional to that of the tree, so call the length of the boy's shadow x, set up a proportion, and solve for x:

$$\frac{24}{18} = \frac{4}{x}$$
$$24x = 72$$
$$x = 3$$

So the boy's shadow is 3 feet long. That's (F).

7. C

You are told that triangle *RST* is a right triangle and that one of its angles is 30°, so *RST* must be a 30-60-90 right triangle, meaning its sides must be in the proportion $x:x\sqrt{3}:2x$. Hypotenuse *RT* is 16, so x must be $\frac{16}{2} = 8$ and *RS* (the longer leg) must be $8\sqrt{3}$. That matches (C).

8. J

To find the length of an arc, you need the measure of the central angle as well as the circumference of the entire circle. In this problem, the central angle is 45° and the circumference of the circle is $2\pi(5) = 10\pi$. Plug these values into the proportion and solve for the length of the arc:

$$\frac{\text{central angle}}{360°} = \frac{\text{length of arc}}{\text{circumference}}$$
$$\frac{45°}{360°} = \frac{\text{length of arc}}{2\pi r}$$
$$\frac{1}{8} = \frac{\text{length of arc}}{10\pi}$$

$$8 \times (\text{length of arc}) = 10\pi$$
$$\text{length of arc} = \frac{10\pi}{8}$$
$$= \frac{5\pi}{4}$$

Choice (J) is correct.

9. B

A square has four right angles and four equal sides. Its diagonal cuts the square into two identical isosceles right triangles. The square in this problem has a side length of 7, so the base and height of each isosceles right triangle is also 7. The sides of an isosceles right triangle are in the proportion $x:x:x\sqrt{2}$, so the length of the diagonal (the hypotenuse of both triangles) is $7\sqrt{2}$. Choice (B) is correct.

10. H

A regular polygon is equilateral, so a regular hexagon is a hexagon with six equal sides. The regular hexagon in the problem has a side of length 11, so its perimeter is $6 \times 11 = 66$.

That's (H).

11. C

The perimeter of the rectangle is 28 and one of its sides is 2.5 times the length of the other, so call the shorter side x. The rectangle now has sides of x and $2.5x$. Draw a figure to help visualize the problem.

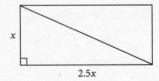

To find x, plug the dimensions into the perimeter formula and solve:

$$\text{Perimeter} = 2l + 2w$$
$$28 = 2(x) + 2(2.5x)$$
$$28 = 2x + 5x$$
$$28 = 7x$$
$$4 = x$$

So $x = 4$, and the dimensions of the rectangle must be $4 \times 1 = 4$ and $2.5 \times 4 = 10$. These values are not parts of a special right triangle, so use the Pythagorean theorem to find the length of the diagonal:

$$\begin{aligned} a^2 + b^2 &= c^2 \\ 4^2 + 10^2 &= c^2 \\ 16 + 100 &= c^2 \\ 116 &= c^2 \\ \sqrt{116} &= c \end{aligned}$$

Because 116 isn't a perfect square, but it lies between $10^2 = 100$ and $11^2 = 121$, $\sqrt{116}$ must be somewhere between 10 and 11. The only choice that fits is (C).

12. **G**

This is a pair of parallel lines cut by a transversal, but this time, there's also a triangle thrown into the mix. Begin with $\overline{AB}$. This is a transversal, so $\angle MAB$ and $\angle ABC$ are alternate interior angles and $\angle MAB = \angle ABC = 55°$. Because triangle ABC is isosceles with $AB = AC$, $\angle ACB$ is also 55°.

That's (G).

13. **D**

The chord is perpendicular to the line segment from the center of the circle, so that line segment must be its perpendicular bisector. This allows you to add more lengths to the figure:

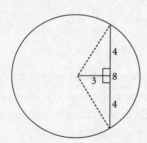

The two right triangles have legs of lengths 3 and 4, so they are both 3-4-5 right triangles with a hypotenuse of length 5. This hypotenuse is also the radius of the circle, so plug that into the area formula and simplify:

$$\begin{aligned} \text{Area} &= \pi r^2 \\ &= \pi(5)^2 \\ &= 25\pi \end{aligned}$$

The correct answer is (D).

14. **J**

Triangle QRS is an isosceles triangle, so its height is also the perpendicular bisector of $\overline{RS}$. Each half of $\overline{RS}$ is $\frac{16}{2} = 8$ units long, so each of the smaller right triangles has a leg of 8 and a hypotenuse of 17. They must therefore be 8-15-17 right triangles, making the height of QRS 15.

Therefore, the area of QRS is

$$\frac{1}{2} \times 16 \times 15 = \frac{1}{2} \times 240 = 120.$$

Choice (J) is correct.

15. **E**

When a square is inscribed within another square as they are in this problem, each of the inner square's vertices bisects one of the outer square's sides, so point Q bisects $\overline{AB}$ and point T bisects $\overline{AD}$. Therefore, $\overline{QA} = \overline{AT}$ and triangle QAT is an isosceles right triangle. Because $QRST$ is a square and $SR = 5$, QR, ST, and TQ do as well, and triangle QAT becomes an isosceles right triangle with a hypotenuse of 5. An isosceles right triangle has sides in the proportion $x:x:x\sqrt{2}$, so $\overline{QA}$ and $\overline{AT}$ each measure $\frac{5}{\sqrt{2}} = \frac{5\sqrt{2}}{2}$.

Plug these into the area formula and simplify:

$$Area = \frac{1}{2}bh$$

$$= \frac{1}{2} \times \left(\frac{5\sqrt{2}}{2}\right)^2$$

$$= \frac{1}{2} \times \frac{25 \times 2}{4}$$

$$= \frac{1}{2} \times \frac{50}{4}$$

$$= \frac{50}{8}$$

$$= 6.25$$

That matches (E).

16. J

Begin by eliminating H, as $50 - 25\pi$ is a negative value, and that cannot be correct. The problem does not provide you with a figure, so draw one to visualize what you are being asked.

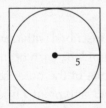

Because the circle is inscribed within the square, its radius is half the side of the square. The circle has a radius of 5, so the square has a side of $2 \times 5 = 10$. Therefore, the area of the circle is $5^2\pi = 25\pi$, and the area of the square is $10^2 = 100$. The difference between the area of the square and that of the circle is thus $100 - 25\pi$. That's (J).

17. E

You are given that the area of the triangle inside the circle is 24 and the length of the chord is 12. $\overline{OH}$ comes from the center and is perpendicular to $\overline{FG}$, so the former must be the perpendicular bisector of the latter and $\overline{FH} = \overline{HG} = 6$. Plug this into the area formula to find the length of $\overline{OH}$:

$$Area = \frac{1}{2}bh$$

$$24 = \frac{1}{2}(6)h$$

$$24 = 3h$$

$$8 = h$$

So $\overline{OH} = 8$, making triangle HOG a 3-4-5 right triangle with legs of length 6 and 8 and a hypotenuse of 10. The latter also happens to be the radius of the circle, so its area is $\pi r^2 = 10^2\pi = 100\pi$. That's (E).

18. G

Sometimes the given figure doesn't contain enough information. Feel free to add to it. Remember that the interior angles of a triangle add up to 180° and that straight angles add up to 180°.

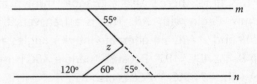

Extend one of the interior line segments to meet line *n*. This creates a triangle in which one angle is $180° - 120° = 60°$, one angle is 55° because it is an alternate interior angle along the transversal, and one angle is $180° - z°$.

$$60° + 55° + 180° - z = 180°$$
$$115° = z$$

That matches (G).

19. D

Use what you know about angles to help you find the missing information. In this case, use the facts that each angle in a rectangle measures 90° and the angles in a triangle add up to 180°.

Because *WZ* and *XY* are parallel, alternate interior angles along the transversal are equal. Angle *WZX* is 60°, so angle *ZXY* is also 60°. Because angle *WXY* is 90°, angle $WXZ = 90° - 60° = 30°$.

Because line segments *WO* and *XO* run from a corner of the rectangle to the center, they are equal in length. This means that triangle *WOX* is isosceles, and angle *WXZ* = angle *XWY* = 30°. Therefore, angle *WOX* = 180° − 30° − 30° = 120°, or (D).

20. **H**

Use the Pythagorean theorem to determine the length of the hypotenuse, which the triangles share, and use the theorem again to determine the length of side $\overline{AD}$.

First triangle (you may have recognized this as a multiple of a 3:4:5 triplet):

$$a^2 + b^2 = c^2$$
$$(8)^2 + (6)^2 = c^2$$
$$64 + 36 = c^2$$
$$100 = c^2$$
$$10 = c$$

Second triangle:

$$a^2 + b^2 = c^2$$
$$(5)^2 + b^2 = (10)^2$$
$$25 + b^2 = 100$$
$$b^2 = 75$$
$$b = \sqrt{75}$$
$$b = \sqrt{25 \times 3}$$
$$b = 5\sqrt{3}$$

Choice (H) is correct.

21. **C**

Remember key triangle rules on Test Day: The sum of the interior angles of a triangle is 180°, and triangles with two equal sides also have two equal angles. $\overline{OA}$ and $\overline{OB}$ are radii of the circle, so they are equal. This means that angles *OAB* and *ABO* are equal as well. The interior angles of the triangle total 180°, and angle *AOB* is 70°. Therefore, *ABO* + *ABO* + 70° = 180°.

$$2ABO = 110°$$

$$ABO = 55°$$

Choice (C) is correct.

22. **H**

The circumference of the shaded region is made up of the sum of the circumference of the bigger circle and the circumference of the smaller circle.

$$\text{Area} = 25\pi = \pi r^2$$
$$r = 5$$
$$\text{Circumference (larger circle)} = 2\pi r = 2\pi(5) = 10\pi$$
$$\text{Radius (smaller circle)} = \frac{1}{2}\text{ radius (larger circle)} = \frac{5}{2}$$
$$\text{Circumference (smaller circle)} = 2\pi r = 2\pi\left(\frac{5}{2}\right) = 5\pi$$

Total circumference of the shaded region = $10\pi + 5\pi = 15\pi$

That's (H).

23. **C**

Use the area of the parallelogram to find its length and then go from there.

$$\text{Area of parallelogram} = b \times h = 14 \times h = 168$$
$$h = 12$$

The height of the parallelogram is also the diameter of the circle, so $r = 6$.

$$\text{Area of circle} = \pi r^2 = 36\pi$$

Choice (C) is correct.

CHAPTER 10

Trigonometry Practice

1. In the triangle below, if cos ∠*BAC* = 0.6 and the length of the hypotenuse of the triangle is 15, what is the length of side *BC*?

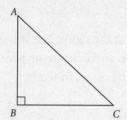

A. 3
B. 5
C. 9
D. 10
E. 12

2. What is the tangent of ∠*EFD* below?

F. $\frac{5}{13}$

G. $\frac{5}{12}$

H. $\frac{12}{13}$

J. $\frac{12}{5}$

K. $\frac{13}{5}$

3. In the triangle below, what is the value of sin ∠QRS?

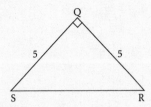

A. $\dfrac{\sqrt{2}}{6}$

B. $\dfrac{\sqrt{2}}{5}$

C. $\dfrac{\sqrt{2}}{2}$

D. $\dfrac{5}{\sqrt{2}}$

E. $5\sqrt{2}$

4. In the right triangle below, $JL = 17$ and $KL = 8$. What is the value of sin ∠JLK?

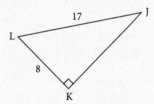

F. $\dfrac{8}{20}$

G. $\dfrac{8}{17}$

H. $\dfrac{8}{15}$

J. $\dfrac{15}{17}$

K. $\dfrac{17}{8}$

5. If sin ∠CAB = $\dfrac{3}{5}$ and $AC = 15$, what is the value of cos ∠CAB?

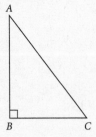

A. $\dfrac{3}{5}$

B. $\dfrac{4}{5}$

C. 1

D. $\dfrac{5}{4}$

E. $\dfrac{15}{4}$

6. The ramp in the figure below has a 20° angle of elevation and a height of 3 feet. What is the length of the ramp, in feet?

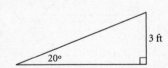

F. $\dfrac{3}{\sin 20°}$

G. $\dfrac{3}{\cos 20°}$

H. $\dfrac{\sin 20°}{3}$

J. $\dfrac{\cos 20°}{3}$

K. $3 \cos 20°$

7. If $\sin A = \dfrac{1}{2}$, which of the following could be the value of $\tan A$?

   A. $\dfrac{\sqrt{3}}{3}$

   B. $\dfrac{1}{3}$

   C. $\dfrac{2}{3}$

   D. $\dfrac{3}{1}$

   E. $\dfrac{5}{3}$

8. A building contractor determines that the angle of elevation from the ground to the top of a small office building is 67°. If the contractor is standing 50 meters from the base of the building when he measures the angle of elevation, what is the height, in meters, of the building?

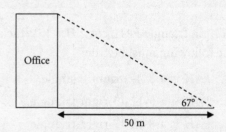

   F. 50 sin 67°
   G. 50 cos 67°
   H. 50 tan 67°
   J. 50 cot 67°
   K. 50 csc 67°

9. In the figure below, which of the following is equivalent to 2?

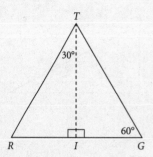

   A. sin ∠TGI
   B. cos ∠RIT
   C. tan ∠RTI
   D. sec ∠TRI
   E. cot ∠ITG

10. If $\tan A = \dfrac{8}{15}$ and $0 \le A \le 90$, which of the following could be a value of $\cos A$?

   F. $\dfrac{8}{17}$

   G. $\dfrac{8}{15}$

   H. $\dfrac{15}{17}$

   J. $\dfrac{17}{15}$

   K. $\dfrac{15}{13}$

11. In the figure below, all of the following are less than 1 EXCEPT:

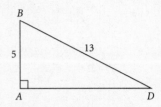

    A. sin ∠D

    B. cos ∠D

    C. tan ∠D

    D. csc ∠B

    E. cot ∠B

12. If sec ∠$A = \dfrac{13}{5}$, which of the following statements must be true?

    F. sin ∠A > cos ∠A

    G. cos ∠A > tan ∠A

    H. tan ∠A < cot ∠A

    J. sec ∠A < csc ∠A

    K. cot ∠A > csc ∠A

13. If $0° < \theta < 90°$ and cot $\theta = \dfrac{4}{5}$, then sec $\theta = ?$

    A. $\dfrac{3}{5}$

    B. $\dfrac{5}{4}$

    C. $\dfrac{\sqrt{41}}{5}$

    D. $\dfrac{\sqrt{41}}{4}$

    E. $\dfrac{5}{3}$

14. George knows that the measure of ∠F in the figure below is 65°. Which of the following additional pieces of information would allow him to determine the length of the hypotenuse?

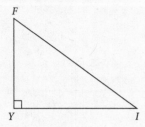

    F. The length of $\overline{YI}$

    G. The value of $\dfrac{\overline{FY}}{\overline{FI}}$

    H. The value of $\dfrac{\overline{YI}}{\overline{FI}}$

    J. The area of triangle FYI

    K. The perimeter of triangle FYI

15. In right triangle FAH, cot ∠$H = 1$. Which of the following must be true?

    A. FAH is a 3-4-5 right triangle.

    B. FAH is a 5-12-13 right triangle.

    C. FAH is a 8-15-17 right triangle.

    D. FAH is a 45-45-90 right triangle.

    E. FAH is a 30-60-90 right triangle.

16. If cos $G = 0.5$, which of the following would also equal 0.5?

   F. $1 - \sin G$

   G. $1 - \tan G$

   H. $1 - \dfrac{1}{\sin G}$

   J. $1 - \dfrac{1}{\sec G}$

   K. $1 - \dfrac{1}{\csc G}$

17. What is the measure of the angle $\dfrac{3\pi}{4}$ expressed in degrees?

   A. $45°$

   B. $60°$

   C. $120°$

   D. $135°$

   E. $270°$

ANSWERS AND EXPLANATIONS

1.	E	10.	H
2.	J	11.	D
3.	C	12.	F
4.	J	13.	D
5.	B	14.	F
6.	F	15.	D
7.	A	16.	J
8.	H	17.	D
9.	D		

1. E

You are given the cosine of $\angle BAC$ and the hypotenuse of the triangle, so begin by using these to find the adjacent side:

$$\cos A = \frac{\text{adjacent}}{\text{hypotenuse}}$$

$$0.6 = \frac{\text{adjacent}}{15}$$

$$\text{adjacent} = 9$$

So the adjacent side, $\overline{AB}$, is 9 and triangle ABC is a right triangle with a leg of length 9 and a hypotenuse of length 15. ABC must therefore be a 3-4-5 right triangle, and $\overline{BC}$ must be 12. That's (E).

2. J

Recall SOHCAHTOA: The tangent of an angle is defined by $\tan A = \frac{\text{opposite}}{\text{adjacent}}$. Because the hypotenuse has length 13 and one of the legs has length 5, this must be a 5-12-13 right triangle. Therefore, the side opposite $\angle EFD$ is 12. The side adjacent to $\angle EFD$ is 5, so $\tan \angle EFD = \frac{12}{5}$. This is (J).

3. C

Because $QS = QR$, triangle QRS must be a 45-45-90 right triangle, and the hypotenuse is $5\sqrt{2}$. Therefore, $\sin \angle QRS = \frac{5}{5\sqrt{2}} = \frac{1}{\sqrt{2}} = \frac{\sqrt{2}}{2}$. That's (C).

4. J

A right triangle with leg 8 and hypotenuse 17 must be an 8-15-17 right triangle, so $JK = 15$. Because JK is opposite $\angle JLK$, $\sin \angle JLK = \frac{15}{17}$. This matches (J).

5. B

You are given the sin of $\angle CAB$ and the hypotenuse of the triangle, so begin by using this information to determine the remaining two sides:

$$\sin \angle CAB = \frac{\text{opposite}}{\text{hypotenuse}}$$

$$\frac{3}{5} = \frac{\text{opposite}}{15}$$

$$5 \times \text{opposite} = 45$$

$$\text{opposite} = 9$$

So the opposite side, $\overline{BC}$, is 9, and triangle ABC is a right triangle with a leg of length 9 and a hypotenuse of length 15. ABC must therefore be a 3-4-5 right triangle, and $\overline{AB}$ must be 12. Now use this information to find the cosine of $\angle CAB$:

$$\cos \angle CAB = \frac{\text{adjacent}}{\text{hypotenuse}}$$

$$= \frac{12}{15}$$

$$= \frac{4}{5}$$

So (B) is correct.

6. **F**

The ramp is the hypotenuse of the right triangle in the figure. Because you only have the side opposite the given angle, use sin 20° to find the hypotenuse:

$$\sin 20° = \frac{\text{opposite}}{\text{hypotenuse}}$$

$$\sin 20° = \frac{3}{\text{hypotenuse}}$$

$$\text{hypotenuse} = \frac{3}{\sin 20°}$$

That's (F).

7. **A**

Whenever you see trig functions, think of right angles. Drawing a figure can go a long way toward making this problem more concrete. Because $\sin A = \frac{1}{2}$, draw a triangle that looks like the following:

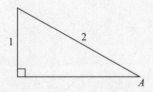

This is a 30-60-90 triangle, and its dimensions must be in the proportion $x : x\sqrt{3} : 2x$, so the length of the longer leg is $\sqrt{3}$. Therefore, $\tan A = \frac{1}{\sqrt{3}} = \frac{\sqrt{3}}{3}$. Choice (A) is the correct answer.

8. **H**

To find the height of the building, you'll need something that establishes a relationship between the known angle (67°), the adjacent side (50 meters), and the opposite side (the height of the building). When dealing with opposite and adjacent, tangent should come to mind:

$$\tan 67° = \frac{\text{opposite}}{\text{adjacent}}$$

$$\tan 67° = \frac{\text{height}}{50}$$

$$50 \tan 67° = \text{height}$$

That's (H).

9. **D**

The two smaller triangles are 30-60-90 triangles whose sides are in the proportion $x : x\sqrt{3} : 2x$, so pick 1 for x and label the relative lengths of the figure:

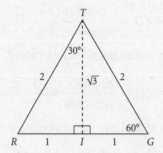

To get a value of 2, you'll need an expression that places the hypotenuse of either of the smaller right triangles over its shorter leg. Of the choices, only (D) does, so it is correct.

10. **H**

Given that $\tan A = \frac{8}{15}$ and A is in the first quadrant, $\cos A$ will also be positive. If you think of this problem as a right triangle with 8 and 15 as the lengths of the opposite and adjacent legs, respectively, you would have an 8-15-17 right triangle with 17 as the length of the hypotenuse. Therefore, $\cos A$ could be $\frac{15}{17}$, which happens to be (H).

11. **D**

A right triangle with a leg of length 5 and a hypotenuse of length 13 is a 5-12-13 right triangle, so the longer leg has length 12. Add this information to the figure:

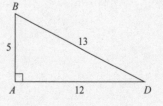

Now, evaluate each choice to find the one that is not less than 1:

A: $\sin \angle D = \dfrac{\text{opposite}}{\text{hypotenuse}} = \dfrac{5}{13}$.

This is less than 1, so eliminate.

B: $\cos \angle D = \dfrac{\text{adjacent}}{\text{hypotenuse}} = \dfrac{12}{13}$.

This is less than 1, so eliminate.

C: $\tan \angle D = \dfrac{\text{opposite}}{\text{adjacent}} = \dfrac{5}{12}$.

This is less than 1, so eliminate.

(D): $\csc \angle B = \dfrac{\text{hypotenuse}}{\text{opposite}} = \dfrac{13}{12}$.

This is greater than 1.

E: $\cot \angle B = \dfrac{\text{adjacent}}{\text{opposite}} = \dfrac{5}{12}$.

This is less than 1, so eliminate.

Only (D) satisfies the criterion, so it must be correct.

12. **F**

When a figure isn't provided, draw your own. Sec $\angle A = \dfrac{13}{5}$, so your triangle must be a 5-12-13 triangle like so:

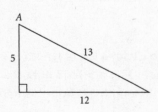

Now, evaluate each choice:

(F): $\dfrac{12}{13} > \dfrac{5}{13}$. This works and on Test Day, you would pick it and move on. For the curious:

G: $\dfrac{5}{13} \not> \dfrac{12}{5}$ Eliminate.

H: $\dfrac{12}{5} \not< \dfrac{5}{12}$ Eliminate.

J: $\dfrac{13}{5} \not< \dfrac{13}{12}$ Eliminate.

K: $\dfrac{5}{12} \not> \dfrac{13}{12}$ Eliminate.

13. **D**

This problem can easily be misread, so be sure to read the question stem carefully. We are looking for a triangle where $\cot \theta = \dfrac{4}{5}$. (Had it said **cos θ**, this would be a 3-4-5 right triangle.) Begin by drawing a figure:

Secant involves the hypotenuse, so use the Pythagorean theorem to find it:

$$\begin{aligned} a^2 + b^2 &= c^2 \\ 4^2 + 5^2 &= c^2 \\ 16 + 25 &= c^2 \\ 41 &= c^2 \\ \sqrt{41} &= c \end{aligned}$$

So the hypotenuse is $\sqrt{41}$. Therefore,

$$\sec \theta = \dfrac{\text{hypotenuse}}{\text{adjacent}} = \dfrac{\sqrt{41}}{4}.$$

That's (D).

14. **F**

George knows one of the angles of the triangle and wants to find the hypotenuse. To use trigonometry, all he would need is the length of one of the legs. Choice (F) provides such a length and is the correct answer. Choices G and H each provide the *ratio* of a leg to the hypotenuse, but a ratio alone cannot give you a specific length. Choices J and K provide information concerning the triangle as a whole but not for one of its sides.

15. **D**

The cotangent of an angle is given by $\dfrac{\text{adjacent}}{\text{opposite}}$. In a right triangle, this gives the ratio of the lengths of its two legs. If cot $\angle H = 1$, the length of the adjacent side must equal the length of the opposite side. That only happens in a 45-45-90 right triangle, which is (D).

16. **J**

While you may not know the values of most of the choices offhand, you don't need to. Every choice subtracts some value from 1, so the correct answer must subtract an expression that is equal to 0.5. Because secant is the reciprocal of cosine, $\cos G = \dfrac{1}{\sec G} = 0.5$, so (J) is correct.

17. **D**

To convert $\dfrac{3\pi}{4}$ into degrees, substitute 180 for π and simplify:

$$\frac{3\pi}{4} = \frac{3 \times 180}{4}$$
$$= \frac{540}{4}$$
$$= 135°$$

Choice (D) is correct.

SECTION THREE

ACT Science

CHAPTER 11

Introduction to ACT Science

The last multiple-choice section of the ACT is Science. The structure of the test is always the same, so you can be certain you'll see six passages, each with six to eight questions, for a total of 40 questions. You'll have 35 minutes to complete the section. That may not sound like a lot of time, but don't panic. This guide will show you how to move quickly and find the correct answers, and you will have plenty of passages on which to practice your new skills.

The passages are drawn from four basic areas of science—biology, chemistry, Earth and space sciences, and physics. You'll have one or two passages on each of these topics. Each passage will display data (Data Representation), describe the results of a set of experiments (Research Summaries), or present different views on a single topic (Conflicting Viewpoints). Most passages will include diagrams, charts, or tables with important information.

The purpose of the Science section is to test your scientific reasoning skills, not your ability to recite specific things you've learned (such as the life cycle of a cell). As long as you have a basic understanding of scientific terms and concepts, you won't have to rely on any outside knowledge to answer the questions. What does this mean specifically? It means *the answer is in the passage*. In fact, not only is the answer in the passage, it's printed right below the question. Your only job is to distinguish the correct answer choice from the three incorrect choices it's mixed in with. We've come up with a method to make that easy (we're Kaplan, it's what we do).

THE KAPLAN METHOD FOR ACT SCIENCE

1. Read the passage, taking notes as you go.

Read the introductory paragraph to get an idea of what you're dealing with, and quickly read the paragraphs explaining experiment setup. Keep note-taking to a minimum—underline or circle key words (they're often already italicized, which makes your job even easier), and underline or circle specific values (this makes them easier to spot if you need them later). Don't spend more than a minute here.

2. Skim the figures.

Ask yourself two questions: "What does this figure show?" and "What are the units of measurement?" Keep your answers general. Figures usually show how one variable changes in relation to another variable. Your job at this point is to simply identify what these variables are and what units they're measured in.

3. Attack the questions.

Ask yourself, "Where in the passage can I find the answer to this question?" Quite often, the question stem will point you to just the paragraph or figure you need. Begin with the questions that look easy to you. If you have no idea where to find the answer to a question, save it for last. You'll develop an increasingly clearer understanding of the passage as you answer the questions.

Now, let's take a closer look at the three types of passages you'll see on the ACT Science test: **Data Representation, Research Summaries**, and **Conflicting Viewpoints**.

DATA REPRESENTATION

These passages do just what they say; they show you data. They will not detail the procedures of multiple experiments nor provide summaries of competing theories. All that's provided is some information on the page for you to see, mostly in the form of graphs and tables. The questions that correspond to these passages are designed to test your ability to find information in these graphs and tables.

Suppose you saw the following graph in a science passage:

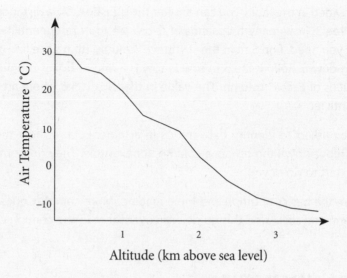

Air Temperature at Various Altitudes at Test Site #3

As you're using the Kaplan Method for ACT Science, you'll stop when you reach this figure. You'll say to yourself, "What does this figure show? What are the units of measurement?" Read the title of the graph: The graph shows air temperature at various altitudes at Test Site #3. Read the axis labels: The temperature is measured in °C (degrees Celsius) and the altitude is measured in kilometers above sea level. Using the information provided in the figure, you can answer the question, "What is the temperature at 2 km above sea level?" To locate the data you need, find 2 km on the x-axis (that's the axis on the bottom). Draw a line straight up until it intersects with the curved line provided in the graph. Then draw a horizontal line from this point to the y-axis (that's the up-and-down axis). The graphs are often pretty small, and eyeballing won't always give you an accurate enough estimate, so using your pencil is important. Here, you can see that the temperature at 2 km is about 5°.

Data will also appear in the form of tables. Suppose you came across this in a passage:

Concentration of *E. coli* in Cooling Pool B	
Distance from Effluent Pipe 3	1,000s of *E. coli* per Centiliter
0 m	0.4
5 m	5.6
10 m	27.6
15 m	14.0
20 m	7.5

Again, you'll identify what the figure (the table) shows and the units of measurement. This table shows the concentration of *E. coli* in Cooling Pool B based on the distance (in meters) from Effluent Pipe 3.

ACT questions based on tables often ask you to identify specific data points. Using the information provided in the table, you can answer the question, "At a distance of five meters from Effluent Pipe 3, how many thousands of *E. coli* are found per centiliter of water?" To locate the data you need, find 5 m in the "Distance" column (that's the line of boxes on the far left, running down). Follow the 5 m row (a row is a line of boxes running horizontally) over to the "1,000s of *E. coli*" column. The value in that box is 5.6. There are 5.6 thousands of *E. coli* per centiliter.

You may also be asked to identify data trends in a table. Look at how the *E. coli* values change as you move down the distance column, for example. They go from 0.4 up to 27.6, and then they start to go down.

These skills form the basics of graph and table reading. More complex questions will build on these fundamental skills, and this guide will cover them in the coming chapters.

RESEARCH SUMMARIES

Research Summary passages contain a short opening description followed by short descriptions of two, three, or four experiments. Similar to Data Representation passages, Research Summaries passages contain graphs and tables. However, they also test your understanding of the purpose, the method, and the results of experiments.

The **purpose** of an experiment is the *why*—the general principle or *hypothesis* (question) being studied. This may be how bacterial growth is affected by the presence of different nutrients, or how friction affects the amount of time it takes a rolling ball to come to a stop. Generally, the opening paragraph of a Research Summaries passage will contain the purpose. You should underline this as you first read the passage.

The **method** of an experiment is the *how*—how the researchers are finding the answer to their question. Usually, this involves changing one variable while the others are held constant. This allows researchers to isolate and study the effect of one variable at a time.

The **results** of the experiment are the *what*—the data that answers the researchers' questions. This is usually presented in tables or graphs. You'll use the same skills to interpret the tables and graphs of a Research Summaries passage that you use in understanding Data Representation passages.

The questions following Research Summaries passages test both your ability to read and interpret figures (they'll be very similar to the figures questions following a Data Representation passage) and your understanding of the scientific method.

These scientific method questions may ask you, for example, to identify which factor in an experiment was varied, to determine how changing the method might affect results, or to explain why researchers included a particular step in their method. They may also ask you to identify the hypothesis that prompted the experiment, to determine whether the results support a particular hypothesis, or to explain how new information might affect the researchers' hypothesis.

Many of the questions following a Research Summaries passage tell you exactly where to find the answer (they'll begin, "According to Experiment 1..."). Take advantage of this and answer these questions first!

In fact, the best way to approach a Research Summaries passage is to read the opening paragraph (or paragraphs) and Experiment 1 first. Then scan through the questions to find those that refer to *only* Experiment 1, and answer those. Next, do the same for Experiment 2 (and Experiments 3 and 4 if there are that many). Finally, answer the questions that don't reference a specific experiment. Research Summaries passages are a lot less confusing when you approach them systematically.

CONFLICTING VIEWPOINTS

The last type of passage on the ACT Science test is called Conflicting Viewpoints. Each Conflicting Viewpoints passage is structured in generally the same way. There is an opening paragraph (or two) that explains some background information about a particular phenomenon. Generally, the opening paragraph is a statement of fact. It often supplies you with a few definitions of key terms.

Next, a scientist will offer one opinion on or possible explanation for the phenomenon. A second scientist will then offer a different opinion on or explanation for the phenomenon. Sometimes, a third or fourth scientist weighs in.

Each scientist will offer different evidence to support a specific viewpoint. Frequently, these viewpoints will contradict each other. That's okay. Your job isn't to decide who is right and who is wrong; it's to recognize each viewpoint and understand how the scientists use evidence to support their viewpoints.

Conflicting Viewpoints passages are different from Research Summaries or Data Representation passages; they are so different, in fact, that they have their own Kaplan Method.

THE KAPLAN METHOD FOR CONFLICTING VIEWPOINTS PASSAGES

1. **Read the introductory text and the first author's viewpoint**, then answer the questions that ask only about the first author's viewpoint.

2. **Read the second author's viewpoint**, then answer questions that ask only about the second author's viewpoint.

3. **Answer the questions that refer to both authors' viewpoints.**[*]
   *When there are three or more author viewpoints, you'll read each viewpoint and answer the related questions before you answer the questions that relate to multiple viewpoints.

As you read the Conflicting Viewpoints passage, try to identify each scientist's hypothesis, and underline it when you find it. Typically, it's right there in the first sentence. The rest of the paragraph will be devoted to providing evidence and supporting details.

Conflicting Viewpoints questions test your ability to follow a scientist's line of reasoning. They may ask you why a scientist included a particular detail in his argument or to apply a scientist's line of reasoning to a different situation. They may ask you to predict, based on what's written, what a scientist might think about a related topic.

Questions may also introduce new information and ask whether the new information weakens or strengthens one or both of the arguments. They may ask you to identify points of agreement between the two scientists (hint: look for answer choices drawn from the opening text). Conversely, a question may ask you to identify a point of disagreement.

Finally, some questions will test your reading comprehension skills. (They'll look something like, "According to the passage, polypeptide molecules are…" or "The passage indicates that…") Answer these questions just as you would a reading question: by going back to the passage to find your answer.

The most important thing to remember as you're taking the ACT Science test is that answering questions, not reading the passage, is how you earn points. Don't get caught up in the details of a complex passage on your first read. Spend your time on the questions. That's where the payoff is.

GUESSING

Do it! There's no wrong answer penalty on the ACT. If you don't know the answer, go ahead and wager your best guess. If you can eliminate any wrong answers, you'll be in even better shape.

CHAPTER 12

Data Representation I

While the ACT doesn't quiz you on science terms or definitions, it does assume that you have completed at least two years of a standard three- or four-year science course of study on the college prep level. You are expected to understand basic scientific terms and methods, and you will be asked to select data from simple and complex presentations, find information in text, and determine how one variable changes in response to another.

BASIC SCIENCE TERMINOLOGY

The ACT primarily uses the International System of Units (SI), which includes meters (m), kilograms (kg), seconds (s), amperes (A), Kelvins (K), moles (mol), as well as liters (L). In addition to these units, you may see feet (ft), pounds (lb), degrees Celsius (°C), and degrees Fahrenheit (°F). You should be comfortable using derivatives of these units as well—milliliters (mL) and grams (g), for example.

You should know what it means to create a *solution* (a mixture in which a *solute* is dissolved into a *solvent*) and to *dilute* (thin or weaken). You should understand terms such as *density* (mass per unit volume) and *force* (that which causes a mass to accelerate).

This is by no means an exhaustive list of terms you may encounter, but it represents the general level of terminology that you will be expected to understand. More specific terms and pieces of equipment (such as a *bomb calorimeter*) will be explained within the text.

FINDING BASIC INFORMATION IN THE PASSAGE

If there's one skill that will earn you more points on the ACT Science test than any other, it's reading data provided in figures and tables. Your pencil is the key. If a question asks you to find a point on a graph, for example, draw a line up from the *x*-axis and over to the *y*-axis.

You can even use a second pencil to keep your lines straight. Often, the graphs are small, and estimating just isn't accurate enough.

Circle a piece of data in a table as a visual aid. The wrong answer choices usually include the values just above or below the correct one, and it's very easy for your eyes to accidentally skip up or down a row in a table.

When you're asked to find two or more pieces of data from within a figure (for example, when a question requires you to order values from least to greatest), use your pencil again. Write your answer in your test booklet, then match it to the correct answer choice.

DETERMINING THE RELATIONSHIP BETWEEN TWO VARIABLES

If locating data within a passage is the number one skill on the ACT Science test, then determining the relationship between two variables is number two.

Determining the relationship between two variables means observing how one variable changes in response to another. A scientist studying the relationship between the number of cars on the highway and the number of accidents, for example, will record the number of cars she counts in one hour along with the number of accidents. Then she'll list her results in a table, with the number of cars in one column and the number of accidents in another. Or perhaps she'll graph them, with the number of cars on the *x*-axis and the number of accidents on the *y*-axis. Suppose these are her results:

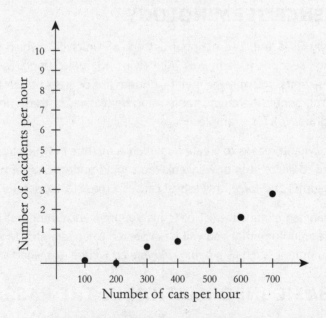

Your job is to determine how the number of accidents changes in relation to how the number of cars changes. In this example, the graph shows that as the number of cars increases, the number of accidents increases. That's all there is to it!

PRACTICE QUESTIONS

PASSAGE I

As the prevalence of wild wetland decreases with human population growth and urban sprawl, efforts have begun to create new wildlife habitats. In particular, scientists have studied the feasibility of conserving amphibian species by creating artificial ponds. Several studies have documented the natural process of *colonization* of artificial ponds with microorganisms (spirogyra and protozoans), aquatic plant life (*Nymphaea odorata* and *Eleocharis acicularis*), and both amphibian larvae and grown amphibians (*Hyla regilla*, *Rama sylvatica*, and *Ambystoma macrodactylum*) over time (see Figure 1).

(Note: Species are organized according to their appearance over time.)

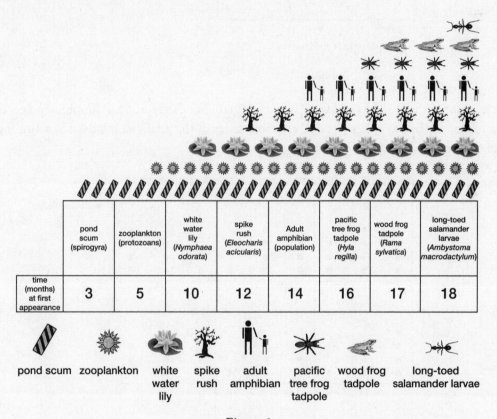

	pond scum (spirogyra)	zooplankton (protozoans)	white water lily (*Nymphaea odorata*)	spike rush (*Eleocharis acicularis*)	Adult amphibian (population)	pacific tree frog tadpole (*Hyla regilla*)	wood frog tadpole (*Rama sylvatica*)	long-toed salamander larvae (*Ambystoma macrodactylum*)
time (months) at first appearance	3	5	10	12	14	16	17	18

pond scum   zooplankton   white water lily   spike rush   adult amphibian   pacific tree frog tadpole   wood frog tadpole   long-toed salamander larvae

Figure 1

As colonization progresses, amphibian larvae gradually displace vegetative cover. The larvae populations were estimated by trapping and counting the number of larvae in 1 m³ of pond water at the edge of the pond and multiplying this number by the circumference of the pond in meters. Percent vegetative cover was estimated using aerial photography to determine the fraction of pond surface area obscured by aquatic plant life. Figure 2 shows the change over time of percent vegetative cover with the presence of tadpoles and larvae in one pond.

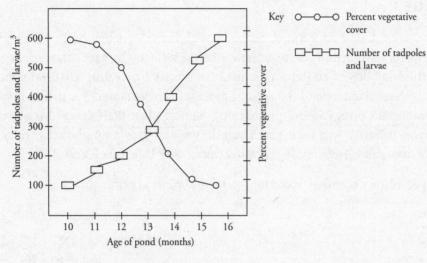

Figure 2

Scientists also wished to determine how the final abundance of adult amphibian species correlated with several key physical characteristics of the artificial ponds. Their findings are summarized in Figure 3 below.

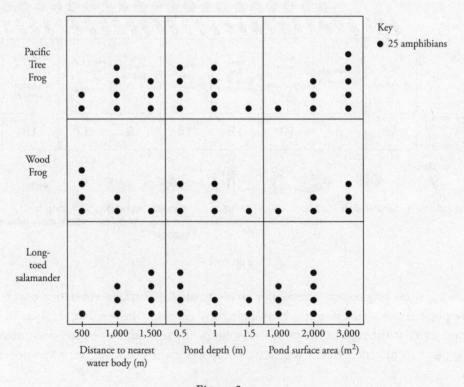

Figure 3

1. According to Figure 1, how would the colonization of an artificial pond best be characterized at 14 months?

    A. The pond would contain a mixture of algae, aquatic plants, and amphibian larvae.

    B. The pond would contain a mixture of aquatic plants and adult amphibians.

    C. The pond would contain a mixture of algae, aquatic plants, and adult amphibians.

    D. The pond would contain a mixture of algae, amphibian larvae, and adult amphibians.

2. Given the information in Figure 3, the greatest numbers of long-toed salamanders can be found in artificial ponds that:

    F. are less than 1,000 m from the nearest water body, are at least 1 m deep, and have a surface area of 2,000 m^2 or greater.

    G. are 1,500 m from the nearest water body, are at least 1 m deep, and have a surface area of 1,000 m^2.

    H. are 500 m from the nearest water body, are 1.5 m deep, and have a surface area greater than 2,000 m^2.

    J. are 1,500 m from the nearest water body, are 0.5 m deep, and have a surface area of 2,000 m^2.

3. The passage defines *colonization* of an artificial pond as:

    A. the natural process by which algae, plant life, adult amphibians, and amphibian larvae appear in an artificial pond.

    B. the introduction by man of algae, plant life, and adult amphibians to an artificial pond.

    C. the natural arrival of predators to an artificial pond.

    D. the introduction by man of predators to an artificial pond.

4. The scientists estimated the number of amphibian larvae present in each artificial pond by:

    F. trapping and counting the number of larvae in one cubic meter of edge pond water, then multiplying by the surface area of the pond in meters squared.

    G. trapping and counting the number of larvae in one cubic meter of edge pond water, then multiplying by the circumference of the pond in meters.

    H. trapping and counting the number of larvae in one cubic meter of central pond water, then multiplying by the circumference of the pond in meters.

    J. trapping and counting the number of larvae in one cubic meter of central pond water, then multiplying by the surface area of the pond in meters squared.

5. According to information that depicts the change in percent vegetative cover and the change in amphibian larvae population, as time passes after the 12th month:

    A. the number of larvae increases and the percent vegetative cover decreases.

    B. the number of larvae increases and the percent vegetative cover increases.

    C. the number of larvae decreases and the percent vegetative cover decreases.

    D. the number of larvae decreases and the percent vegetative cover increases.

6. On the basis of the data presented in Figure 1, which of the following conclusions concerning colonization is correct?

    F. Colonization of an artificial pond usually begins in winter.

    G. The final state of colonization is characterized by the presence of only plant life in the pond.

    H. Colonization of an artificial pond is characterized by an increasing diversity of plant and animal species.

    J. As colonization progresses, there is a decrease in the diversity of plant and animal species present in the pond.

7. According to the information in Figure 3, the number of pacific tree frogs:

    A. stays constant as pond surface area increases.

    B. decreases as pond surface area increases.

    C. increases as pond surface area decreases.

    D. increases as pond surface area increases.

ANSWERS AND EXPLANATIONS

1.	C	5.	A
2.	J	6.	H
3.	A	7.	D
4.	G		

PASSAGE I

1. C

Figure 1 shows that algae (pond scum and zooplankton) appear between 3 and 5 months, aquatic plants (white water lily and spike rush) show up after 10 to 12 months, and an adult amphibian population is present at 14 months. That fits (C).

Choices A and D are incorrect because they state that amphibian larvae are present in the pond at 14 months. Figure 1 shows that the first amphibian larvae (pacific tree frog tadpoles) don't appear until 16 months. Choice B doesn't include algae, so it is likewise incorrect.

2. J

This question asks you to put together two or more pieces of data from a single data presentation. You're directed to Figure 3 and asked to detemine which artificial ponds have the most long-toed salamanders. The key next to Figure 3 explains that each dot represents 25 amphibians.

Long-toed salamanders are represented by the bottom row in Figure 3. There are 25 in artificial ponds that are 500 m from the nearest water body, 75 in artificial ponds 1,000 m from the nearest water body, and 100 in ponds that are 1,500 m from the nearest water body. Similarly, the maximum number of long-toed salamanders are found in ponds that are 0.5 m deep and have a surface area of 2,000 m^2. This fits (J).

3. A

This question is testing your ability to find the definition of a term used in the written part of the passage. The term *colonization* first appears in the first paragraph: "Several studies have documented the natural process of *colonization* of artificial ponds with microorganisms (spirogyra and protozoans), aquatic plant life (*Nymphaea odorata* and *Eleocharis acicularis*), and both amphibian larvae and grown amphibians (*Hyla regilla*, *Rama sylvatica*, and *Ambystoma macrodactylum*) over time..." You can infer from this sentence that the process of colonization refers to the natural process by which plants and animals come to inhabit a pond. This best matches (A).

You could have approached this question by examining the answer choices. Choice B is incorrect because it states that colonization is the "introduction by man" of plants and animals into the pond, which contradicts the information in the first paragraph. Choice C mentions predators, which aren't mentioned anywhere in the passage, and D mentions both introduction by man and predators, neither of which is supported by the passage. That leaves (A), which is correct.

4. G

This question is testing your ability to understand the scientists' methods. Numbers of amphibian larvae are shown on the left-hand axis of Figure 2, so start by reading the paragraph just before Figure 2. It states, "The larvae populations were estimated by trapping and counting the larvae in 1 m^3 of pond water at the edge of the pond and multiplying this number by the circumference of the pond in meters." This is correctly restated in (G).

Notice that only subtle changes make the other choices incorrect. Choice F substitutes surface area for circumference, H substitutes central pond water for edge water, and J makes both substitutions.

5. **A**

This kind of question asks you to determine how the variables change over time. Figure 2 shows how the percentage of vegetative cover changes in relation to how the population of larvae changes, so that's where you should look for your answer. Notice that the question stem restricts you to "after 12 months." This makes your job easier, because from 10 to 11 months, the percentage of vegetative cover is steady. After 12 months, as the population of larvae increases, the percentage of vegetative cover steadily drops. This matches (A).

6. **H**

This question asks you to draw a "conclusion" based on information presented in the figure. In this case, that is really just another way of asking you to state what the figure shows. So what does Figure 1 show? It shows that as time goes by, an artificial pond becomes home to first algae and plankton, then plant life, and finally animals. The icons accompanying Figure 1 indicate that none of these species entirely replaces another. Rather, the pond becomes home to an increasingly diverse population of plants and animals. This matches (H).

There is no information in the entire passage that supports F. Choice G is incorrect because the final state of colonization is characterized by the presence of both plant and animal life, and J is incorrect because it states the opposite of what Figure 1 shows.

7. **D**

You can save yourself a little time by looking at the answer choices before you answer this question. The question asks you to compare the pacific tree frog population to pond surface area. That's depicted in the third column of the first row of Figure 3. You can see that the number of frogs increases as the surface area of the pond increases. Choice (D) is correct.

CHAPTER 13

Data Representation II

Answering the questions that accompany Data Representation passages will also require more complex skills that build upon the ones covered in the previous chapter. To do well on these questions, you must be able to locate data within complex presentations; compare or combine data points from one, two, or more presentations that may be simple or complex; translate information into a table, graph, or diagram; interpolate between data points; and determine the relationship between two variables in a complex presentation.

LOCATING DATA WITHIN COMPLEX PRESENTATIONS

Nothing is more intimidating than a science chart with lots of lines heading in all different directions, except for maybe a chart with lots of lines and a few random Greek letters. But keep your cool! The ACT doesn't expect that you have a Ph.D. The complex charts and graphs might look scary, but remind yourself that they're designed to be comprehensible to ACT test takers.

As you read a passage containing a complex figure, focus on the paragraph preceding the figure for an explanation of what it shows. Also be sure to read the axis labels and any legend or key accompanying the figure.

When a question specifically asks about a complex figure, the key is to focus only on the relevant information. Often, complex figures are only complex because they show how more than one variable behaves at a time. Use your pencil to trace only the curve that applies to the question at hand, and you'll often find that the figure isn't really as complex as it first appeared.

COMPARING AND COMBINING DATA

You may be asked to combine or compare data. To compare data, circle all of the relevant points in the graph or table. This will help you to focus on only what is being asked and exclude any possible distractions. Most often, questions of this type ask you to order something from least to greatest (or vice versa). For example, you may be given a table that lists the viscosity of several liquids, and a question may ask you to order them from most to least viscous.

Other questions will ask you to combine data. Usually, this just means adding or subtracting the numbers. These questions will often ask you to find the difference between two data points.

Remember that on a first read, some questions (like some figures) will appear rather complicated. Remind yourself that complicated questions are really just questions that ask you to put together a few straightforward steps. Take the question one step at a time, and you'll find that getting the answer is much easier than it first appeared.

TRANSLATING INFORMATION INTO A TABLE, GRAPH, OR DIAGRAM

Occasionally, a question will require generating your own graph, table, or diagram. This sounds difficult, but these are usually among the easier questions related to a passage. That is because: a) the graphs are usually very straightforward, and b) you don't actually have to generate them. You just have to pick the correct plot out of four choices.

To identify the correct plot, first check the general trend of the data by looking at the slope of the curve or the values in the table. Is the data you're looking for increasing, decreasing, or staying the same? Match the shape you're looking for to the answer choices.

If more than one choice remains after you check the trends, check individual values. If the information in the passage indicates that the data should have a specific value at a given point, look for the choice that contains that point. It shouldn't take more than a few seconds to backtrack your way to the correct answer.

INTERPOLATING POINTS

Interpolation simply means "reading between." Some questions will ask you to approximate values that fall between those actually shown in a figure or table. Again, this is a lot easier than it sounds.

Suppose, for example, a passage contains a table—one that lists distance in multiples of 10 cm in one column and corresponding intensity in the other—and you're asked to find the intensity at 15 cm. Your job would be to look for the values of intensity at 10 cm and 20 cm and make a guess halfway between those. That's certainly manageable.

If you're looking at a graph, the task is similar. Look for the values just below and above the point in question, and make a guess somewhere in the middle. Then match your guess to the answer choices.

UNDERSTANDING THE RELATIONSHIP BETWEEN VARIABLES IN A COMPLEX DATA PRESENTATION

Understanding the relationship between variables will remain a primary skill in passages with complex data presentations. The basic methods are the same as those discussed in the previous chapter of this guide, but you'll have to focus harder to zero in on the variables you need and ignore the ones you don't.

Work methodically. Locate one variable at a time (check the axes of a graph or the left-most column and top row of a table for the variables). If you can't find them, look for the appropriate units. For example, if a question asks for temperature, look for °C or °F.

When you find the correct variables, ask yourself how one changes as the other increases or decreases. Then match your answer to the choices. Again, you can often break down complex questions into a few straightforward steps. Take them one step at a time, and you'll find that getting the answer is far more manageable than it looks!

PRACTICE QUESTIONS

PASSAGE II

Table 1 below shows the percent composition of Earth's atmosphere by volume[*], temperature, and pressure according to elevation above a specific location. An elevation of 0 kilometers represents sea level, the surface of Earth. Temperature is given in degrees Celsius and pressure in mmHg.

Table 1

Atmospheric layer	Elevation (km)	Pressure (mmHg)	Temperature (°C)	Concentration (% by volume)				
				N_2	O_2	He	H_2	H_2O
Troposphere	0	1000	25	78.08	20.94	0.0005	0.00005	2
	15	100	−45	78.08	20.94	0.0005	0.00005	1
Stratosphere	50	10	−5	78.08	20.94	0.0005	0.00005	0.25
Mesosphere	90	3	−90	78.08	20.94	0.0005	0.00005	0
Thermosphere	200	1	900	0	99.78	0.0005	0.00005	0
	500	0.001	1,500	0	0	99.6	0.4	0
Exosphere	1,000	0.0001	0	0	0	0.05	99.05	0

[*]Percent composition by volume is equivalent to the mole-fraction of the gas—i.e., the number of molecules of the gas over the total number of molecules in any given volume.

The temperature in Earth's atmosphere does not vary continuously with elevation. Figure 1, which follows, shows the complex relationship between elevation and temperature in Earth's atmosphere.

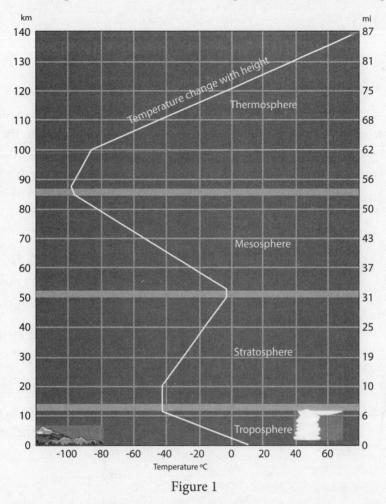

Figure 1

Image and associated databases are available from the National Oceanic and Atmospheric Administration, U.S. Department of Commerce, www.srh.noaa.gov/srh/jetstream/atmos/atmprofile.htm.

1. According to the data presented in Table 1, what is the percent concentration of O_2 at 50 km?

    A. 99.78

    B. 78.08

    C. 20.94

    D. 0.005

2. The percent concentrations of which of the following gases are constant below an elevation of 90 km?

    F. N_2, O_2, and H_2O only

    G. H_2O only

    H. N_2, O_2, and H_2 only

    J. N_2, O_2, He, and H_2 only

3. Which of the following graphs represents the relationship between O_2 concentration and elevation?

    A.

    B.
    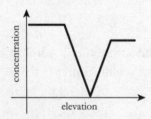

    C.
    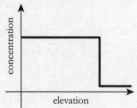

    D.
    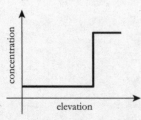

4. According to Table 1, the concentration of H_2O at 10 km is approximately:

    F. 1%.

    G. 1.5%.

    H. 2%.

    J. 10%.

5. A group of researchers sent up a weather balloon from their lab. The balloon sent back a temperature reading of −20°C. Given the information in Figure 1, which of the following elevations could possibly be the elevation of the balloon?

    A. 7 km only

    B. 23 km, 59 km, or 116 km

    C. 3 km, 23 km, 35 km, or 72 km

    D. 7 km, 37 km, 59 km, or 116 km

6. Suppose a rocket malfunctioned on its way into outer space. The last readings the rocket sent to Earth before the malfunction were an atmospheric O_2 concentration of 33.75%, an H_2O concentration of 0%, and a temperature of −85°C. At what elevation did the rocket likely malfunction?

    F. 10 km

    G. 50 km

    H. 100 km

    J. 500 km

7. According to the information in Figure 1, for distances between 50 and 85 km, as elevation increases:

    A. temperature increases.

    B. temperature decreases.

    C. temperature decreases, then increases.

    D. temperature increases, then decreases.

ANSWERS AND EXPLANATIONS

1. C
2. J
3. A
4. G
5. D
6. H
7. B

PASSAGE II

1. C

This question asks you to locate data within a complex data presentation. Table 1 has nine rows and nine columns, and it contains a lot of information. To locate the data you need, start by looking for the column labeled "O_2." (Hint: it's the sixth column.) Then use a pencil to follow that column down until you reach the 50 km row. Underline the 50 km row, and see where it meets the line down the O_2 row. It's easy for your eyes to accidentally skip up, down, or over a row in such a big table, so marking the row and column you need will help avoid unnecessary errors. The correct answer is (C), 20.94%.

2. J

Here's a great example of a comparing-data question. The question asks you to list the gases that have constant concentrations below an elevation of 90 km. To answer this question, underline the 90 km row in Table 1. (Again, this makes it easier to see.) Which gases have concentrations that don't change above this line? N_2, O_2, He, and H_2. The correct answer is (J).

Choice G is a trap. Notice that the question stem asks for concentrations "below an elevation of 90 km." Below 90 km means less than 90 km, which in the table means the rows *above* 90 km. If you had interpreted the word "below" literally, you might have picked G.

3. A

This question asks you to translate information from a table into a graph. Looking at the answer choices, you can see that the graph you are asked to generate doesn't contain specific points (in fact, there are no numbers at all on any of the graphs in the answer choices), merely an approximation of the general shape of the curve you would expect to see.

To find the correct graph, look at the column for O_2 concentration in Table 1. (That's the sixth column from the left.) From 0 to 90 km, the concentration remains constant at 20.94%. Then, at 200 km, it spikes to 99.8%. Above 500 km, it's 0%.

Choice (A) best represents this trend. Notice that D shows the steep rise, but not the drop back down to zero, and C shows the drop down to zero, but not the steep rise to 99.8%.

4. G

This question tests your ability to interpolate between two data points. Table 1 doesn't directly give you a value for percent H_2O concentration at 10 km, but it does give you the percent concentrations at 0 and 15 km (which are 2% and 1%, respectively). Your job is to read between the lines. Because 10 km is between 0 and 15 km, you need to find the answer choice between 2% and 1%. That's (G).

5. D

This question tests your ability to combine data. The curve in Figure 1 zigzags a few times and crosses over −20°C more than once. To identify the elevations that could measure −20°C, you should draw a line up from −20°C on the *x*-axis. This line will intersect the curve four times. Tracing left to the *y*-axis from those four points of intersection, you would find that possible elevations at −20°C are approximately 7 km, 37 km, 59 km, and 116 km. That matches (D).

Choice C is a trap designed to catch you if you weren't paying attention to units and the presence of a double *y*-axis. The left *y*-axis is in km and the right one is in miles. Choice C provides the correct values for this question in miles, but labels the answer in km, so it appears consistent with all the other choices.

6. H

This question requires you to combine information from two complex sources: Table 1 and Figure 1. Your first clue is that the O_2 concentration at the elevation of malfunction is 33.75%. The O_2 concentration is given in Table 1, so start by checking there; it shows that above 90 km, the O_2 concentration rises from 20.94%. Without going any further, you know you can eliminate F and G. The rocket is definitely above these elevations.

Your next clues are an H_2O concentration of 0% and a temperature of $-85°C$. The H_2O concentration is 0% for any elevation above 90 km, so that doesn't give any additional information. Temperature, however, is an important clue. According to Figure 1, a temperature of $-85°C$ only occurs above 90 km once, at about 100 km. Choice (H), then, is the correct answer.

7. B

To answer this question, you need to determine the relationship between the variables in a complex data presentation. Fortunately, this question makes your job a bit easier by restricting elevation to between 50 and 85 km. Start by drawing a horizontal line through 50 km and a second horizontal line through 85 km. Now you can focus on only the relevant section of the curve.

You can see that between your two lines, as elevation increases, temperature decreases. This is stated in (B).

CHAPTER 14

Data Representation III

The final set of skills required for Data Representation passages includes selecting hypotheses, conclusions, or predictions supported by the data, comparing and combining data from complex sources, identifying and using complex relationships between variables, and extrapolating from the data. Don't let the sound of these terms scare you—as usual, they break down into a few relatively straightforward steps. Zeroing in on exactly which variables are addressed in the question stem and finding the relevant part of the passage will take you 90 percent of the way to the correct answer.

SELECTING HYPOTHESES, CONCLUSIONS, OR PREDICTIONS FROM THE DATA

Recall from an earlier chapter that a hypothesis is the question a researcher is testing, and a conclusion is the answer to that question that is suggested by the data. Occasionally, you'll be asked to identify valid hypotheses or conclusions based on what is written in the passage. Remember, the answer is always right there in the passage. Your job is never more complicated than distinguishing the correct choice from among three choices that *must* be wrong.

To answer these questions, first go back to the passage. Find where it discusses the information in the question stem, and see if you can predict the answer. If you can, great. If you can't, see if you can identify the fatal flaw in each of the incorrect answer choices. Often, these flaws will be glaring: some choices might reference information that has absolutely nothing to do with the passage; some might reference an unrelated detail from another part of the passage; still others might reverse the correct relationship.

COMPARING AND COMBINING DATA FROM COMPLEX PRESENTATIONS

Comparing and combining data from complex presentations isn't any more difficult than comparing and combining data from simple presentations. What is harder is *finding* the appropriate data in the midst of a complex presentation.

What makes a presentation complex? Usually, it's the presence of lots of variables in one giant table, a bizarre-looking graph with three or more lines, at least one of which is zigzagging, and some kind of strange units, such as angstroms (Å) (that's 0.1 nanometer, usually used for wavelength).

To find the right data in a complex presentation, use the skills discussed in the previous chapter: Examine to the axes to find units (which can point you to the correct variable), and use your pencil to highlight only the appropriate data. Then focus on the precise calculation the question stem is asking for. Sometimes you'll need to figure out which piece of data is bigger, or rank the data from least to greatest. Sometimes you'll have to add or subtract values. The actual calculations you make will be very simple. Calculators aren't allowed on the Science test, and you'll never need them.

IDENTIFYING AND USING COMPLEX RELATIONSHIPS BETWEEN VARIABLES

ACT Science focuses on the relationship between variables. Often, those relationships are fairly straightforward; as one variable increases by a constant amount, the other increases or decreases by a constant amount. Occasionally, the relationship will be more complex; as one variable doubles, for example, the other will quadruple. This is known as an *exponential* relationship, and is the most common sort of complex relationship on the ACT.

If you suspect a complex relationship between variables, be sure to look at more than one data point. For example, if doubling the concentration of a solution quadruples the reaction time, check to see what tripling the concentration does (it should multiply the reaction time by nine). Take those extra few seconds to make sure you're seeing what you think you're seeing.

Using complex relationships is simpler than identifying them. Once you've identified the relationship between the two variables, use that to predict what will happen next.

EXTRAPOLATING FROM DATA

Extrapolating from data means predicting the value of a data point that falls beyond the range given. The rule here is to assume the data trend continues as indicated by the table or graph. If you see a line graphed as straight from 1 to 100 on the *x*-axis, assume it continues to be straight. If it's a curve that's beginning to level off at the top, assume that leveling trend continues. If a table shows a variable steadily increasing, you can safely assume that subsequent values not shown in the table also continue to increase.

PRACTICE QUESTIONS

PASSAGE III

Sunburn is a serious reaction to excess sun exposure. Figure 1 shows effects of sunburn on the body from the time of exposure until six days post-exposure. Following excess sun exposure, the body begins to produce melanin, a pigment that helps protect skin cells from further sun damage. Figure 1 shows levels of melanin production in the skin following sun exposure.

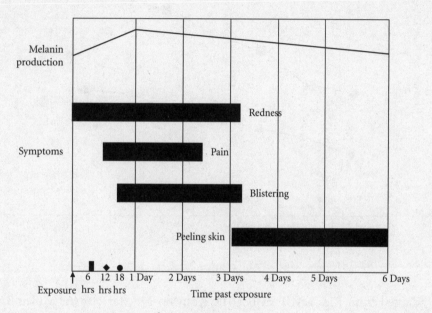

Figure 1

Sun exposure is also suspected to contribute to melanoma, a particularly deadly form of skin cancer. The frequency of melanoma reported in the United States among populations that have suffered at least one blistering sunburn (squares) and among populations that have never suffered a blistering sunburn (circles) are shown as follows in Figure 2.

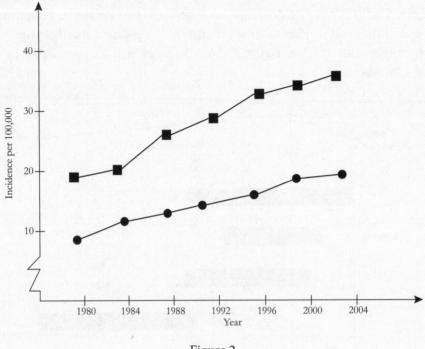

Figure 2

Figure adapted from Ries LAG, Melbert D, Krapcho M, Mariotto A, Miller BA, Feuer EJ, Clegg L, Horner MJ, Howlader N, Eisner MP, Reichman M, Edwards BK (eds). SEER Cancer Statistics Review, 1975–2004, National Cancer Institute. Bethesda, MD, www.seer.cancer.gov/csr/1975_2004/, based on November 2006 SEER data submission, posted to the SEER website, 2007.

1. One day after a serious sunburn is sustained, one could conclude that the major symptoms would be:

   A. redness, pain, and peeling skin.

   B. redness, pain, and blistering.

   C. pain, blistering, and peeling skin.

   D. redness, pain, blistering, and peeling skin.

2. Suffering just one blistering sunburn has which of the following effects, if any, on the chance of developing melanoma?

   F. It doubles an individual's chance of developing melanoma.

   G. It triples an individual's chance of developing melanoma.

   H. It decreases an individual's chance of developing melanoma by one-half.

   J. It has no effect on an individual's chance of developing melanoma.

3. Several students spent a day at the beach and suffered serious sunburns. According to the information in the passage and in Figure 1, should these students expect to be fully recovered within four days?

   A. Yes, all symptoms will have resolved within four days.

   B. Yes, only blistering is expected to remain at four days.

   C. No, the students will still be suffering from blisters and peeling skin at four days.

   D. No, the students will still be suffering from peeling skin only at four days.

4. Based on the data in Figure 1, one can conclude that the body's production of melanin peaks:

   F. 6 hours after sun exposure.

   G. 12 hours after sun exposure.

   H. 24 hours after sun exposure.

   J. 2 days after sun exposure.

5. Assume that sustaining a blistering burn causes people to develop melanoma within one year. If approximately 1 in 100 people who develop a blistering burn also develop melanoma, approximately what was the incidence of blistering burns per 100,000 people in 1980?

   A. 2,000

   B. 1,000

   C. 200

   D. 100

6. Which years showed the sharpest increase in incidence of melanoma among populations who have experienced a blistering sunburn?

   F. 2000–2004

   G. 1996–2000

   H. 1988–1992

   J. 1984–1988

7. According to the information in Figure 2, approximately what incidence of melanoma can be expected in 2008 among people who have never experienced a blistering sunburn?

   A. 35/100,000

   B. 19/100,000

   C. 35/10,000

   D. 19/10,000

ANSWERS AND EXPLANATIONS

1.	**B**	**5.**	**A**
2.	**F**	**6.**	**J**
3.	**D**	**7.**	**B**
4.	**H**		

PASSAGE III

1. B

To find the conclusion that is supported by the passage, go back to where sunburn symptoms are discussed in the passage. A quick glance tells you that's Figure 1.

According to the figure, redness, pain, and blistering will be present after one day. This matches choice (B). Note that peeling skin doesn't occur until after three days, making A, C, and D incorrect.

2. F

This is a comparing and combining data question. To determine the effect that experiencing a blistering sunburn has on a person's chances of developing melanoma, check Figure 2. Specifically, compare the odds of a person getting melanoma after experiencing a blistering sunburn to the odds of a person getting melanoma after never experiencing a blistering sunburn. To do that, compare a few points on the graph.

Look, for example, at the odds of a person getting melanoma in 1984. People who did not suffer a blistering sunburn have about a 11/100,000 chance of getting melanoma, versus a 21/100,000 chance for those who did get a blistering sunburn. That means one blistering sunburn about doubles the chances for getting melanoma. If you check a few other points, you can see that this relationship is consistent through 2004. That matches (F).

Note that you can't answer this question by simply eyeballing the answer. That's because the y-axis of Figure 2 may not start at 0. Researchers often "edit

out" a section of one axis (most often the y-axis) to save space. That's what makes this figure complex, and that's why it's so important that you actually compare the values at a few different points. You should check the values at 1984, rather than 1980, because the values for 1984 are beyond the section where the y-axis has been "shortened." This avoids the uncertainty associated with the labeling in this region.

3. D

This question asks you to make a prediction based on the information presented in one of the figures. Because symptoms are shown in Figure 1, start there.

Figure 1 tells you that four days after a serious sunburn, only peeling skin should remain as a symptom. This matches (D).

4. H

Melanin production is shown near the top of Figure 1. The curve rises steeply until one day after the initial exposure, then begins to taper off gradually. This tells you that the body's production of melanin peaks 24 hours after exposure, (H).

5. A

Mathematically speaking, this is as complex as questions get on the ACT Science test. By working through it slowly and methodically, finding the answer is not exceptionally difficult.

First, make sure you understand exactly what the question is asking you to do—find the incidence of *blistering burns* per 100,000 people in 1980. Figure 2 gives the incidence of *melanoma* per 100,000 people who have previously suffered a blistering burn in 1980 (about 20/100,000). How can you get from the incidence of *melanoma* to the incidence of *blistering burns* themselves? The question stem provides the link. It says that about 1 in 100 people who develop a blistering burn also develop melanoma that year. If 20 in 100,000 people developed melanoma, the incidence of blistering burns must have been 100 times higher, or 2,000 per 100,000. Choice (A) is correct.

6. J

To answer this question, use the answer choices to limit the amount of work you need to do. The sharpest increase must occur between 2000 to 2004, 1996 to 2000, 1988 to 1992, or 1984 to 1988. Figure 2 shows a fairly slow increase in melanoma rates for the blistering sunburn population (represented with squares) from 1996 to 2004, so eliminate F and G. The line segment from 1984 to 1988 has a slope that is obviously steeper than the segment that connects 1988 to 1992, so (J) is correct.

7. B

This question requires you to extrapolate from a graph. Values for 2008 aren't plotted in Figure 2, so you have to use the information in the figure to wager your best guess.

The most important thing to do is make sure you're looking at the right line on the graph. The question is asking for incidence among those who have "never experienced a blistering sunburn," a population represented by the line with circles. The figure shows a value of about 18/100,000 in 2000 and a little more than that—maybe 18.5/100,000—in 2004. You can guess, then, that (B), 19/100,000, will be the rate in 2008.

CHAPTER 15

Research Summaries I

Research Summaries passages build on the skills tested by Data Representation passages. In addition to testing your ability to find and interpret data, Research Summary passages also test your understanding of and ability to use the scientific method and to identify the purpose, methods, and results of an experiment. These question types will be thoroughly explained in the next three chapters.

UNDERSTANDING THE TOOLS IN AN EXPERIMENT

As explained in chapter eleven, Research Summaries passages will typically consist of an opening paragraph explaining some background information and the basic setup of the experiments, followed by two or more short descriptions of the specifics of the experiments researchers conducted along with their results. The opening paragraph or experiment descriptions might contain a complex-looking diagram, like this one:

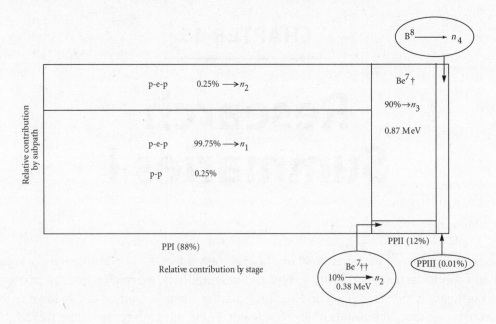

The experiment descriptions will always be followed by the results that particular experiment achieved.

Some of the experimental setup diagrams will look pretty unfamiliar and complicated. Don't let that intimidate you! Treat them like you would any other figure on the ACT: Summarize the gist of it, and then move on. You can look at it in more detail when you come to a question that relates to it. That's all!

Other tools involved in an experiment are the same tools you've used in your high school science labs. Remember, the ACT assumes you've taken at least two years of college-prep-level science, and that usually includes a certain amount of lab work. You should be familiar with what it means to measure temperature, measure quantities of liquids, record masses, record physical observations, and carry out many of the other routine tasks required by a high school science lab.

UNDERSTANDING AN EXPERIMENTAL DESIGN

Some of the questions following a Research Summaries passage will test your ability to understand how an experiment was put together. Typically, this will require you to understand the researchers' methods. Ask yourself, "How did they set up their experiment? Which variables in the experiment did the researchers change? Which variables changed in response? What kinds of results did the researchers record?"

Remember that experiments on the ACT, while they may look complicated at first glance, are usually very straightforward. Some basic equipment will be set up and a few trials will be run. Only a handful of variables will be manipulated. Remind yourself that the passages usually look much harder than they really are.

IDENTIFYING A CONTROL IN AN EXPERIMENT

There are two terms that are very important when tackling Research Summary passages: independent and dependent variables. **Independent variables** are the variables that are deliberately manipulated by researchers. **Dependent variables** are those that change in response to the independent variables.

When researchers are trying to figure out the effects of the independent variables they're studying, they try to limit effects caused by other variables they're not thinking of or measuring—the humidity of the room if it's a chemistry experiment, the amount of rainfall this year if it's an outdoor biology experiment, or any other factor that's not in their control or on their radar.

To do this, they usually run what is called a *control*. Think of this as a "blank" experiment of sorts; the experiment run without manipulating any of the independent variables. For example, if researchers are measuring how three patches of sunflowers grow when they're fed three different kinds of nutrient solutions, they may plant a fourth patch of sunflowers that are fed only water. The water-fed patch would be the experiment's control.

IDENTIFYING SIMILARITIES AND DIFFERENCES AMONG EXPERIMENTS

Questions asking you to identify the similarities and differences among experiments are a guarantee on the ACT, but luckily, finding the answer is almost always straightforward. Usually, it's printed in the paragraph below the *Experiment 1, 2,* or *3* headings. Often, you'll see an explanation of Experiment 2 along the lines of "Scientists repeated Experiment 1, but increased the temperature of the solution to 25°C." Then, looking at the description of Experiment 1, you'll find that the scientists conducted that experiment at 15°C.

You should also look quickly at the results the scientists achieved with each experiment. These are almost always printed in graphs and/or tables in Research Summaries passages. Independent variables are usually found on the horizontal axis of a graph or in the leftmost columns of a table, while dependent variables are usually found on the vertical axis or in the rightmost columns.

PRACTICE QUESTIONS

PASSAGE I

Friction is the force acting between two surfaces in contact with each other. Students wished to determine the amount of force required to overcome the force of friction between two stationary objects. To conduct their experiment, they rested a block of wood on top of a larger wooden plane. Next, they set a weight atop the wooden block. Then, a tangential force was applied to the wooden block and measured with a spring scale, as diagrammed in Figure 1. The tangential force required to cause the wooden block to begin sliding across the plane was recorded.

(Note: Assume the mass of the spring scale is insignificant.)

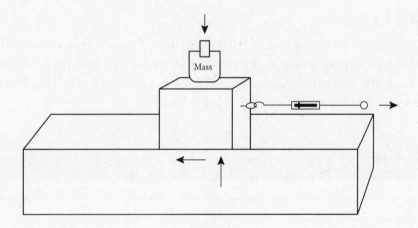

Figure 1

EXPERIMENT 1

Students placed a wooden block with a surface area of 100 cm^2 in the apparatus depicted above. A tangential force was applied, and the force at which the block began to slip was recorded in newtons (N). The experiment was repeated using weights of differing masses atop the wooden block. The forces required to cause the wooden block to slip were recorded in Table 1.

Table 1

Mass of Weight (kg)	Tangential Force (N)
0.5	6.1
1.0	12.3
1.5	18.8
2.0	24.5
2.5	30.6

EXPERIMENT 2

Experiment 1 was repeated using a wooden block of the same mass, but a surface area of 200 cm². The mass of the weight atop the block was varied as in Experiment 1, and the tangential forces required to cause the block to slip were recorded in Table 2 below.

Table 2

Mass of Weight (kg)	Tangential Force (N)
0.5	6.1
1.0	12.3
1.5	18.8
2.0	24.5
2.5	30.6

EXPERIMENT 3

In this experiment, the coefficient of friction between the wooden block and the plane was lowered by applying grease to the plane. The procedures from Experiment 1 were repeated using the wooden block and weights from Experiment 1. The forces required to cause the wooden block to slip were recorded in Table 3.

Table 3

Mass of Weight (kg)	Tangential Force (N)
0.5	3.6
1.0	6.2
1.5	9.4
2.0	12.3
2.5	15.3

1. Which of the following statements best explains why the students varied the mass of the weight atop the wooden block in Experiment 1? The students wanted to:

   A. measure the amount of downward force between the wooden block and the plane.

   B. vary the amount of tangential force on the wooden block.

   C. study how different materials affect friction.

   D. vary the downward force between the wooden block and the plane.

2. In Experiment 1, students varied which of the following factors?

   F. The surface area of the wooden block

   G. The mass atop the wooden block

   H. The length of the plane

   J. The composition of the plane

3. How is the design of Experiment 2 different from that of Experiment 1?

   A. The surface area of the block is twice as large in Experiment 2 as it is in Experiment 1.

   B. The surface area of the block is half as large in Experiment 2 as it is in Experiment 1.

   C. The surface of the plane is modified by adding grease in Experiment 2.

   D. The mass of the weights atop the wooden block ranges from 0.5 to 2.5 kg.

4. If Experiment 2 is continued using a 3 kg mass atop the wooden block, the tangential force required for the block to slip will be approximately:

   F. 18.5 N.

   G. 24.5 N.

   H. 30.6 N.

   J. 36.8 N.

5. Several students were asked to push a box to the other end of their classroom. Based on the results of the experiments, how could the students minimize the amount of force necessary to push the box? The students could:

   A. rotate the box so that the side with the smallest area is in contact with the ground.

   B. decrease the coefficient of friction between the box and the floor by painting the bottom of the box with grease.

   C. pull the box instead of pushing the box.

   D. increase the mass of the box by adding weights.

6. Which of the following factors was NOT directly controlled by the students in Experiment 3?

   F. The tangential force required to move the wooden block

   G. The mass of the weight atop the wooden block

   H. The coefficient of friction between the wooden block and the plane

   J. The surface area of the side of the wooden block in contact with the plane

7. If Experiment 3 was repeated using a wooden block of the same mass but a larger surface area, how would the tangential force required to cause the block to slip be affected? The force required would be:

   A. higher than the force recorded in Experiment 3.

   B. lower than the force recorded in Experiment 3.

   C. the same as the force recorded in Experiment 3.

   D. higher when heavier masses are used, and lower when lighter masses are used.

ANSWERS AND EXPLANATIONS

1.	D	5.	B
2.	G	6.	F
3.	A	7.	C
4.	J		

PASSAGE I

1. D

This question tests your understanding of the experimental design. What was the point of putting weights on top of the wooden block? The students wanted to be able to increase the downward force of the wooden block without changing the wooden block itself. By placing different masses on top of the same wooden block, they could increase or decrease the downward force between the wooden block and the plane. Choice (D) is correct.

2. G

Look at Table 1. It shows how the tangential force required to cause the block to slip changes as the mass of the weight atop the block increases. This information is also in the paragraph before Table 1. ("The experiment was repeated using weights of differing masses atop the wooden block.") Choice (G) is correct.

3. A

This question tests your ability to distinguish the differences between two experiments. To answer, first look at the paragraph underneath the Experiment 2 heading. It reads, "Experiment 1 was repeated using a wooden block of the same mass, but a surface area of 200 cm^2." The surface area of the wooden block in contact with the plane was 100 cm^2 in Experiment 1, so the correct choice is (A). Note that B reverses the relationship between surface area in the experiments and is therefore incorrect.

Choice C is incorrect because it states the difference between Experiments 1 and 3, and D is incorrect because both experiments varied the mass of the weights atop the wooden blocks. It represents something that is the same in both experiments, not different.

4. J

Research Summaries passages will have a few of the same question types that you see in Data Representation passages. Here, you're asked to extrapolate from Table 2. The mass of the weight in Table 2 varies from 0.5 to 2.5 kg, and the tangential force increases steadily as the mass increases. Because the tangential force required to move a block with a 2.5 kg mass atop it is 30.6 N, and each 0.5 kg increase results in an approximately 6 N increase in tangential force, you need an answer choice close to 36 N. Only (J) is close enough to be correct.

5. B

According to the passage, which factors lead to a decrease in the amount of tangential force required to move a block (or, in this case, a box)? Experiment 1 shows that decreasing the downward force on the box will make it easier to push, but this doesn't match any of the answer choices. In fact, D states the exact opposite of this idea, so you can eliminate it.

Experiment 2 shows that, contrary to what your intuition might tell you, increasing the surface area does not increase the amount of tangential force required to move the box. Rotating the box so that the side with the smallest area is in contact with the ground, then, won't make it any easier to push, and you can eliminate A.

Experiment 3 shows that decreasing the coefficient of friction between two surfaces by covering one with grease decreases the tangential force needed to push a block. Covering one side of the box with grease, however messy, would then make the box easier to push, so (B) is correct.

Note that the passage never mentioned the difference between a push and a pull, so C must be incorrect as well.

6. **F**

The key to answering this question correctly is the phrase "directly controlled" in the question stem. While all of the answer choices represent variables that changed over the course of Experiment 3, only one was NOT "directly" controlled by the students: the tangential force required to move the wooden block, (F).

The precise terms here are *dependent* and *independent* variables. Independent variables are deliberately manipulated by researchers. Dependent variables are those that change in response to the independent variables.

In Experiment 3, the mass of the weight atop the wooden block is an independent variable, directly manipulated by the students. The coefficient of friction between the wooden block and the plane was directly manipulated by the students through the application of grease, and the surface area of the wooden block in contact with the plane was also directly controlled by the students. Note that these variables don't need to be changed over the course of the experiment to be directly controlled by the experimenters—holding something constant is also a way of controlling it.

Only the tangential force required to move the wooden block changed in response to another variable (the mass of the weight atop the block), so only this was not directly controlled by the students. The tangential force was a dependent variable, and so (F) is correct.

7. **C**

This question requires you to apply the results of one experiment to another. To determine how surface area affects tangential force, review the results of Experiments 1 and 2. Tables 1 and 2 show that when the students doubled the area of contact between the wooden block and the plane, the tangential force remained unchanged.

You would expect, then, that increasing the area of contact in Experiment 3 would have no effect on tangential force. Choice (C) is correct.

CHAPTER 16

Research Summaries II

Research Summaries passages test your ability to understand the purpose, method, and results of experiments in many different ways. In this chapter, you'll learn how to identify additional trials or experiments that can be performed to enhance or evaluate experimental results; predict the results of these additional trials or measurements; predict how modifying the design of an experiment could affect its results; determine which experimental conditions would produce specific results; and analyze the given information.

ADDITIONAL TRIALS AND EXPERIMENTS

Research Summaries passages may include questions that ask you to identify trials or experiments that would give researchers more information. To answer these questions correctly, you must understand how experiments are set up and recognize the independent and dependent variables. As you learned in the previous chapter, independent variables are the variables deliberately manipulated by researchers. Dependent variables are those that change in response to the independent variables.

The best way to answer questions about new trials or experiments is to determine, from the question stem, exactly what the researchers need to know that they don't already. Then all you have to do is identify which answer choice addresses only this variable, while keeping the others constant. Watch out for answer choices that identify variables already explored in previous trials or experiments—don't let them confuse you just because they look familiar. Students are drawn to wrong answers that look familiar. Also avoid answer choices that stray too far from the topic. Students are likewise drawn to answer choices that look confusing and complicated.

Other questions will ask you to predict the results of a hypothetical new trial or experiment. These questions are very similar to the questions that accompany Data Representation passages because they require you to recognize patterns and trends in the data. The key in these is to identify *which* patterns and trends apply to the question at hand. The question stem will give you clues to figure out where you need to look.

RESULTS

You may see a question that asks you to do the reverse of the above; that is, instead of asking what results come from new conditions, it'll ask which conditions would yield new results. The method for answering is the same: Identify the relevant part of the passage (each experiment is set up to test a different variable, so look for the experiment that manipulates that variable), find the trend in the data, and use that trend to identify the conditions that will give you the desired result.

The key to additional trials and experiments questions is the fact that the answer is in the passage. Additional trials and experiments must conform to the data you've already been given. Be wary of answer choices that stray too far from the passage.

NEW INFORMATION

Sometimes, the questions following a Research Summaries passage will introduce new information. That information may come in the form of text in the question stem or a new table or graph. Your job with these questions is to figure out how this new information relates to what's already stated in the passage and find an answer choice that works with both the passage and the new information.

Questions with new information can be confusing on first read. One strategy is to save these questions for last. That gives you a chance to become more familiar with the data in the passage before you try to integrate additional data.

PRACTICE QUESTIONS

PASSAGE II

Before it can be considered safe to drink, water pumped from the ground or piped from an above-ground reservoir must be treated to remove contaminants. Suspended particles of organic and inorganic matter, parasites such as *Giardia*, bacteria, and toxic metal ions, especially lead (Pb), arsenic (As), and copper (Cu), must be removed before the water is considered safe for consumption. Several methods of water filtration were investigated by scientists.

METHOD 1

Water was filtered through a rapid sand filter. The funnel-shaped filter is constructed from an upper layer of activated carbon and a thicker layer of sand below that, as shown in Figure 1. Particles too large to pass through the spaces between grains of sand are trapped in the upper layers of sand, and smaller particles are trapped in pore spaces or adhere to the sand. Rapid sand filters do not remove a significant number of dissolved metal ions. The efficiency of particle removal was tested using filters of different depths. The results are shown in Table 1.

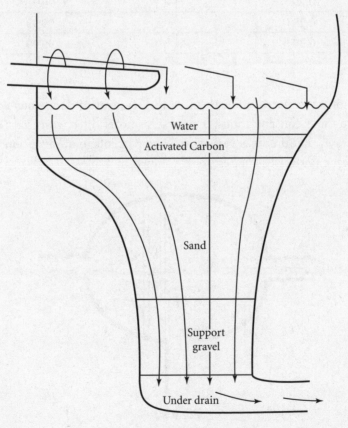

Figure 1

Table 1

Filter Depth (m)	Particles Removed (%)	Time for Filter Efficiency to Drop to 50% (hours)
0.6	95	2
0.8	98	3
1.0	99	4
1.2	99	5

METHOD 2

After filtering through the rapid sand filter, the water was pumped into storage tanks and treated with various concentrations of chlorine to destroy parasites and bacteria. The results are shown in Table 2.

Table 2

Chlorine Concentration (parts per million)	Parasites and Bacteria Killed (%)
0.10	75
0.15	98
0.20	99.9
0.25	99.99
0.30	99.99

METHOD 3

The final step of water purification involves the removal of dangerous dissolved metal ions. To reduce these concentrations, a reverse osmosis filter was used (Figure 2). The size of the pores in the filter was varied, and the percentage of metal ions removed was calculated for each pore size. The results are shown in Table 3.

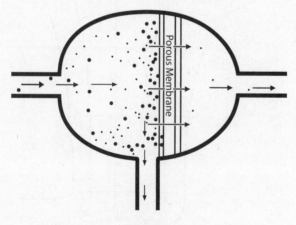

Figure 2

Table 3

Pore Size (angstroms)	As Removed (%)	Cu Removed (%)	Pb Removed (%)
2	62	78	85
3	59	77	98
4	51	68	92
5	47	62	84
6	30	56	70

1. How would the results of chlorinating water be affected, if at all, if excess chlorine-free water was accidentally added to the water in the chlorination tanks?

   A. The percentage of bacteria and parasites killed would decrease.

   B. The percentage of bacteria and parasites killed would increase.

   C. The percentage of bacteria and parasites killed would not change.

   D. The percentage of bacteria killed would increase while the percentage of parasites killed would decrease.

2. Which of the following statements about pore size is consistent with the results of Method 3?

   F. Increasing pore size results in an increased amount of Cu removed.

   G. Increasing pore size results in an increased amount of bacteria and parasites killed.

   H. Decreasing pore size results in an increased amount of As removed.

   J. Decreasing pore size requires more time to filter heavy metal ions.

3. Which of the following procedures would yield more information about the effects of chlorine concentration on parasites and bacteria in water?

   A. Determining what percentage of parasites and bacteria are killed by ultraviolet light exposure

   B. Determining what percentage of parasites and bacteria are killed during the process of rapid sand filtering

   C. Determining how the nutrient content of the water changes with chlorine concentration

   D. Determining what percentages of individual types of parasites and bacteria are killed at different chlorine concentrations

4. Scientists were concerned that sediment particles contained within the rapid sand filters were contributing to the amount of particles in the water exiting the filter. Which of the following procedures would best allow the scientists to investigate this issue?

    F. Filtering pure water through the filter and measuring the amount of particles in the water exiting the filter

    G. Testing the amount of dissolved metal ions in the water before and after filtering

    H. Adding additional sediment particles to the water before filtering

    J. Measuring the amount of sediment in the water before and after filtering

5. Over-chlorinating water can result in the production of potentially harmful chemical by-products called trihalomethanes (THMs). If chlorine concentrations above 0.25 parts per million (ppm) can result in dangerous levels of THMs in the water supply, which of the following chlorine concentrations should scientists recommend to best purify drinking water of parasites and bacteria?

    A. 0.15 ppm

    B. 0.20 ppm

    C. 0.25 ppm

    D. 0.30 ppm

6. Molecules of zinc (Zn) are very close in size to molecules of Cu. If the percentage of Zn ions in water was also measured before and after reverse osmosis filtration, what relationship, if any, would scientists expect to see between pore size and percent Zn removal?

    F. As pore size increased, the percentage of Zn removed would increase.

    G. As pore size increased, the percentage of Zn removed would decrease.

    H. As pore size increased, the percentage of Zn removed would first increase, then decrease.

    J. As pore size increased, the percentage of Zn removed would first decrease, then increase.

7. If a city's water contained equal amounts of As, Cu, and Pb, a reverse osmosis filter with which pore size would remove the most heavy metals?

    A. 2 angstroms

    B. 3 angstroms

    C. 4 angstroms

    D. 5 angstroms

ANSWERS AND EXPLANATIONS

1.	A	5.	C
2.	H	6.	G
3.	D	7.	B
4.	F		

PASSAGE II

1. **A**

You're asked to predict how the accidental addition of chlorine-free water would affect the results obtained in Method 2. To answer, consider the relationship between the chlorine concentration and the percentage of bacteria and parasites killed. A quick glance back at Table 2 shows that as chlorine concentration increases, the percentage of bacteria and parasites killed also increases.

What, then, will happen to the concentration of chlorine in the water if chlorine-free water is introduced? Chlorine-free water would dilute the concentration of chlorine. This decrease in concentration would decrease the percentage of bacteria and parasites killed, so (A) is correct.

2. **H**

Remember that some Research Summaries questions will ask you to evaluate the experimental results, just like Data Representation questions. Here, you're asked to examine the relationship between pore size and heavy metal removal.

Begin by reviewing Table 3. For both As and Cu, an increase in pore size results in a decrease in the percentage of heavy metal removed. For Pb, percentage removed increases when pore size is increased from 2 to 3 angstroms, and then decreases as pore size increases further.

Only (H) correctly describes this relationship (though the wording is a bit confusing—"decreasing" pore size means you have to read the table from bottom to top, which isn't how one usually approaches a table). Note that G and J don't relate to Method 3 at all and that F gets the relationship backward.

3. **D**

The effect of chlorine concentration on parasites and bacteria was examined in Method 2. Further investigations about chlorine concentration should test something not already studied in Method 2. Only (D) identifies an experiment that would give additional information about the effects of chlorine concentration on parasites and bacteria; it suggests finding out how specific kinds of parasites and bacteria each respond to chlorine concentration.

Choices A and B are incorrect because they do not offer any additional information about how *chlorine concentration* affects parasites and bacteria. Choice C is incorrect because you're asked to find out more about the effects of chlorine concentration on parasites and bacteria, not nutrient content.

4. **F**

The question stem indicates that the scientists are concerned that sediment contained within the rapid sand filter itself is contributing to the amount of particles in the water exiting the filter. To test whether this is an issue, it would make sense for the scientists to run pure, particle-free water through the filter and see if any sediment particles are contained in the water exiting the filter, as in (F).

Choice G is irrelevant to the issue of particles within the filter. Heavy metal ions are discussed in Method 3, but have no bearing on the particle-removing rapid sand filters discussed in Method 1.

Choice H is also incorrect. Adding additional sediment to the water before filtering won't tell scientists anything about whether the filter itself is adding particles to the water. Choice J is likewise incorrect. The amount of sediment in the water is already being measured before and after filtering, to determine the percentage of sediment removed by each filter. It offers no additional information.

5. C

Many questions on the ACT Science test will introduce some additional information, either in the form of information within the question stem, as is the case here, or in the form of additional tables or graphs. Your task in this situation is to integrate this new information with the information already in the passage.

To answer this question, look back at Table 2. It shows that chlorine concentrations of 0.10 ppm kill 75% of parasites and bacteria, concentrations of 0.15 ppm kill 98% of parasites and bacteria, concentrations of 0.20 ppm kill 99.9% of parasites and bacteria, and concentrations of 0.25 ppm and higher kill 99.99% of parasites and bacteria. The question stem, however, tells you that concentrations *over* 0.25 ppm result in dangerously high THM levels. Scientists should recommend that water be treated with the lowest chlorine concentration that still kills the highest percentage of parasites and bacteria. That would be 0.25 ppm, as in (C).

6. G

Here, you are asked to predict the results of an additional measurement in an experiment. Scientists didn't measure the percentage of zinc removed by reverse osmosis filtration. However, you're told in the question stem that zinc molecules are about the same size as copper molecules. Table 3 shows that pore size is directly related to the percentage of each metal that is removed. You can infer, then, that if zinc is about the same size as copper, it'll be removed by the reverse osmosis filter similarly to copper.

How is the percentage of copper removed related to pore size? You should resist the temptation to speed through this question by answering from memory. It only takes a few more seconds to double-check the table!

A quick scan down the Cu column shows that as pore size increases, the percentage of copper removed decreases. This matches (G).

7. B

This question asks you to determine which set of experimental conditions gives a specified result. In this case, the specified result is the greatest amount of heavy metals removed.

To investigate, look back to Table 3. It shows that as pore size increases, the percent of As and Cu removed decreases. However, notice that Pb doesn't exactly follow this trend. At 2 angstroms, 85% of Pb is removed. But at 3 angstroms, 98% of Pb is removed. That's a different trend and underscores the importance of evaluating *all* of the data before answering a question. Once pore size is increased over 3 angstroms, the percentage of Pb removed sharply plummets.

To determine whether 2 angstroms or 3 angstroms is the best answer, you must evaluate, approximately, which size results in more total metal removed. You're told that the city's water contains equal amounts of As, Cu, and Pb. The decrease in the percent of As and Cu removed when pore size is increased from 2 to 3 angstroms is tiny (3% and 1% respectively), but the increase in Pb removed is significant (13%). You don't need a calculator (and aren't allowed one anyway!) to realize that more metal is removed at 3 angstroms than at 2 angstroms. Choice (B) is correct.

CHAPTER 17

Research Summaries III

The final set of skills needed to approach Research Summaries passages with confidence is the ability to understand complex experiments, to understand the hypothesis for an experiment, and to integrate new, complex information into the information already given in the passage. The good news about complex Research Summaries passages is that the ACT likes to keep passages at approximately the same level of difficulty. That means that complicated passages will usually be balanced with more manageable questions. So don't let all of the complicated-looking figures and diagrams scare you off!

UNDERSTANDING COMPLEX EXPERIMENTS

You'll be able to recognize a complex Research Summaries passage instantly: They're the ones with scary-looking diagrams of experimental apparatus, lots of text, and a combination of tables and graphs. The multiple experiments will likely be quite different from each other, which is a big contrast to the majority of Research Summaries passages. Most Research Summaries passages contain experiments that are very similar variations of each other, changing just one variable at a time and using the same experimental design.

If you get a really complex Research Summaries passage on your ACT, read carefully through it one time, but do not get bogged down in the details. Remember, you don't get points for understanding the passage, just for answering questions (correctly, of course!). Stay focused. Don't let your mind wander as you're reading, but don't worry about the details until a question sends you there.

The same is true for the complex apparatus diagrams. On your first read, note their existence. You can puzzle them out in detail when (or if) a question directs you to.

Finally, treat a complex data presentation like a combination of straightforward ones. Look carefully for the variable in question (search for the appropriate units if you're having trouble finding it), and identify any trends in the data. The information you need is there, but it's surrounded by data you don't need. Work methodically and you won't be thrown by it.

UNDERSTANDING HYPOTHESES

While you'll likely never be asked to directly identify the hypothesis of an experiment, the ACT does test your understanding of hypotheses in general. So what is a hypothesis? It's the scientists' guess at the answer to the question they're asking (reread chapter eleven for more on hypotheses and the scientific method in general).

Usually, the ACT makes it clear when they're giving you a hypothesis, so you don't have to worry about ever coming up with one on your own. You'll see clues when you're dealing with a hypothesis question, such as "If Scientist 1's hypothesis is correct…" Your job is to determine how to support or weaken that hypothesis or how to further test it.

Supporting a hypothesis means backing it up with more information that fits the hypothesis. Weakening it means finding information that contradicts the hypothesis. To further test a hypothesis, you'll need to ask yourself what else you need to know to confirm or disprove this guess.

NEW INFORMATION

As you may recall from chapter sixteen, you should be prepared for questions that introduce new information, be it in the form of text, graphs, tables, or diagrams. Your job on these questions is to find an answer choice that combines both the new information and the information already presented in the passage. Questions that introduce new information are generally not as hard as they seem, but can be time-consuming. It's a good idea to save these for last.

PRACTICE QUESTIONS

PASSAGE III

The *reaction rate* of a chemical reaction is defined as how quickly the reaction takes place. Students wished to investigate the reaction rate, burn time, and peak burn temperature for three common combustion reactions.

STUDY 1

Students used a computer model to simulate the combustion of pure hydrogen (H_2), methane (CH_4), and propane (C_3H_8). The model was created with the assumption that the reactions took place in a closed system of constant volume. Figure 1 shows the computer-generated reaction rates as a function of temperature for hydrogen, methane, and propane gas.

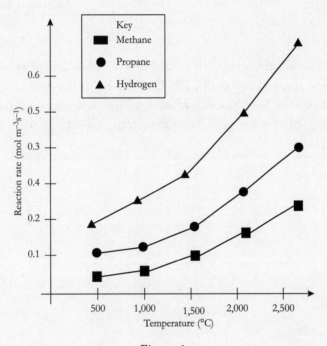

Figure 1

STUDY 2

Students next measured the *burn time* (the length of time for which a fixed amount of each fuel will burn), *ignition temperature* (the temperature at which the fuel spontaneously ignites), and the peak temperature of each fuel by igniting a fixed amount of each fuel and a fixed amount of air inside a closed, heat-proof, insulated container. A *thermocouple* (high-temperature electronic thermometer) was affixed to the container to measure temperature. The results of the study are shown in Table 1.

Table 1

Fuel	Amount (g)	Burn Time (s)	Ignition Temperature (°C)	Peak Temperature (°C)
Hydrogen	0.1	103	585	2,318
Propane	0.1	256	487	2,385
Methane	0.1	321	540	2,148

STUDY 3

Students learned that of ten fires in their county this year, four had peak burn temperatures over 1,500°C: Fire 1 reached a peak of 2,295°C, fire 4 reached a peak of 2,100°C, and fires 5 and 9 reached a peak of 2,400°C (note: all of these temperatures have an error of ±50°C). Students hypothesized that the peak burn temperature could be used to determine the cause of these fires.

1. If the hypothesis made by the students in Study 3 is correct, which fuel would most likely have been the cause of fire 9?

   A. Hydrogen

   B. Methane

   C. Propane

   D. Kerosene

2. A student hypothesized that the higher the molecular weight of the fuel, the longer its burn time. Do the results of Study 2 and all of the information in the table below support this hypothesis?

Gas	Molecular Weight (grams per mole)
Hydrogen	1.008
Propane	44.096
Methane	16.043

   F. Yes; propane has the highest molecular weight of the gasses and the longest burn time.

   G. Yes; propane has a higher molecular weight than hydrogen and a longer burn time.

   H. No; the higher the molecular weight of a fuel, the shorter its burn time.

   J. No; there is no clear relationship between molecular weight and burn time.

3. A student hypothesized that increasing the pressure on a fuel would increase its reaction rate. The best way to verify this hypothesis would be to repeat Study 1 and use:

   A. changing pressures instead of changing temperatures.

   B. higher temperatures.

   C. different fuels.

   D. greater initial amounts of the fuels.

4. According to Study 1, which of the following statements best describes the relationship, if any, between temperature and reaction rate?

   F. As temperature increases, reaction rate increases.

   G. As temperature increases, reaction rate decreases.

   H. As temperature increases, reaction rate first increases, then decreases.

   J. There is no apparent relationship between temperature and reaction rate.

5. Which of the following graphs best represents the relationship between the ignition temperature and burn time of the three fuels?

   A.
   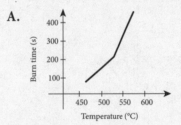

   B.
   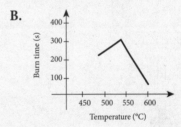

   C.
   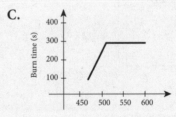

   D.
   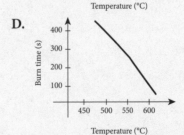

6. Which of the following figures best illustrates the apparatus described in Study 2?

F.

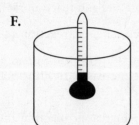

G.

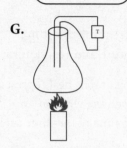

H.

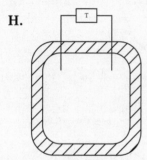

J.

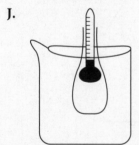

7. Which of the following best explains why students used a fixed amount of fuel to determine burn times in Study 2? The students wanted to be certain that:

A. variations in burn time were not due to varying starting amounts of fuel.

B. the fuel had enough time to burn completely.

C. the temperature in the container was hot enough to ignite the fuel.

D. no other fuels were present in the container.

ANSWERS AND EXPLANATIONS

1.	C	5.	B
2.	J	6.	H
3.	A	7.	A
4.	F		

PASSAGE III

1. C

To answer this question, you must first determine exactly what hypothesis the students made in Study 3. Going back to the passage, you see that "students hypothesized that the peak burn temperature could be used to determine the cause of these fires." In other words, the students believed that the peak burn temperature of the fire would correlate to the peak burn temperature of the fuel that caused it. The questions to ask are what was the peak temperature of fire 9 and which fuel has a peak burn temperature closest to that?

According to Study 3, the peak temperature of fire 9 was 2,400°C, plus or minus 50°C. You can find peak burn temperatures for each fuel in Table 1. According to the table, the only fuel with a peak burn temperature in the range of 2,350°C to 2,450°C is propane, with a peak burn temperature of 2,385°C. The correct answer is (C).

2. J

This is a rather complex question. It requires you to combine information from two sources to come up with an answer. Remember, though, that complex doesn't have to mean difficult. Even the most complex questions usually break down into a series of manageable steps. Your job is to keep your cool and work methodically, one step at a time.

The first step is to understand the hypothesis presented in the question stem. A student is saying that higher molecular weights of fuels mean longer burn times. It's a good idea to underline that first sentence, or at least circle the words "higher molecular weight" and "longer burn time," to avoid making an error.

Molecular weight isn't mentioned anywhere in the passage, but this missing information is provided in a table for us just below the question stem. It shows that hydrogen has the lowest molecular weight, followed by methane, then propane. Because the fuels aren't listed in increasing or decreasing order of molecular weight, it's a good idea to jot "hydrogen, methane, propane" (the fuels in order of increasing molecular weight) down next to the table to avoid confusion.

Next, order the fuels according to burn time. According to Table 1, hydrogen has the shortest burn time, then propane, then methane. Finally, you can determine whether a higher molecular weight means a longer burn time. It doesn't. While hydrogen has the shortest burn time and lowest molecular weight, the pattern breaks with propane and methane. The correct answer, then, is (J).

While it's true that propane has a higher molecular weight than hydrogen and a longer burn time, as in G, that's not enough to make the entire hypothesis true, so G is incorrect.

3. A

To verify a hypothesis, you need data. To acquire this data, the student would have to modify Study 1 so that it provided information about the effects of pressure on reaction rate. Only (A) mentions pressure, and it is correct.

Choice B is incorrect. If Study 1 were repeated with higher temperatures, you'd learn more about the effect of temperature on reaction rate, but nothing about pressure. Because the question stem asks you to verify a hypothesis concerning the effects of pressure on reaction rate, the correct answer choice must incorporate pressure.

Choices C and D are again incorrect because they don't test pressure. Repeating the study with different fuels, or differing initial amounts of fuel, would tell you nothing about pressure and reaction rate. If you are familiar with the ideal gas law, it might occur to you that increasing the amount of fuel (choice D) in a fixed-volume container will ultimately increase pressure in the system. However, this is definitely not the *best* way to study the effect of pressure on rate because you would be unsure if a rate increase was caused by the greater amount of fuel or the increase in pressure. Good experimental design means only one variable is changed at a time.

4. F

To find the relationship between temperature and reaction rate, look back to Figure 1. Temperature is given on the *x*-axis and reaction rate on the *y*-axis. Temperature is increasing left to right from 500°C to 2,500°C, and reaction rate increases from 0.1 to 0.6 mol m^{-3}s^{-1} as you move up. Now, look at the shape of the curve itself; it points up. This means that as temperature increases, the reaction rate increases. Choice (F) is correct.

Note that G states the opposite relationship, and H gives a relationship that is more complicated than the one shown by the figure. If the reaction rate first increased, then decreased, you'd expect to see a curve that looks like a hill—going up and then going back down. That's not what Figure 1 shows.

5. B

This question asks you to plot the relationship between two variables. You saw these kinds of questions in chapter thirteen (look back to the section titled "Translating Information into a Table, Graph, or Diagram" for a quick review). These kinds of questions can appear anywhere in the Science test, even in the Conflicting Viewpoints passage.

Remember, you don't actually have to generate your own plot here, just pick the correct one from the four answer choices. So what should the correct plot look like?

The correct plot should list ignition temperature in degrees Celsius on the *x*-axis and burn time in seconds on the *y*-axis. It should also plot the values shown in Table 1. Propane's ignition temperature is 487°C and its burn time is 256 seconds, methane's ignition temperature is 540°C and its burn time is 321 seconds, and hydrogen's ignition temperature is 585°C and its burn time is 103 seconds.

Only the graph shown in (B) correctly plots these values, so it is correct.

6. H

Study 2 explains the experimental setup by stating that students ignited "a fixed amount of each fuel and a fixed amount of air inside a closed, heat-proof, insulated container. A *thermocouple* (high-temperature electronic thermometer) was affixed to the container to measure temperature."

Find the answer choice that shows a closed, heat-proof container containing fuel and air, with a thermocouple affixed. Only (H) comes close, and it is correct.

Choice F is incorrect because it shows the ignition taking place in an open container. Furthermore, it shows a standard thermometer inside the container, which contradicts the explanation given.

Choice G is incorrect because it shows a flask held over a Bunsen burner, which likewise contradicts the given explanation.

Choice J is incorrect because it shows a flask with a standard thermometer inside another container, which again contradicts the given explanation.

7. A

This question requires an understanding of the methods of a complex experiment. What was the purpose of using a fixed amount of fuel in determining burn time?

Usually, the easiest way to answer "which of the following" kinds of questions is to go straight to the answer choices. Of course, if the correct answer pops into your head first, even better. But failing that, it's generally fastest to scan the answer choices for the correct answer.

Choice (A) states that the students wanted to be certain that "variations in burn time were not due to varying starting amounts of fuel." This is certainly true—it's reasonable to assume that the more fuel you start with, the longer it will burn. So if you're going to compare burn times directly, you have to be certain you're controlling all of the other variables. You can check the other choices to be certain you have the correct one.

Choice B states that the students wanted to be certain that "the fuel had enough time to burn completely." This doesn't make sense. Students were measuring the burn time, but using equal amounts of starting fuel each time wouldn't help students be certain that they had allowed the fuel to burn completely. Cross this one off.

Choice C states that the students wanted to be certain that "the temperature in the container was hot enough to ignite the fuel." This likewise doesn't make sense. There's no connection between ignition temperature and using equal starting amounts of fuel. Cross this one off, too.

Choice D states that the students wanted to be certain that "no other fuels were present in the container." While students would certainly want to be sure they were burning one fuel at a time when measuring burn times, using fixed amounts of each fuel wouldn't accomplish this. Cross it off. Choice (A), then, is correct.

CHAPTER 18

Conflicting Viewpoints I

Conflicting Viewpoints passages are a bit of an oddball in the ACT Science test. While there are typically two Data Representation and three Research Summaries passages, you'll likely have just one Conflicting Viewpoints passage. And while Data Representation and Research Summaries passages present information from one or more experiments, Conflicting Viewpoints passages by and large contain only some opening information and the viewpoints of two, three, or four authors. They are far less likely to contain any figures (although it does happen), and they resemble ACT Reading test passages as much as they do Science passages.

Conflicting Viewpoints passages focus on a different set of skills than Data Representation and Research Summaries passages. While the latter two test your ability to locate and analyze data, Conflicting Viewpoints passages primarily test your ability to understand hypotheses (you can consider each viewpoint to be one author's hypothesis). They'll also test your ability to determine whether additional information would support or weaken a viewpoint and to identify similarities and differences between the viewpoints.

UNDERSTANDING THE VIEWPOINTS

Some of the questions that accompany a Conflicting Viewpoints passage will simply test your ability to identify key issues or assumptions in each viewpoint. They're typically rather straightforward but almost without fail contain wrong answer choices that draw from other viewpoints, which may tempt you into selecting the wrong answer. That's why it's important to follow the Kaplan Method for Conflicting Viewpoints passages, which was introduced in chapter eleven.

THE KAPLAN METHOD FOR CONFLICTING VIEWPOINTS PASSAGES

1. Read the introductory text and the first author's viewpoint, then answer the questions that ask only about the first author's viewpoint.

2. Read the second author's viewpoint, then answer the questions that ask only about the second author's viewpoint.

3. Answer the questions that refer to both authors' viewpoints.*

(*When there are three or more author viewpoints, you'll read each viewpoint and answer the related questions before you answer the questions that relate to multiple viewpoints.)

This will require that you skip around in the questions. That's fine. There's no rule saying you have to answer the questions in the order in which they appear. And you'll find that it's much, much easier to deal with Conflicting Viewpoints passages when you minimize your opportunities to confuse the two (or three, or four) viewpoints.

SUPPORTING AND WEAKENING VIEWPOINTS

It is common to see at least one question asking you to identify whether information supports or weakens the view of one or both authors. Typically, these questions will introduce new information or data, and your job will be to determine whether the data supports or weakens a viewpoint.

Remember that information supports a viewpoint if it offers data that makes the viewpoint even more likely to be true. Conversely, information weakens a viewpoint if it contradicts some part of it. The information doesn't have to prove a viewpoint to be true in order to support it, and it doesn't have to prove it false to weaken it. You're simply supposed to judge whether the information adds to or detracts from any part of the viewpoint.

IDENTIFYING SIMILARITIES AND DIFFERENCES BETWEEN VIEWPOINTS

You may see questions that ask you to identify a point on which the authors agree or a point on which they disagree. It's helpful in answering these questions to underline each author's main point as you're reading. Then when it's time to identify similarities and differences, you'll have the essence of their viewpoints highlighted.

Answer these questions immediately after you answer the questions that deal with just one viewpoint at a time. Questions that ask you to identify similarities or differences are typically among the easier questions in the passage, so they're great for giving you a chance to make sure you have determined the subtleties of each viewpoint.

PRACTICE QUESTIONS

PASSAGE I

Acids are chemical compounds that are today most fundamentally defined as having a pH less than 7.0—that is, a hydrogen ion activity greater than pure water—when dissolved in water and that are capable of donating a proton. Bases are chemical compounds with a pH higher than 7.0 and are commonly understood as compounds that can accept protons. Before chemists developed the modern model of acid and base behavior, there were many attempts to explain their observable qualities; especially the observable qualities of acids (sour taste, an ability to turn blue vegetable dye red, an ability to conduct an electric current, and, most notably, an ability to react with certain compounds and give off hydrogen or carbon dioxide gas). Two historical definitions of acids are presented below.

LAVOISIER DEFINITION

Lavoisier believed that all acids showed the same observable properties because they contained one common substance. Because he primarily knew of strong acids such as HNO_3 and H_2SO_4, he believed that this common substance was oxygen (O). He believed that a metal, reacting with an acid, would combine with the oxygen in the acid to produce a *calx* (a metal oxide). The calx would then further react with the acid to form a salt. Hydrogen, the other component of water, was released as a gas. The reaction could also run in reverse. Water, combined with a salt and calx, would produce an acid. Oxygen, according to Lavoisier, was the driving force behind the reaction. In fact, Lavoisier himself actually recognized and named the element oxygen, deriving the word from the Greek for "acid-producer."

Lavoisier further believed that the other portions of the acidic compound were responsible for the acid's specific individual properties, and called this other portion the "acidifiable base." He had no further understanding of bases.

LIEBIG DEFINITION

Liebig primarily studied organic acids such as CH_3COOH (acetic acid) and $C_3H_6O_3$ (lactic acid). According to the Liebig definition, it was not oxygen but rather hydrogen (H) that was responsible for the common properties of an acid. However, not all compounds containing hydrogen could be considered acidic; Liebig defined an acid as a hydrogen-containing compound in which the hydrogen could be replaced by a metal. The hydrogen would then be released as a gas. Liebig recognized a base as a substance capable of *neutralizing* an acid (neutralization is the process by which an acid is transformed into a salt and water and no longer displays the common characteristics of an acid), but had no theoretical understanding of how and why bases worked.

1. Which of the following assumptions was made by Lavoisier?

    A. All acids combine with oxygen as they react.

    B. Acids cannot be broken down into other compounds.

    C. All acids contain oxygen.

    D. Any compound that contains oxygen is an acid.

2. Adherents of which definition, if either, would predict that hydrochloric acid (HCl) has a pH of less than 7?

    F. The Lavoisier definition only

    G. The Liebig definition only

    H. Both the Lavoisier definition and the Liebig definition

    J. Neither the Lavoisier definition nor the Liebig definition

3. According to the Lavoisier definition, the combination of a salt, hydrogen gas, and a metal oxide would produce a(n):

    A. calx.

    B. neutral salt.

    C. base.

    D. acid.

4. According to the Liebig definition of acids, the gas produced from the reaction of an acid:

    F. is flammable.

    G. is composed of hydrogen.

    H. is suitable for breathing.

    J. is composed of oxygen.

5. When phosphoric acid (H_3PO_4) reacts with sodium bromide (NaBr), monosodium phosphate (NaH_2PO_4) and hydrobromic acid (HBr) are produced according to the equation NaBr + H_3PO_4 → NaH_2PO_4 + HBr. What conclusion would adherents of each definition draw about the mechanics of this reaction?

    A. Both adherents of the Lavoisier definition and adherents of the Liebig definition would conclude that the reaction is driven by the oxygen contained in the phosphoric acid.

    B. Both adherents of the Lavoisier definition and adherents of the Liebig definition would conclude that the reaction is driven by the hydrogen contained in the phosphoric acid.

    C. Adherents of the Lavoisier definition would conclude that the reaction is driven by the oxygen contained in the phosphoric acid; adherents of the Liebig definition would conclude that the reaction is driven by the hydrogen contained in the phosphoric acid.

    D. Adherents of the Lavoisier definition would conclude that the reaction is driven by the hydrogen contained in the phosphoric acid; adherents of the Liebig definition would conclude that the reaction is driven by the oxygen contained in the phosphoric acid.

6. Suppose an acid is allowed to react completely with a metal. Which of the following statements about acids is most consistent with the information presented in the passage?

  F. If the Lavoisier definition is correct, the oxygen in the acid will have donated a proton in the reaction.

  G. If the Lavoisier definition is correct, the hydrogen in the acid will have accepted a proton in the reaction.

  H. If the Liebig definition is correct, the oxygen in the acid will have donated a proton in the reaction.

  J. If the Liebig definition is correct, the hydrogen in the acid will have accepted a proton in the reaction.

7. The Liebig definition states that the hydrogen in an acid is capable of being replaced by a metal. Which of the following findings, if true, could be used to *counter* this definition?

  A. Acids, when reacted, release pure hydrogen gas.

  B. Acids, when neutralized, form solutions with a pH equal to that of pure water.

  C. In the reaction of sulfuric acid (H_2SO_4) with sodium hydroxide (NaOH), sodium sulfate (Na_2SO_4) and water are formed.

  D. Some acids, when completely reacted, form pure metals in water.

8. The Lavoisier definition differs from the Liebig definition in that only the Liebig definition states that a base:

  F. is capable of neutralizing an acid.

  G. has a pH of greater than 7.

  H. is responsible for an acid's specific qualities.

  J. could be formed by combining a calx with a metal oxide.

ANSWERS AND EXPLANATIONS

1.	C	5.	C
2.	G	6.	F
3.	D	7.	D
4.	G	8.	F

PASSAGE I

1. C

The passage states that "Lavoisier believed that all acids showed the same observable properties because they contained one common substance." In the next sentence, the passage goes on to state that "he believed that this common substance was oxygen (O)." So according to Lavoisier, all acids contain one common substance, and that substance is oxygen. Choice (C) is correct.

Choice A is incorrect because it very subtly twists the relationship in the passage. Acids don't *combine* with oxygen, they actually *contain* oxygen. If you were to rush through this question because it looks pretty easy, you might fall into this wrong answer trap.

Choice B is incorrect. The Lavoisier definition never states that acids cannot be broken down into other compounds. Quite the opposite—the definition states that acids can react with metals to produce hydrogen gas and a salt.

Choice D is incorrect because it also subtly twists the definition given in the passage. The Lavoisier definition states that all acids contain oxygen, but you cannot conclude from this statement that all compounds containing oxygen are acids.

2. G

The opening paragraph states that "Acids are chemical compounds that are today most fundamentally defined as having a pH less than 7.0." So this question stem is really asking you to decide who would predict that HCl is an acid.

According to the Lavoisier definition, all acids contain oxygen (O). According to the Liebig definition, all acids contain hydrogen (H). Hydrochloric acid, which contains hydrogen and not oxygen, would only be expected to be an acid according to the Liebig definition. Choice (G) is correct.

3. D

Go back to the passage to answer this question. The details about how salts, gases, and metal oxides react in this definition are too complicated to recall by memory. The Lavoisier definition states in the first paragraph that Lavoisier "believed that a metal, reacting with an acid, would combine with the oxygen in the acid to produce a *calx* (a metal oxide). The calx would then further react with the acid to form a salt. Hydrogen, the other component of water, was released as a gas. The reaction could also run in reverse. Water, combined with a salt and calx, would produce an acid."

The question stem is asking about the combination of a salt, hydrogen gas, and a metal oxide. The first two sentences quoted above describe how an acid can break a metal down into a salt and hydrogen gas (a component of water). The last sentence tells you that the reaction also works in reverse—water can combine with a salt and calx to produce an acid. The question stem asks what will happen when you combine hydrogen gas with a salt and a metal oxide. The passage tells you that hydrogen gas is a component of water, and a metal oxide is also known as calx. So the question stem is merely asking what happens when you combine water with a salt and a calx. According to the last sentence quoted, the answer is an acid. Choice (D) is correct.

4. G

You should approach this question by making sure you understand exactly what the Liebig definition says about the gas produced from the reaction of an acid. Go back to the passage.

The passage states that "Liebig defined an acid as a hydrogen-containing compound in which the hydrogen could be replaced by a metal. The hydrogen could then be released as a gas." So look for an answer choice that matches this statement. Choice (G) fits best.

Choice F is incorrect. The passage doesn't say anything about whether hydrogen gas is flammable (it is, though).

Choice H is also incorrect; hydrogen gas is not suitable for breathing. No gas's suitability is mentioned in the passage, so you can eliminate this answer choice anyway. Very rarely will questions rely on your knowledge of science facts, but it does happen. The ACT very occasionally assumes that you know basic things such as which gases are safe for breathing.

Choice J is incorrect. The Liebig definition does not state that oxygen is given off when an acid is reacted.

5. **C**

This question manages to make a very straightforward idea look extraordinarily complicated. By working methodically and not panicking, you'll see that getting to the answer is far easier than it first appears.

The question stem gives the equation for a reaction between an acid (H_3PO_4) and a metal-containing compound (NaBr). You're asked to decide who would conclude what based on this reaction.

Make sure you understand exactly what each definition states. According to the Lavoisier definition, acids are acids because they contain oxygen. According to the Liebig definition, acids are acids because they contain hydrogen. Because H_3PO_4 contains both hydrogen and oxygen, you can expect that adherents of the Lavoisier definition would conclude that the oxygen in H_3PO_4 is responsible for the reaction, while adherents of the Liebig definition would conclude that the hydrogen is responsible for the reaction. This matches (C).

Note that D gets the definitions backward and is therefore incorrect.

6. **F**

This question requires you to combine information in the passage with information in the question stem to support one of the viewpoints. As is typical with "which of the following" type questions, it's easiest to begin this one by moving straight to the answer choices.

Choice (F) reads, "If the Lavoisier definition is correct, the oxygen in the acid will have donated a proton in the reaction." To know whether this is true, you need to know what the passage says about protons. That information appears in the opening paragraph. The first sentence says that "Acids are chemical compounds…capable of donating a proton." So if acids donate protons, and a hypothetical acid has reacted with a metal, is it consistent with the Lavoisier definition to state that the oxygen in the acid donated a proton? Yes. The Lavoisier definition maintains that oxygen is the component of an acid responsible for the characteristic behavior of an acid. Therefore, if the definition is correct, then the oxygen will have donated the proton.

Note that G makes two mistakes. It attributes the acid-like behavior to the hydrogen in the acid, which is consistent with the Liebig definition and not the Lavoisier definition, and it states that the acid will have *accepted* a proton, which according to the passage, is characteristic of a base, not an acid.

Choices H and J similarly confuse the main tenets of each definition and the definitions of an acid and a base given in the first paragraph.

7. **D**

To counter an argument means to go against it. So you're looking for the finding that contradicts Liebig's definition. This is a "which of the following" question, so go straight to evaluating the answer choices.

Choice A states, "Acids, when reacted, release pure hydrogen gas." Does this contradict the idea that the hydrogen in an acid is capable of being replaced by a metal? No. If anything, it strengthens it, because it supports the idea that the hydrogen in an acid is released (though it doesn't speak to what that hydrogen might be replaced by). Eliminate this one.

Choice B states, "Acids, when neutralized, form solutions with a pH equal to that of pure water." Does this counter the given definition of acids? No. The pH of neutralized acids doesn't tell you whether the hydrogen has been replaced by a metal. Eliminate it.

Choice C describes the reaction of sulfuric acid and sodium hydroxide. The product is sodium sulfate. This reaction supports the Liebig definition; the hydrogen in the acid has been replaced in the final product by a metal (Na). This supports the Liebig definition, so it is certainly not correct.

Choice (D) states, "Some acids, when completely reacted, form pure metals in water." This would counter the Liebig definition. The hydrogen in an acid can't be replaced by a metal if pure metals in water are formed. It is correct.

8. **F**

This question asks you to identify one difference between the definitions. This is the first question in the passage to ask about a base, so make sure you understand what each definition states about them.

According to the first viewpoint, Lavoisier believed that "the other portions of the acidic compound were responsible for the acid's specific individual properties, and called this other portion the 'acidifiable base.'"

According to the second viewpoint, Liebig believed that a base was "a substance capable of neutralizing an acid." This matches (F) perfectly, and it is correct.

Note that G pulls from the introductory material. This is tricky because while it is true, it's not specific to the Liebig definition. Choice H is incorrect because it matches the Lavoisier definition of bases. While J might be tempting because it uses terms from the passage, there is nothing in the Liebig definition to suggest how a base might be formed.

CHAPTER 19

Conflicting Viewpoints II

Even though you're likely to see just one Conflicting Viewpoints passage on your ACT Science test, as opposed to two Data Representation and three Research Summaries passages, the questions that accompany Conflicting Viewpoints passages are often as varied as those that accompany the other two passage types. Some Conflicting Viewpoints passages and their questions are more technical, focusing on the mechanics of a specific scientific question (like the passage that accompanies this chapter), while others are more theoretical (like the passages that accompany chapters eighteen and twenty). You should practice with a variety of Conflicting Viewpoints passages so you'll be prepared for whichever passage you get.

Some passages will contain questions focused on strengthening and weakening the viewpoints. These questions will incorporate new information in the question stems. Some questions may ask you to make predictions about seemingly unrelated subjects based on the viewpoints, and some may ask you to find hypotheses, conclusions, or predictions that are supported by two viewpoints or pieces of information.

MORE ON STRENGTHENING AND WEAKENING VIEWPOINTS

Some questions will introduce new information, and then ask you whether this information supports or weakens one or both viewpoints. Some questions may even ask you *why* the information supports or weakens a viewpoint.

The questions that introduce new information sometimes look confusing. The information may look overly complex or appear at first to have nothing to do with the rest of the passage. Remember that the questions typically don't rely on any outside knowledge, so you *do* have the information you need to answer them. Break these questions down into a series of manageable steps, and always start by making sure you understand each viewpoint. Remember that information supports a viewpoint if it makes it more likely to be true and weakens it if it makes the viewpoint less likely to be true.

MAKING PREDICTIONS BASED ON VIEWPOINTS

Questions that ask you to make predictions based on viewpoints are especially common with the more technical passages. These questions often require you to apply a viewpoint to a specific process or a hypothetical situation, and to make a prediction about the outcome.

Confusion is the biggest source of error in these situations. Make sure you minimize potential confusion by reading one viewpoint at a time, and then answering the questions that apply to just that viewpoint.

HYPOTHESES, CONCLUSIONS, AND PREDICTIONS SUPPORTED BY TWO VIEWPOINTS

Another complicated question type will ask you to find hypotheses, conclusions, and predictions that are supported by two or more viewpoints or pieces of information. Again, your job is to find the answer choice that is compatible with both viewpoints or with one viewpoint and some additional information. Make sure you understand exactly what each viewpoint states, or how the viewpoint in question relates to the additional information in the question stem, before you answer the question. A big source of error on the ACT Science test is moving to the answer choices before you know what the question is truly asking. Make sure you have a guess in mind before reading the answer choices. For questions that ask "which of the following," it's generally quicker to evaluate the answer choices without making a guess, but you need to be certain that you know how to tell whether each choice is correct before you start.

PRACTICE QUESTIONS

PASSAGE II

The *molecular geometry* of a molecule is the three-dimensional appearance of that molecule. Molecular geometry determines many of the properties of a substance, most notably its phase of matter, reactivity, polarity, and magnetism. While molecular geometries can be determined absolutely through X-ray crystallography or quantum mechanical modeling, they can also be predicted using the *valence shell electron-pair repulsion* (VSEPR) theory.

The VSEPR theory uses Lewis structures (diagrams that show the bonding between atoms and lone electron pairs in a molecule) to predict the molecular geometry of covalently bonded molecules. VSEPR theory states that the shape of a molecule is determined by the repulsions between the bonded and nonbonded pairs of electrons in the outer shell of the molecule's central atom. These electron pairs arrange themselves as far apart as possible to minimize repulsion.

There are five basic molecular geometries:

1. *Linear.* In a linear molecule, two electron pairs are arranged in a line around a central atom. The bond angle between the electron pairs is 180°.

2. *Trigonal planar.* A molecule with trigonal planar geometry has three electron pairs arranged around a central atom, with a bond angle of 120° between electron pairs.

3. *Tetrahedral.* In a molecule with tetrahedral geometry, the central atom is surrounded by four electron pairs, each with bond angles of 109.5°.

4. *Trigonal bipyramidal.* Molecules with this geometry contain five electron pairs. The bond angles are 90°, 120°, and 180°.

5. *Octahedral.* Molecules with octahedral geometry contain six electron pairs around one central atom. Bond angles are 90° and 180°.

Two students discuss the molecular geometry of water (H_2O).

STUDENT 1

Predicting the shape of a molecule is relatively straightforward. A molecule's shape will always be determined by the number of electron pairs around the central atom. The number of electron pairs corresponds to the number of atoms that are bonded to the central atom of the molecule. For example, water contains two hydrogen atoms bonded to one atom of oxygen, giving the molecule a linear geometry.

STUDENT 2

The geometry of a molecule is dependent on the number of atoms bonded to the central atom of that molecule. However, the geometry of a molecule is also dependent on the presence of electron pairs around the central atom that are not bonded to any atoms. These nonbonded electron pairs also have a negative charge and therefore equally repel the electrons that are bonded to atoms, influencing the shape of a molecule. Because one molecule of water contains two nonbonded electron pairs around a central oxygen atom and two electron pairs bonded to hydrogen atoms, it has a tetrahedral geometry.

1. According to the passage, electron pairs tend to:

   A. attract each other.

   B. prefer a linear geometry.

   C. arrange themselves close to each other.

   D. arrange themselves as far apart as possible.

2. Based on Student 2's explanation, the south (negatively charged) end of a magnet will repel the south end of another magnet because:

   F. positive charges always attract each other.

   G. negative charges always repel each other.

   H. a positive and negative charge always attract each other.

   J. a positive and a negative charge always repel each other.

3. Suppose a molecule of sulfur dioxide (SO_2) was found to have a bond angle of 120° and contain one nonbonded electron pair. If true, this finding would most likely support the viewpoint(s) of:

   A. Student 1 only.

   B. Student 2 only.

   C. both Student 1 and Student 2.

   D. neither Student 1 nor Student 2.

4. Quantum mechanical models suggest that a molecule of sulfur hexafluoride (SF_6) contains no nonbonded electron pairs. Both students would predict that a molecule of SF_6 would have:

   F. an octahedral geometry.

   G. a trigonal bipyramidal geometry.

   H. bond angles of 90°, 120°, and 180°.

   J. a bond angle of 109.5°.

5. Suppose that the model presented by Student 1 is correct. Based on the information provided, what would be the bond angle in a molecule of perchlorate ion (ClO_4^-)?

   A. 90°

   B. 109.5°

   C. 120°

   D. 180°

6. Based on the model presented by Student 1, a molecule of ammonia (NH_3) would have a:

   F. trigonal planar geometry, because three atoms of H are bonded to one central atom of N.

   G. trigonal planar geometry, because three atoms of H and one nonbonded electron pair are bonded to one central atom of N.

   H. tetrahedral geometry, because three atoms of H are bonded to one central atom of N.

   J. tetrahedral geometry, because three atoms of H and one nonbonded electron pair are bonded to one central atom of N.

7. Which of the following diagrams showing the relationship between an electron pair's bonded status and the relative strength of the repulsive force is consistent with Student 2's assertions about the shape of a water molecule, but is NOT consistent with Student 1's assertions about the shape of a water molecule?

A.

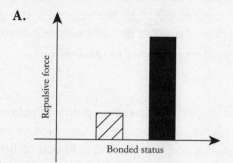

B.

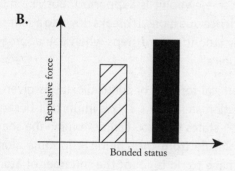

C.

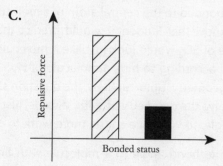

D.

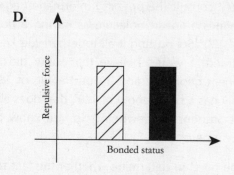

8. Measurements made through X-ray crystallography show that the repulsive force of nonbonded electron pairs is greater than the repulsive force of bonded electron pairs. Based on the information provided, this finding would most likely weaken the viewpoint(s) of:

F. Student 1 only.

G. Student 2 only.

H. both Student 1 and Student 2.

J. neither Student 1 nor Student 2.

ANSWERS AND EXPLANATIONS

1.	D	5.	B
2.	G	6.	F
3.	B	7.	D
4.	F	8.	H

PASSAGE II

1. D

Remember that Conflicting Viewpoints passages can look a lot like Reading Comprehension passages. This question tests your ability to locate and comprehend information from within the passage (information that just happens to be scientific, of course!). This is also a great question to answer first, because it doesn't require you to read either viewpoint yet.

To answer, scan back through the opening information in the passage to find where it discusses electron pairs. That's in the second half of the second paragraph, which reads, "VSEPR theory states that the shape of a molecule is determined by the repulsions between the bonded and nonbonded pairs of electrons in the outer shell of the molecule's central atom. These electron pairs arrange themselves as far apart as possible to minimize repulsion."

The last sentence perfectly matches (D), which is correct.

Notice that A and C get it exactly backward. This is a common wrong answer format. Choice B brings up a detail from later in the passage and uses it inappropriately, another common wrong answer format.

2. G

To answer this question, you must apply what you know about charge from Student 2's explanation to a new situation (namely, magnets). What does Student 2 believe about negative charges? Go back to the passage to find out.

Student 2 states that "nonbonded electron pairs also have a negative charge and therefore equally repel the electrons that are bonded to atoms." This means that both bonded and nonbonded electron pairs have a negative charge and that the two repel. You can conclude that negative charges repel each other.

Based on this theory, the south (negatively charged) end of a magnet will repel the negatively charged end of another magnet because negative charges always repel each other. This matches (G).

3. B

This kind of question can seem pretty complicated at first glance. It introduces a new type of molecule, gives a bond angle, and asks you to figure out whose viewpoint is supported. But like most other Science questions, it breaks down to a series of relatively strightforward steps when you approach it methodically.

The chemical formula of sulfur dioxide is given in the question stem (SO_2). This is important because Student 1 states that you can determine the shape of a molecule from the chemical formula alone. All you have to do is count the "number of atoms that are bonded to the central atom of a molecule." So you know that Student 1 would predict that a molecule of SO_2 would look just like a molecule of H_2O, and according to him, a molecule of H_2O has a linear geometry. Unfortunately, the question stem doesn't give the geometry of sulfur dioxide, just the bond angle, so you have a little more work to do.

To find the bond angle in a molecule with linear geometry, consult the opening information again. That defines a linear molecule as having a bond angle of 180°. So, putting it all together, you know that Student 1 would believe that sulfur dioxide has a linear geometry and a bond angle of 180°. If it really has a bond angle of 120°, that does NOT support Student 1's viewpoint. So you know the answer isn't A or C.

Now you need to determine whether the fact that sulfur dioxide has a bond angle of 120° supports the viewpoint of Student 2. Student 2 states that "the geometry of a molecule is also dependent on the presence of electron pairs around the central atom that are not bonded to any atoms" and that because water contains two nonbonded pairs of electrons and two atoms bonded to the central atom, it has a tetrahedral geometry.

If a molecule of sulfur dioxide contains one nonbonded electron pair and two bonded electron pairs, Student 2 would predict that it would have a trigonal planar geometry. According to the passage, this geometry has a bond angle of 120°. This is consistent with Student 2's viewpoint, so (B) is correct.

4. F

This question is asking you to make another prediction about molecular shape. This time, however, you're dealing with a situation in which Student 1 and Student 2 are in agreement. That is because according to the question stem, sulfur hexafluoride (SF_6) doesn't contain any nonbonded electron pairs, and these nonbonded electron pairs are the source of the two students' disagreement.

The question, then, is fairly straightforward. You have a central atom (S) with six atoms (F) bonded to it. According to the information in the passage, molecules that contain six electron pairs around a central atom will have an octahedral geometry. Choice (F) is correct.

5. B

It would be wise to answer this question before reading Student 2's viewpoint, to minimize confusion.

Student 1 believes that "a molecule's shape will always be determined by the number of electron pairs around the central atom" and that "the number of electron pairs corresponds to the number of atoms that are bonded to the central atom of the molecule." Because a molecule of perchlorate ion (ClO_4^-) contains four O atoms around one central Cl atom, it should have a tetrahedral geometry.

The names of the possible shapes aren't given in the answer choices, just bond angles. According to the information in the passage, molecules with a tetrahedral geometry have bond angles of 109.5°, and (B) is correct.

6. F

Here is another question you should answer before going on to read Student 2's viewpoint. Student 1 believes that the number of atoms bonded to the central atom of a molecule determines the number of electron pairs around the central atom, and thus the shape of the molecule. If ammonia (NH_3) contains three H molecules around one N molecule, it should have a trigonal planar geometry. Choice (F) is correct.

If you look at G and J, you can see that the wrong choices are drawn from either Student 2's viewpoint alone or a combination of Student 1's and Student 2's viewpoints. Choice H incorrectly applies Student 1's viewpoint to predicting the geometry of NH_3. Answering this question before you go on to read Student 2's viewpoint will help you avoid a lot of potential confusion.

7. D

The first step in answering this question is to make sure you know exactly what Student 1 and Student 2 assert about the strength of the repulsive force of bonded and nonbonded electron pairs.

Student 1 claims that nonbonded electron pairs don't factor into the shape of a molecule at all. That's why water has a linear geometry, he says. Student 2 claims that nonbonded electron pairs exert a repulsive force equal in magnitude to the repulsive force exerted by bonded electron pairs. That's why he believes a molecule of water, which contains two bonded and two nonbonded electron pairs, should have a tetrahedral geometry.

So you're looking for a diagram that shows the relative strength of the repulsive force to be equal for bonded and nonbonded electron pairs. This is consistent with (D), which shows the two bars to be of equal length. Choice (D), then, is correct.

Note that the question stem asks you to determine which answer choice is consistent with Student 2's viewpoint and not Student 1's. It doesn't matter what you might know about the repulsive force of bonded and nonbonded electrons or which student is actually right in this case (if either). Your job is simply to identify the diagram that is consistent with the viewpoints in the passage.

8. **H**

According to the passage, Student 1 believes that nonbonded electron pairs don't factor into the shape of a molecule. Measurements that prove non-bonded electron pairs *do* show a repulsive force would weaken his viewpoint, because this would allow them to influence the shape of the molecule. You can eliminate G and J.

You're not done, though. You still need to determine whether this information weakens the viewpoint of Student 2. He believes that nonbonded electron pairs behave just like bonded electron pairs in determining the shape of a molecule. If measurements were to show that nonbonded electron pairs exerted a repulsive force *greater* than that of bonded electron pairs, this viewpoint would also be weakened—nonbonded electron pairs could not be counted on to behave just like bonded electron pairs and would have an even greater influence on the shape of a molecule than he believes. His viewpoint would also be weakened, so (H) is correct.

CHAPTER 20

Conflicting Viewpoints III

COMPLEX PASSAGES

Some passages are more complex than others. A passage may seem complex to some people but not to others because the subject matter isn't their favorite science topic. While you're more likely to see some relatively straightforward (for you) and some relatively complex (for you) topics among the Data Representation or Research Summaries passages simply because there are more of those, you might get a Conflicting Viewpoints passage that seems really complex, or just isn't your favorite. That's okay; don't let it psych you out. There are ways to deal with complex Conflicting Viewpoints passages.

If you come across a Conflicting Viewpoints passage that immediately strikes you as really confusing, relax, take a deep breath, and remember that you're prepared for this test and you're going to do well (it's the power of positive thinking, and it really works). Then start reading attentively; nothing gets your mind wandering like a topic that just doesn't interest you. The passages are short, though, so stay focused. Be on the lookout for the authors' main points—they usually sum up their beliefs in one neat thesis sentence. Underline it when you see it. If you finish reading a viewpoint without finding a thesis sentence, check the first and last paragraphs of the viewpoint again. They're most often there, but unfortunately not always. If you don't see a thesis statement, don't worry. Move on to the questions and refer back to individual details of the viewpoint as necessary.

When you finish reading one viewpoint, skip ahead to the questions and answer the ones that relate to only that viewpoint. This will give you the chance to solidify your understanding of that viewpoint before reading any others. It'll also break up the passage into even smaller chunks, making it less likely that you'll zone out and waste time reading without comprehending. With only six to eight questions in the passage, though, not every viewpoint may have corresponding questions. If you do get some, take advantage!

Sometimes, especially if the opening information is rather long, you'll get a question or two that relates to only that paragraph. It's worth it to check for questions that relate solely to the opening information (look for ones that start with a vague, "According to the passage…").

When you're done answering any questions that relate to only the first viewpoint, go ahead and read the second, and answer the questions that relate to just that viewpoint (repeat this process if there are more than two viewpoints). Finally, you can answer the questions that relate to multiple viewpoints. Remember that you can answer questions in any order you'd like, so skip the really difficult-looking ones on the first pass.

COMPLEX QUESTIONS

Questions can also look complex. Most often, the ACT makes a question complex by introducing seemingly unrelated concepts, asking you to make a prediction or draw a conclusion that feels pretty abstract, or asking you to apply a scientist's reasoning to an entirely new situation.

Complex questions ARE answerable. Start by making sure you understand exactly what the author is saying. Reread the relevant parts of the passage and summarize the main idea in your own words (to yourself, of course!). Then go straight to the answer choices. Three will contain fatal flaws. Look for these. The most common kind of wrong answer choice uses ideas drawn from the wrong parts of the passage (for example, ideas in the opening information or another author's viewpoint). Also look out for answer choices that appear to use all the right words, but actually state the exact opposite of what is true. For example, watch for choices that use keywords from the passage but aren't relevant to the question at hand (these are always tempting because they look so good), and for choices that are outside the scope of the passage. An answer choice doesn't have to be one you'd come up with on your own to be correct—it just has to be true according to the passage.

Finally, remember that you do have a time limit. Don't spend too much time on any one question. Guess, circle it in your test booklet, and come back for another look if you have time when you finish the section.

PRACTICE QUESTIONS

PASSAGE III

Scientists discuss three possible explanations for the mechanism by which water (H_2O) conducts electric current.

SCIENTIST 1

Pure water is incapable of conducting an electric current. Electric current in this sense is defined as the flow of electrically charged particles, known as electrons, through a medium. Because the electrons in a water molecule are tightly bound to the molecule, they will not flow freely and cannot create an electric current.

Pure water will, however, conduct ions—that is, charged particles within the water will physically move from one place to another when an electric charge is applied. However, the physical flow of ions from one location to another is technically different from the flow of electric current, so it cannot be accurately stated that water conducts electric current.

SCIENTIST 2

Water itself is capable of conducting electricity, but it does so very poorly. The hydrogen and oxygen in water molecules continuously *dissociate* (come apart) into H^+ and OH^- ions and reform into H_2O. As the bonds in water are continuously broken and reformed, electrons are temporarily freed. When an electric field or current is then applied to water, the electrons that constitute the electric current are able to flow from one end to another, constituting a feeble electric current. The individual H^+ and OH^- ions are not required to move in this process, so the result is the flow of a true electric current.

When a substance that suppresses the formation of ions is dissolved in water, it will no longer conduct electricity. So when sugar, for example, is stirred into water, the production of ions is disrupted, and the water will cease to conduct electricity. When a substance that increases ion concentration is added to water, the water will conduct a greater amount of electricity.

SCIENTIST 3

Water is capable of conducting electricity, but only because it contains impurities. Most water contains dissolved salts, such as sodium chloride (NaCl), minerals such as calcium (Ca) and copper (Cu), and gases such as carbon dioxide (CO_2). The impurities dissolved in water are capable of conducting an electric current because they contain unpaired electrons in the atoms' outer valence shells. It is these unpaired valence electrons that permit the flow of electric current. The number of electrons available to transmit charge is proportional to the amount of electricity the water can conduct.

Seawater is a perfect example of how impurities are responsible for the electrical conductivity of water. Seawater primarily contains dissolved sodium (Na^+), chloride (Cl^-), sulfate (SO_4^{2-}), magnesium (Mg^{2+}), calcium (Ca^{2+}), and potassium (K^+) ions. While pure water can conduct approximately 0.000055 $S\cdot m^{-1}$ (Siemens per meter) of electricity, seawater can conduct 5 $S\cdot m^{-1}$. The concentration of dissolved impurities is also directly proportional to the conductivity of water; the saltier the water, the more electricity it can conduct.

1. Based on Scientist 3's explanation, if the number of unpaired electrons in the valence shells of the impurities in water were decreased, which of the following quantities would simultaneously decrease?

   A. The resistivity of water

   B. The electrical conductivity of the water

   C. The H^+ concentration of the water

   D. The density of the water

2. Which scientist(s), if any, would predict that water will not truly conduct an electric current if it is 100% pure?

   F. Scientist 1, but not Scientist 2 or Scientist 3

   G. Scientist 2, but not Scientist 1 or Scientist 3

   H. Both Scientist 1 and Scientist 2, but not Scientist 3

   J. Both Scientist 1 and Scientist 3, but not Scientist 2

3. City A's drinking water is typically heavily chlorinated, resulting in the formation of hypochlorite ions (ClO^-). Scientist 3 would most likely predict that, compared to less-chlorinated water, City A's water would show:

   A. an increase in the amount of electricity the water is capable of conducting.

   B. a decrease in the amount of electricity the water is capable of conducting.

   C. no change in the amount of electricity the water is capable of conducting.

   D. a decrease in the formation of H^+ ions.

4. When a sample of zinc (Zn) is doped with chromium (Cr), the electrical conductivity at room temperature is increased fivefold. Based on Scientist 3's explanation, the reason that the conductivity increases is most likely that:

   F. the addition of Cr allows for the formation of ions within the sample.

   G. Cr atoms contain fewer electrons in their outer valence shells than Zn atoms.

   H. Cr atoms contain more electrons in their outer valence shells than Zn atoms.

   J. Cr is a better conductor of electricity because it is a metal.

5. Salt (NaCl) in water ionizes to form Na^+ and Cl^- ions. When an electric current is passed through a dish of pure water mixed with salt, a current is conducted. Scientist 1 would most likely argue that:

   A. an ion flow, not a true electric current, has been conducted.

   B. a true electric current, not an ion flow, has been conducted.

   C. the salt encourages H^+ and OH^- ions to form, resulting in an electric current.

   D. the salt disrupts H^+ and OH^- ion formation, reducing the maximum possible electric current.

6. Suppose it is found that the more water is heated, the more readily it dissociates into H^+ and OH^- ions. Scientist 2 would most likely state that a greater electric current could be conducted by:

   F. cooler water than by warmer water.

   G. warmer water than by cooler water.

   H. ice than liquid water.

   J. a mixture of ice and water than by liquid water.

7. Scientist 2 and Scientist 3 would most likely agree that water will conduct more electricity under which of the following conditions?

   A. The water is purified of any impurities.

   B. The water is polarized.

   C. Sugar is added to the water.

   D. Ion-producing impurities are added to the water.

8. Suppose it is discovered that in the presence of an electric current, water readily ionizes into H^+ and OH^- ions. This finding would support the viewpoint of:

   F. both Scientist 1 and Scientist 2.

   G. both Scientist 1 and Scientist 3.

   H. Scientist 2 only.

   J. neither Scientist 1, nor Scientist 2, nor Scientist 3.

ANSWERS AND EXPLANATIONS

1. B   5. A
2. J   6. G
3. A   7. D
4. H   8. J

PASSAGE III

1. B

To answer this question, make sure you understand what Scientist 3 has to say about unpaired electrons in the valence shells of impurities in water. In the first paragraph, Scientist 3 states, "The impurities dissolved in water are capable of conducting an electric current because they contain unpaired electrons in the atoms' outer valence shells. It is these unpaired valence electrons that permit the flow of electric current. The number of electrons available to transmit charge is proportional to the amount of electricity the water can conduct."

This is complicated, so it is crucial that you make sure you really understand her point. Scientist 3 is saying that unpaired electrons in the outer shells of impurities in water conduct electricity. Furthermore, Scientist 3 is saying that the more electrons you have available, the more electricity you can conduct.

If more valence electrons mean more electricity can be conducted, then what would happen if there were fewer valence electrons? A smaller electric current could be conducted. That is, if the number of available electrons decreased, the electrical conductivity of the water would decrease. Choice (B) is correct.

2. J

This question is asking you to predict what each scientist would believe about the ability of pure water to truly conduct electricity. Work methodically, starting with Scientist 1.

Scientist 1 states in the first sentence that "Pure water is incapable of conducting an electric current." Scientist 1 does, however, concede that pure water is capable of conducting ions, but he clarifies at the end that "the physical flow of ions from one location to another is technically different from the flow of electric current, so it cannot be accurately stated that water conducts electric current." Scientist 1 agrees that water will not truly conduct an electric current if it is 100% pure, so eliminate G.

Scientist 2 believes that pure water is capable of conducting electricity: "Water itself is capable of conducting electricity, but it does so very poorly." Scientist 2 also states that the flow is "a true electric current." Because Scientist 2 disagrees with the statement in the question stem, eliminate H.

Scientist 3 believes that water is capable of conducting electricity, but "only because it contains impurities." Pure water, then, would not conduct electricity in her opinion. Scientist 3 agrees with the statement in the question stem, so (J) is correct.

3. A

You have to make sure you understand Scientist 3's viewpoint before you attempt to answer this question. Scientist 3 writes that "the impurities dissolved in water are capable of conducting an electric current because they contain unpaired electrons in the atoms' outer valence shells" and that "the number of electrons available to transmit charge is proportional to the amount of electricity the water can conduct." Scientist 3 also ends the discussion of why seawater conducts electricity by noting that "the concentration of dissolved impurities is also directly proportional to the conductivity of water."

It makes sense, then, that Scientist 3 would believe that City A's water, with its greater concentration of hypochlorite ions, would conduct more electricity than less heavily chlorinated water. Choice (A) is correct.

Note that B gets the relationship backward, and D introduces a detail from Scientist 2's argument. Both of these answer choices might be tempting if you didn't take the time to review Scientist 3's viewpoint.

4. **H**

Again, you need to start answering this question by making sure you understand Scientist 3's viewpoint. Scientist 3 writes that "unpaired valence electrons… permit the flow of electric current. The number of electrons available to transmit charge is proportional to the amount of electricity the water can conduct."

Basically, this means that electric current flows in water because of the availability of valence electrons. The more electrons you have, the more electricity you can conduct.

Now, apply this same line of reasoning to the example in the question stem. When chromium is added to a sample of zinc, the sample can conduct five times more electricity. According to Scientist 3's line of reasoning, this must be because chromium has more valence electrons free to transmit charge. This matches (H).

Choice F is incorrect. It is true that Scientist 3 does indicate that water conducts electricity because impurities are present, and these impurities are likely ionized. However, F provides no real explanation for a difference in conductivity, because zinc would likely form ions in solution too. More importantly, Scientist 3's focus is on the number of valence electrons present in the impurities, rather than the impurities themselves, so (H) is a better answer.

Choice G is incorrect because it states the exact opposite of what is true.

Choice J is incorrect because it is unrelated to the passage.

5. **A**

This is a great question to answer first, before reading Scientist 2 or Scientist 3's viewpoints.

Scientist 1 states that "pure water will…conduct ions—that is, charged particles within the water will physically move from one place to another when an electric charge is applied. However, the physical flow of ions from one location to another is technically different from the flow of electric current, so it cannot be accurately stated that water conducts electric current."

The question stem tells you that salt forms ions in water. You should assume, then, that Scientist 1 would think that, while perhaps more current is conducted, it is still technically an ion flow, and not a true current. Choice (A) is correct.

Note that B states the exact opposite of what you would expect Scientist 1 to say, and C and D are drawn from Scientist 2's viewpoint.

6. **G**

First, review what Scientist 2 has to say about ions. Scientist 2 states that "the hydrogen and oxygen in water molecules continuously *dissociate* (come apart) into H^+ and OH^- ions and reform into H_2O. As the bonds in water are continuously broken and reformed, electrons are temporarily freed. When an electric field or current is then applied to water, the electrons that constitute the electric current are able to flow from one end to another, constituting a feeble electric current." The result, Scientist 2 says, is "the flow of a true electric current."

The question stem tells you that the more water is heated, the more H^+ and OH^- ions it forms. Put these two pieces of information together: Warmer water has more ions and therefore must conduct more electricity. (According to Scientist 2 and the information in the question, anyway. Remember, it doesn't matter which viewpoint is true as long as you're using the one the question asks for!) Choice (G) is correct.

7. **D**

This is a "which of the following" question. Your protocol here is to make sure you understand each viewpoint, then move directly to the answer choices.

Scientist 2 believes that the ions in water free the electrons that are responsible for conducting electricity. Because water essentially self-ionizes into H^+ and OH^- ions, it can conduct electricity even when no other ions are present.

Scientist 3 believes that only the presence of added ions, in the form of impurities, makes water able to conduct electricity. The two disagree over whether pure water can conduct electricity, but not over the fact that ions are essential for electricity conduction in water. Now the answer choices:

Choice A is incorrect. Scientist 3 specifically states that impurities are necessary for the conduction of an electric current.

Choice B is incorrect. No one ever mentions polarization. This is outside the scope of the passage.

Choice C is incorrect. Scientist 2 actually states that adding sugar to water will make it unable to conduct electricity. You're looking for something that will make water conduct *more* electricity.

Choice (D) is correct. Both scientists agree that ions are responsible for electricity conduction in water. More ions would therefore lead to more electricity conducted.

8. J

By now you can readily recognize that H^+ and OH^- ion formation is the key to Scientist 2's viewpoint. Scientist 2 believes that the ability of water to self-ionize into H^+ and OH^- ions is responsible for its ability to conduct electricity. However, while Scientist 2 believes that H^+ and OH^- are crucial for electrical conduction, Scientist 2 makes no claims about the effect of an electrical field on the dissociation of water. Such an effect is not implied either. If the question had said that additional dissociation caused by an electric field resulted in an increase in current, then this viewpoint would be supported. Just stating that an electric field results in dissociation does not bolster Scientist 2's argument. The viewpoints of Scientists 1 and 3 would not be supported either, because neither discusses H^+ and OH^- ions. Choice (J) is correct.

Additional Resources

100 KEY MATH CONCEPTS FOR THE ACT

NUMBER PROPERTIES

1. UNDEFINED

On the ACT, *undefined* almost always means **division by zero.** The expression $\dfrac{a}{bc}$ is undefined if either b or c equals 0.

2. REAL/IMAGINARY

A real number is a number that has a **location on the number line.** Imaginary numbers are numbers that involve the square root of a negative number. For example, $\sqrt{-4}$ is an imaginary number.

3. INTEGER/NONINTEGER

Integers are **whole numbers;** they include negative whole numbers and zero.

4. RATIONAL/IRRATIONAL

A **rational number** is a number that can be expressed as a **ratio of two integers. Irrational numbers** are real numbers—they have locations on the number line—they just **can't be expressed precisely as a fraction or decimal.** For the purposes of the ACT, the most important **irrational numbers** are $\sqrt{2}, \sqrt{3},$ and π.

5. ADDING/SUBTRACTING SIGNED NUMBERS

To **add a positive and a negative number,** first ignore the signs and find the positive difference between the number parts. Then attach the sign of the larger original number part. For example, to add 23 and −34, first ignore the minus sign and find the positive difference between 23 and 34—that's 11. Then attach the sign of the number with the larger number part—in this case, it's the minus sign from the −34. So 23 + (−34) = −11.

Make **subtraction** situations simpler by turning them into addition. For example, you can think of −17 − (−21) as −17 + (+21) or −17 − 21 as −17 + (−21).

To **add or subtract a string of positives and negatives,** first turn everything into addition. Then combine the positives and negatives so that the string is reduced to the sum of a single positive number and a single negative number.

6. MULTIPLYING/DIVIDING SIGNED NUMBERS

To multiply and/or divide positives and negatives, treat the number parts as usual and **attach a negative sign if there were originally an odd number of negatives.** For example, to multiply –2, –3, and –5, first multiply the number parts: $2 \times 3 \times 5 = 30$. Then go back and note that there were *three*—an *odd* number—negatives, so the product is negative: $(-2) \times (-3) \times (-5) = -30$.

7. PEMDAS

When performing multiple operations, remember to perform them in the right order. **PEMDAS**, which stands for **Parentheses** first, then **Exponents,** then **Multiplication** and **Division** (left to right), and lastly **Addition** and **Subtraction** (left to right).

In the expression $9 - 2 \times (5 - 3)^2 + 6 \div 3$, begin with the parentheses: $(5 - 3) = 2$. Then do the exponent: $2^2 = 4$. Now the expression is: $9 - 2 \times 4 + 6 \div 3$. Next do the multiplication and division to get: $9 - 8 + 2$, which equals 3. If you have difficulty remembering PEMDAS, use this sentence to recall it: Please Excuse My Dear Aunt Sally.

8. ABSOLUTE VALUE

The absolute value of a number is its **distance from zero** on the number line. To calculate the absolute value of a positive number, leave it alone. To calculate the absolute value of a negative number, make it positive.

Treat absolute value signs like **parentheses.** Do what's inside them first and then take the absolute value of the result. Don't take the absolute value of each piece between the bars before calculating. In order to calculate $|(-12) + 5 - (-4)| - |5 + (-10)|$, first do what's inside the bars to get: $|-3| - |-5|$, which is $3 - 5$, or -2.

9. COUNTING CONSECUTIVE INTEGERS

To count the number of consecutive integers between two given numbers, **subtract the smallest from the largest and add 1.** To count the number of integers from 13 through 31, subtract: $31 - 13 = 18$. Then add 1: $18 + 1 = 19$.

DIVISIBILITY

10. FACTOR/MULTIPLE

The **factors** of an integer, n, are the positive integers that divide into n with no remainder. The **multiples** of n are the integers that n divides into with no remainder. For example, 6 is a factor of 12, and 24 is a multiple of 12. And 12 is both a factor and a multiple of itself, since $12 \times 1 = 12$ and $12 \div 1 = 12$.

11. PRIME FACTORIZATION

A **prime number** is a positive integer that has exactly two positive integer factors: 1 and the integer itself. The first eight prime numbers are 2, 3, 5, 7, 11, 13, 17, and 19.

To find the prime factorization of an integer, continue factoring until **all the factors are prime.** For example, to factor 36:

$$36 = 4 \times 9 = 2 \times 2 \times 3 \times 3$$

12. RELATIVE PRIMES

Relative primes are integers that have no common factor other than 1. To determine whether two integers are relative primes, break them both down into their prime factorizations. For example: $35 = 5 \times 7$, and $54 = 2 \times 3 \times 3 \times 3$. They have **no prime factors in common,** so 35 and 54 are relative primes.

13. COMMON MULTIPLE

A common multiple is a number that is a multiple of two or more integers. You can always get a common multiple of two numbers by **multiplying** them, but unless the two numbers are relative primes, the product will not be the *least* common multiple. For example, to find a common multiple for 12 and 15, you could just multiply: $12 \times 15 = 180$.

14. LEAST COMMON MULTIPLE (LCM)

To find the **least common multiple,** examine the **multiples of the larger integer** until you find one that's **also a multiple of the smaller.** To find the LCM of 12 and 15, begin by taking the multiples of 15: 15 is not divisible by 12; 30 is not; nor is 45. But the next multiple of 15, 60, *is* divisible by 12, so it's the LCM.

15. GREATEST COMMON FACTOR (GCF)

To find the greatest common factor, break down the integers into their prime factorizations, and **multiply all the prime factors they have in common.** For example, $36 = 2 \times 2 \times 3 \times 3$, and $48 = 2 \times 2 \times 2 \times 2 \times 3$. These integers have a 2×2 and a 3 in common, so the GCF is $2 \times 2 \times 3 = 12$.

16. EVEN/ODD

To predict whether a sum, difference, or product will be even or odd, just **take simple numbers like 1 and 2 and see what happens.** There are rules—"odd times even is even," for example—but there's no need to memorize them. What happens with one set of numbers generally happens with all similar sets.

17. MULTIPLES OF 2 AND 4

An integer is divisible by 2 (even) if the **last digit is even.** An integer is divisible by 4 if the **last two digits form a multiple of 4.** The last digit of 562 is 2, which is even, so 562 is a multiple of 2. The last two digits form 62, which is *not* divisible by 4, so 562 is not a multiple of 4. The integer 512, however, is divisible by 4 because the last two digits form 12, which is a multiple of 4.

18. MULTIPLES OF 3 AND 9

An integer is divisible by 3 if the **sum of its digits is divisible by 3.** An integer is divisible by 9 if the **sum of its digits is divisible by 9.** The sum of the digits in 957 is 21, which is divisible by 3 but not by 9, so 957 is divisible by 3 but not by 9.

19. MULTIPLES OF 5 AND 10

An integer is divisible by 5 if the **last digit is 5 or 0.** An integer is divisible by 10 if the **last digit is 0.** The last digit of 665 is 5, so 665 is a multiple of 5 but *not* a multiple of 10.

20. REMAINDERS

The remainder is the **whole number left over after division.** For example, 487 is 2 more than 485, which is a multiple of 5, so when 487 is divided by 5, the remainder is 2.

FRACTIONS AND DECIMALS

21. REDUCING FRACTIONS

To reduce a fraction to lowest terms, **factor out and cancel** all factors the numerator and denominator have in common.

$$\frac{28}{36} = \frac{4 \times 7}{4 \times 9} = \frac{7}{9}$$

22. ADDING/SUBTRACTING FRACTIONS

To add or subtract fractions, first find a **common denominator,** and then add or subtract the numerators.

$$\frac{2}{15} + \frac{3}{10} = \frac{4}{30} + \frac{9}{30} = \frac{4+9}{30} = \frac{13}{30}$$

23. MULTIPLYING FRACTIONS

To multiply fractions, **multiply** the numerators and **multiply** the denominators.

$$\frac{5}{7} \times \frac{3}{4} = \frac{5 \times 3}{7 \times 4} = \frac{15}{28}$$

24. DIVIDING FRACTIONS

To divide fractions, **invert** the second one and **multiply.**

$$\frac{1}{2} \div \frac{3}{5} = \frac{1}{2} \times \frac{5}{3} = \frac{1 \times 5}{2 \times 3} = \frac{5}{6}$$

25. CONVERTING A MIXED NUMBER TO AN IMPROPER FRACTION

To convert a mixed number to an improper fraction, **multiply** the whole number part by the denominator, then **add** the numerator. The result is the new numerator (over the same denominator). To convert $7\frac{1}{3}$, first multiply 7 by 3, then add 1, to get the new numerator of 22. Put that over the same denominator, 3, to get $\frac{22}{3}$.

26. CONVERTING AN IMPROPER FRACTION TO A MIXED NUMBER

To convert an improper fraction to a mixed number, **divide** the denominator into the numerator to get a **whole number quotient with a remainder.** The quotient becomes the whole number part of the mixed number, and the remainder becomes the new numerator—with the same denominator. For example, to convert $\frac{108}{5}$, first divide 5 into 108, which yields 21 with a remainder of 3. Therefore, $\frac{108}{5} = 21\frac{3}{5}$.

27. RECIPROCAL

To find the reciprocal of a fraction, **switch the numerator and the denominator**. The reciprocal of $\frac{3}{7}$ is $\frac{7}{3}$. The reciprocal of 5 is $\frac{1}{5}$. The product of a number and its reciprocal is 1.

28. COMPARING FRACTIONS

One way to compare fractions is to **re-express them with a common denominator.**

$\frac{3}{4} = \frac{21}{28}$ and $\frac{5}{7} = \frac{20}{28}$. Because $\frac{21}{28}$ is greater than $\frac{20}{28}$, $\frac{3}{4}$ is greater than $\frac{5}{7}$.

Another way to compare fractions is to **convert them both to decimals.** For example, $\frac{3}{4}$ converts to 0.75, and $\frac{5}{7}$ converts to approximately 0.714, and 0.75 is greater than 0.714.

29. CONVERTING FRACTIONS TO DECIMALS

To convert a fraction to a decimal, **divide the bottom into the top.** To convert $\frac{5}{8}$, divide 8 into 5, yielding 0.625.

30. REPEATING DECIMALS

To find a particular digit in a repeating decimal, note the **number of digits in the cluster that repeats.** If there are two digits in that cluster, then every second digit is the same. If there are three digits in that cluster, then every third digit is the same, and so on. For example, the decimal equivalent of $\frac{1}{27}$ is 0.037037037…, which is best written as $0.\overline{037}$.

There are three digits in the repeating cluster, so every third digit is the same: 7. To find the 50th digit, look for the multiple of 3 just less than 50—that's 48. The 48th digit is 7, and with the 49th digit, the pattern repeats beginning with 0 again. The 50th digit is 3.

31. IDENTIFYING THE PARTS AND THE WHOLE

The key to solving most fraction and percent word problems is to identify the part and the whole. Usually, you'll find the **part** associated with the verb *is/are* and the **whole** associated with the word *of.* In the sentence "Half of the boys are blonds," the whole is the boys ("*of* the boys"), and the part is the blonds ("*are* blonds").

PERCENTS

32. PERCENT FORMULA

Whether you need to find the part, the whole, or the percent, use the same formula:

Part = Percent × Whole

Example: What is 12% of 25?
Setup: Part = 0.12 × 25

Example: 15 is 3% of what number?
Setup: 15 = 0.03 × Whole

Example: 45 is what percent of 9?
Setup: 45 = Percent × 9

33. PERCENT INCREASE AND DECREASE

To increase a number by a percent, **add the percent to 100%,** convert to a decimal, and multiply. To increase 40 by 25%, add 25% to 100%, convert 125% to 1.25, and multiply by 40 to get 1.25 × 40 = 50.

34. FINDING THE ORIGINAL WHOLE

To find the **original whole before a percent increase or decrease,** set up an equation. Think of a 15% increase over x as $1.15x$.

Example: After a 5% increase, the population was 59,346. What was the population
               before the increase?
Setup: 1.05x = 59,346

35. COMBINED PERCENT INCREASE AND DECREASE

To determine the combined effect of multiple percents increase and/or decrease, **start with 100 and see what happens.**

Example: A price went up 10% one year, and the new price went up 20% the next year. What was the combined percent increase?

Setup: First year: 100 + (10% of 100) = 110. Second year: 110 + (20% of 110) = 132. That's a combined 32% increase.

RATIOS, PROPORTIONS, AND RATES

36. SETTING UP A RATIO

To find a ratio, put the number associated with the word **of on top** and the quantity associated with the word **to on the bottom** and reduce. The ratio of 20 oranges to 12 apples is $\frac{20}{12}$, which reduces to $\frac{5}{3}$.

37. PART-TO-PART AND PART-TO-WHOLE RATIOS

If the parts add up to the whole, a part-to-part ratio can be turned into two part-to-whole ratios by putting **each number in the original ratio over the sum of the numbers.** If the ratio of males to females is 1 to 2, then the males-to-people ratio is $\frac{1}{1+2} = \frac{1}{3}$, and the females-to-people ratio is $\frac{2}{1+2} = \frac{2}{3}$. In other words, $\frac{2}{3}$ of all the people are female.

38. SOLVING A PROPORTION

To solve a proportion, **cross multiply:**

$$\frac{x}{5} = \frac{3}{4}$$
$$4x = 5 \times 3$$
$$x = \frac{15}{4} = 3.75$$

39. RATE

To solve a rate problem, **use the units** to keep things straight.

Example: If snow is falling at the rate of 1 foot every 4 hours, how many inches of snow will fall in 7 hours?

Setup:

$$\frac{1 \text{ foot}}{4 \text{ hours}} = \frac{x \text{ inches}}{7 \text{ hours}}$$
$$\frac{12 \text{ inches}}{4 \text{ hours}} = \frac{x \text{ inches}}{7 \text{ hours}}$$
$$4x = 12 \times 7$$
$$x = 21$$

40. AVERAGE RATE

Average rate is *not* simply the average of the rates.

$$\text{Average } A \text{ per } B = \frac{\text{Total } A}{\text{Total } B}$$

$$\text{Average Speed} = \frac{\text{Total distance}}{\text{Total time}}$$

To find the average speed for 120 miles at 40 mph and 120 miles at 60 mph, **don't just average the two speeds.** First, figure out the total distance and the total time. The total distance is $120 + 120 = 240$ miles. The times are 3 hours for the first leg and 2 hours for the second leg, or 5 hours total. The average speed, then, is $\frac{240}{5} = 48$ miles per hour.

AVERAGES

41. AVERAGE FORMULA

To find the average of a set of numbers, **add the terms and divide by the number of terms.**

$$\text{Average} = \frac{\text{Sum of the terms}}{\text{Number of terms}}$$

To find the average of the five numbers 12, 15, 23, 40, and 40, first add them: $12 + 15 + 23 + 40 + 40 = 130$. Then divide the sum by 5 to get $130 \div 5 = 26$.

42. AVERAGE OF EVENLY SPACED NUMBERS

To find the average of evenly spaced numbers, just **average the smallest and the largest.** The average of all the integers from 13 through 77 is the same as the average of 13 and 77.

$$\frac{13 + 77}{2} = \frac{90}{2} = 45$$

43. USING THE AVERAGE TO FIND THE SUM

Sum = (Average) × (Number of terms)

If the average of ten numbers is 50, then they add up to 10×50, or 500.

44. FINDING THE MISSING NUMBER

To find a missing number when you're given the average, **use the sum.** If the average of four numbers is 7, then the sum of those four numbers is 4×7, or 28. Suppose that three of the numbers are 3, 5, and 8. These three numbers add up to 16 of that 28, which leaves 12 for the 4th number.

POSSIBILITIES AND PROBABILITY

45. COUNTING THE POSSIBILITIES

The fundamental counting principle: if there are *m* **ways** one event can happen and *n* **ways** a second event can happen, then there are *m* × *n* **ways** for the two events to happen. For example, with 5 shirts and 7 pairs of pants to choose from, you can put together 5 × 7 = 35 different outfits.

46. PROBABILITY

$$\text{Probability} = \frac{\text{Favorable outcomes}}{\text{Total possible outcomes}}$$

If you have 12 shirts in a drawer and 9 of them are white, the probability of picking a white shirt at random is $\frac{9}{12} = \frac{3}{4}$. This probability can also be expressed as 0.75 or 75%.

POWERS AND ROOTS

47. MULTIPLYING AND DIVIDING POWERS

To multiply powers of the same base, **add the exponents and keep the same base:**

$$x^3 \times x^4 = x^{3+4} = x^7$$

To divide powers of the same base, **subtract the exponents and keep the same base:**

$$y^{13} \div y^8 = y^{13-8} = y^5$$

48. RAISING POWERS TO POWERS

To raise a power to an exponent, **multiply the exponents:**

$$(x^3)^4 = x^{3 \times 4} = x^{12}$$

49. SIMPLIFYING SQUARE ROOTS

To simplify a square root, **factor out the perfect squares** under the radical, unsquare them, and write the result in front:

$$\sqrt{12} = \sqrt{4 \times 3} = \sqrt{4} \times \sqrt{3} = 2\sqrt{3}$$

50. ADDING AND SUBTRACTING ROOTS

You can add or subtract radical expressions only if the part under the radicals is the same:

$$2\sqrt{3} + 3\sqrt{3} = 5\sqrt{3}$$

Don't try to add or subtract when the radical parts are different. There's not much you can do with an expression like:

$$3\sqrt{5} + 3\sqrt{7}$$

51. MULTIPLYING AND DIVIDING ROOTS

The product of square roots is equal to the **square root of the product:**

$$\sqrt{3} \times \sqrt{5} = \sqrt{3 \times 5} = \sqrt{15}$$

The quotient of square roots is equal to the **square root of the quotient:**

$$\frac{\sqrt{6}}{\sqrt{3}} = \sqrt{\frac{6}{3}} = \sqrt{2}$$

ALGEBRAIC EXPRESSIONS

52. EVALUATING AN EXPRESSION

To evaluate an algebraic expression, **plug in** the given values for the unknowns and calculate according to **PEMDAS**. To find the value of $x^2 + 5x - 6$ when $x = -2$, plug in -2 for x:

$$(-2)^2 + 5(-2) - 6 = 4 - 10 - 6 = -12$$

53. ADDING AND SUBTRACTING MONOMIALS

To combine like terms, **keep the variable part unchanged while adding or subtracting the coefficients:** $2a + 3a = (2 + 3)a = 5a$

54. ADDING AND SUBTRACTING POLYNOMIALS

To add or subtract polynomials, **combine like terms:**

$$(3x^2 + 5x - 7) - (x^2 + 12) =$$

$$(3x^2 - x^2) + 5x + (-7) - 12 = 2x^2 + 5x - 19$$

55. MULTIPLYING MONOMIALS

To multiply monomials, **multiply the coefficients and the variables separately:**

$$2a \times 3a = (2 \times 3)(a \times a) = 6a^2$$

56. MULTIPLYING BINOMIALS—FOIL

To multiply binomials, use **FOIL.** To multiply $(x + 3)$ by $(x + 4)$, first multiply the **F**irst terms: $x \times x = x^2$. Next, the **O**uter terms: $x \times 4 = 4x$. Then, the **I**nner terms: $3 \times x = 3x$. And finally, the **L**ast terms: $3 \times 4 = 12$. Then combine like terms:

$$x^2 + 4x + 3x + 12 = x^2 + 7x + 12$$

57. MULTIPLYING OTHER POLYNOMIALS

FOIL works only when you want to multiply two binomials. If you want to multiply polynomials with more than two terms, make sure you **multiply each term in the first polynomial by each term in the second:**

$$(x^2 + 3x + 4)(x + 5)$$

$$= x^2(x + 5) + 3x(x + 5) + 4(x + 5)$$

$$= x^3 + 5x^2 + 3x^2 + 15x + 4x + 20$$

$$= x^3 + 8x^2 + 19x + 20$$

After multiplying two polynomials together, the number of terms in your expression should equal the number of terms in one polynomial multiplied by the number of terms in the second. In the example, you should have $3 \times 2 = 6$ terms in the product before you combine like terms.

FACTORING ALGEBRAIC EXPRESSIONS

58. FACTORING OUT A COMMON DIVISOR

A factor common to all terms of a polynomial can be **factored out.** All three terms in the polynomial $3x^3 + 12x^2 - 6x$ contain a factor of $3x$. Pulling out the common factor yields $3x(x^2 + 4x - 2)$.

59. FACTORING THE DIFFERENCE OF SQUARES

One of the test maker's favorite factorables is the **difference of squares**.

$$a^2 - b^2 = (a - b)(a + b)$$

$x^2 - 9$, for example, factors to $(x - 3)(x + 3)$.

60. FACTORING THE SQUARE OF A BINOMIAL

Learn to recognize polynomials that are squares of binomials:

$$a^2 + 2ab + b^2 = (a + b)^2$$
$$a^2 - 2ab + b^2 = (a - b)^2$$

For example, $4x^2 + 12x + 9$ factors to $(2x + 3)^2$, and $n^2 - 10n + 25$ factors to $(n - 5)^2$.

61. FACTORING OTHER POLYNOMIALS—FOIL IN REVERSE

To factor a quadratic expression, **think about what binomials you could use FOIL on to get that quadratic expression.** To factor $x^2 - 5x + 6$, think about what **F**irst terms will produce x^2, what **L**ast terms will produce $+6$, and what **O**uter and **I**nner terms will produce $-5x$. Some common sense—and a little trial and error—lead you to $(x - 2)(x - 3)$.

62. SIMPLIFYING AN ALGEBRAIC FRACTION

Simplifying an algebraic fraction is a lot like simplifying a numerical fraction. The general idea is to **find factors common to the numerator and denominator and cancel them.** Thus, simplifying an algebraic fraction begins with factoring.

To simplify $\dfrac{x^2 - x - 12}{x^2 - 9}$, first factor the numerator and denominator:

$$\frac{x^2 - x - 12}{x^2 - 9} = \frac{(x - 4)(x + 3)}{(x - 3)(x + 3)}$$

Canceling $x + 3$ from the numerator and denominator leaves $\dfrac{x - 4}{x - 3}$.

SOLVING EQUATIONS

63. SOLVING A LINEAR EQUATION

To solve an equation, do whatever is necessary to **isolate the variable.** Be sure to follow the rule: Whatever you do to one side of an equation, you must do to the other side. To solve the equation $5x - 12 = -2x + 9$, first get all the xs on one side by adding $2x$ to both sides: $7x - 12 = 9$. Then add 12 to both sides: $7x = 21$. Then divide both sides by 7 to arrive at $x = 3$.

64. SOLVING "IN TERMS OF"

To solve an equation for one variable **in terms of** another means to **isolate the one variable on one side of the equation,** leaving an expression containing the other variable on the other side of the equation. To solve $3x - 10y = -5x + 6y$ for x in terms of y, isolate x:

$$
\begin{aligned}
3x - 10y &= -5x + 6y \\
3x + 5x &= 6y + 10y \\
8x &= 16y \\
x &= 2y
\end{aligned}
$$

65. TRANSLATING FROM ENGLISH INTO ALGEBRA

To translate from English into algebra, look for key words and systematically turn phrases into algebraic expressions and sentences into equations. Be careful about order, especially when subtraction is called for.

Example: Celine and Remi play tennis. Last year, Celine won 3 more than twice the number of matches that Remi won. If Celine won 11 more matches than Remi, how many matches did Celine win?

Set Up: You are given two sets of information. One way to solve this is to write a system of equations–one equation for each set of information. Use variables that relate well with what they represent. For example, use r to represent Remi's winning matches. Use c to represent Celine's winning matches. The phrase "Celine won 3 more than twice Remi" can be written as $c = 2r + 3$. The phrase "Celine won 11 more matches than Remi" can be written as $c = r + 11$.

INTERMEDIATE ALGEBRA

66. SOLVING A QUADRATIC EQUATION

To solve a quadratic equation, put it in the form $ax^2 + bx + c = 0$, **factor** the left side (if you can), and set each factor equal to 0 separately to get the two solutions. To solve $x^2 + 12 = 7x$, first rewrite it as $x^2 - 7x + 12 = 0$. Then factor the left side:

$$(x - 3)(x - 4) = 0$$

$$x - 3 = 0 \quad \text{or} \quad x - 4 = 0$$

$$x = 3 \quad \text{or} \quad x = 4$$

Sometimes, the left side might not be obviously factorable, or not factorable at all. You can always use the **quadratic formula.** Just plug in the coefficients a, b, and c from $ax^2 + bx + c = 0$ into the formula:

$$x = \frac{-b \pm \sqrt{b^2 - 4ac}}{2a}$$

To solve $x^2 + 4x + 2 = 0$, plug $a = 1$, $b = 4$, and $c = 2$ into the formula:

$$x = \frac{-4 \pm \sqrt{4^2 - 4 \times 1 \times 2}}{2 \times 1}$$

$$= \frac{-4 \pm \sqrt{8}}{2} = -2 \pm \sqrt{2}$$

67. SOLVING A SYSTEM OF EQUATIONS

You can solve for two variables only if you have two distinct equations. Two forms of the same equation will not be adequate. **Combine the equations in such a way that one of the variables cancels out.** To solve the two equations $4x + 3y = 8$ and $x + y = 3$, multiply both sides of the second equation by -3 to get: $-3x - 3y = -9$. Then add the equations; the $3y$ and the $-3y$ cancel out, leaving $x = -1$. Plug that back into either of the original equations, and you'll find that $y = 4$.

68. SOLVING AN EQUATION THAT INCLUDES ABSOLUTE VALUE SIGNS

To solve an equation that includes absolute value signs, **remember that there are two different solutions.** For example, to solve the equation $|x - 12| = 3$, think of it as two equations:

$$x - 12 = 3 \quad \text{or} \quad x - 12 = -3$$
$$x = 15 \qquad\qquad x = 9$$

69. SOLVING AN INEQUALITY

Solve an inequality the same way you would an equation. Just remember that if you **multiply or divide both sides by a negative number,** you must **reverse the inequality symbol.** To solve $-5x + 7 < -3$, subtract 7 from both sides to get $-5x < -10$. Then divide both sides by -5, remembering to reverse the sign: $x > 2$.

70. GRAPHING INEQUALITIES

To graph a range of values, use a thick black line over the number line, and at the end(s) of the range, use a **solid circle** if the point **is included** or an **open circle** if the point is **not included.** The figure here shows the graph of $-3 < x \le 5$.

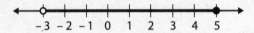

COORDINATE GEOMETRY

71. FINDING THE DISTANCE BETWEEN TWO POINTS

To find the distance between two coordinate points, **use the Pythagorean theorem or special right triangles.** The difference between the xs is one leg and the difference between the ys is the other leg.

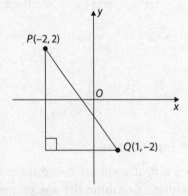

In the figure above, $\overline{PQ}$ is the hypotenuse of a 3-4-5 triangle, so $PQ = 5$.

You can also use the **Distance formula:**

$$d = \sqrt{(x_2 - x_1)^2 + (y_2 - y_1)^2}$$

To find the distance between $R(3,6)$ and $S(5,-2)$:

$$\begin{aligned} d &= \sqrt{(5-3)^2 + (-2-6)^2} \\ &= \sqrt{(2)^2 + (-8)^2} \\ &= \sqrt{68} = 2\sqrt{17} \end{aligned}$$

72. USING TWO POINTS TO FIND THE SLOPE

In mathematics, the slope of a line is often represented by m.

$$\text{Slope} = m = \frac{\text{Change in } y}{\text{Change in } x} = \frac{\text{Rise}}{\text{Run}}$$

The slope of the line that contains the points $A(2,3)$ and $B(0,-1)$ is:

$$\frac{Y_B - Y_A}{X_B - X_A} = \frac{-1 - 3}{0 - 2} = \frac{-4}{-2} = 2$$

73. USING AN EQUATION TO FIND THE SLOPE

To find the slope of a line from an equation, put the equation into **slope-intercept** form:

$$y = mx + b$$

The **slope is m.** To find the slope of the equation $3x + 2y = 4$, rearrange it:

$$\begin{aligned} 3x + 2y &= 4 \\ 2y &= -3x + 4 \\ y &= -\frac{3}{2}x + 2 \end{aligned}$$

The slope is $-\dfrac{3}{2}$.

74. USING AN EQUATION TO FIND AN INTERCEPT

To find the y-intercept, you can either put the equation into $y = mx + b$ **(slope-intercept)** form—in which case b **is the y-intercept**—or you can just **plug $x = 0$ into the equation** and **solve for y.** To find the x-intercept, plug $y = 0$ into the equation and **solve for x.**

75. EQUATION FOR A CIRCLE

The equation for a circle of radius r centered at (h,k) is:

$$(x - h)^2 + (y - k)^2 = r^2$$

The following figure shows the graph of the equation $(x - 2)^2 + (y + 1)^2 = 25$:

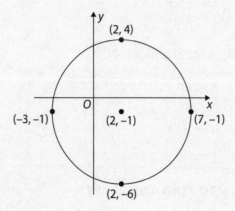

76. EQUATION FOR A PARABOLA

The graph of an equation in the form $y = ax^2 + bx + c$ is a parabola (a U-shaped curve). The figure below shows the graph of seven pairs of numbers that satisfy the equation $y = x^2 - 4x + 3$:

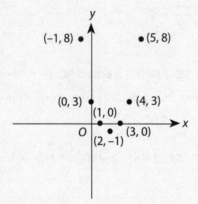

77. EQUATION FOR AN ELLIPSE

The graph of an equation in the form

$$\frac{x^2}{a^2} + \frac{y^2}{b^2} = 1$$

is an ellipse with $2a$ as the sum of the focal radii and with foci on the x-axis at $(0, -c)$ and $(0, c)$, where $c = \sqrt{a^2 - b^2}$. The following figure shows the graph of $\frac{x^2}{25} + \frac{y^2}{16} = 1$:

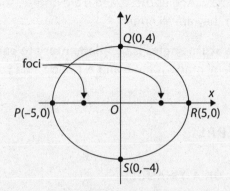

The foci are at (–3,0) and (3,0). $\overline{PR}$ is the **major axis,** and $\overline{QS}$ is the **minor axis.** This ellipse is symmetrical about both the *x*- and *y*-axes.

LINES AND ANGLES

78. INTERSECTING LINES

When two lines intersect, **adjacent angles are supplementary** and **vertical angles are equal.**

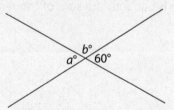

In the figure above, the angles marked *a°* and *b°* are adjacent and supplementary, so $a + b = 180$. Furthermore, the angles marked *a°* and 60° are vertical and have equal measures, so $a = 60$.

79. PARALLEL LINES AND TRANSVERSALS

A transversal across parallel lines forms **four equal acute angles and four equal obtuse angles.** If the transversal meets the lines at a right angle, then all eight angles are right angles.

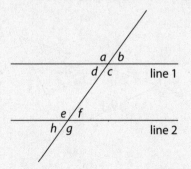

Here, line 1 is parallel to line 2. Angles *a*, *c*, *e*, and *g* are obtuse, so they are all equal. Angles *b*, *d*, *f*, and *h* are acute, so they are all equal.

Furthermore, **each of the acute angles is supplementary to each of the obtuse angles.** Angles *a* and *h* are supplementary, as are *b* and *e*, *c* and *f*, and so on.

TRIANGLES—GENERAL

80. INTERIOR ANGLES OF A TRIANGLE

The three angles of any triangle **add up to 180°.**

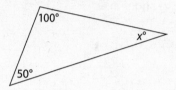

In the figure above, $x + 50 + 100 = 180$, so $x = 30$.

81. EXTERIOR ANGLES OF A TRIANGLE

An exterior angle of a triangle is equal to the sum of the remote interior angles.

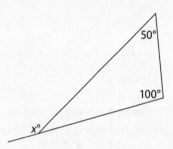

In the figure above, the exterior angle labeled x° is equal to the sum of the remote interior angles:

$x = 50 + 100 = 150$

The three exterior angles of any triangle add up to 360°.

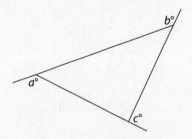

In the figure above, $a + b + c = 360$.

82. SIMILAR TRIANGLES

Similar triangles have the same shape: corresponding angles are equal and corresponding sides are proportional.

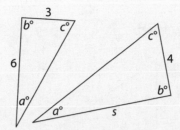

The triangles are similar because they have the same angles. The side of length 3 in the first triangle corresponds to the side of length 4 in the second triangle, and the side of length 6 in the first corresponds to the side of length s in the second.

$$\frac{3}{4} = \frac{6}{s}$$
$$3s = 24$$
$$s = 8$$

83. AREA OF A TRIANGLE

$$\text{Area of Triangle} = \frac{1}{2}(\text{base})(\text{height})$$

The height is the perpendicular distance between the side that's chosen as the base and the opposite vertex.

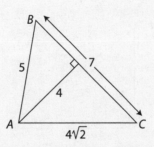

In the triangle above, 4 is the height when side BC is chosen as the base.

$$\text{Area} = \frac{1}{2}bh = \frac{1}{2}(7)(4) = 14$$

RIGHT TRIANGLES

84. PYTHAGOREAN THEOREM

For all right triangles:

$$(\text{leg}_1)^2 + (\text{leg}_2)^2 = (\text{hypotenuse})^2$$

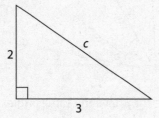

If one leg has length 2 and the other leg has length 3, then:

$$2^2 + 3^2 = c^2$$
$$c^2 = 4 + 9$$
$$c = \sqrt{13}$$

85. SPECIAL RIGHT TRIANGLES

• 3-4-5

If a right triangle's leg-to-leg ratio is 3:4, or if the leg-to-hypotenuse ratio is 3:5 or 4:5, then it's a 3-4-5 triangle and you don't need to use the Pythagorean theorem to find the third side. Just figure out what multiple of 3-4-5 it is.

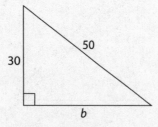

In the right triangle above, one leg has length 30 and the hypotenuse has length 50. This is 10 times 3-4-5, so the other leg has length 40.

• 5-12-13

If a right triangle's leg-to-leg ratio is 5:12, or if the leg-to-hypotenuse ratio is 5:13 or 12:13, then it's a 5-12-13 triangle and you don't need to use the Pythagorean theorem to find the third side. Just figure out what multiple of 5-12-13 it is.

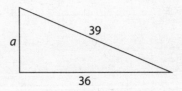

Here, one leg has length 36 and the hypotenuse has length 39. This is 3 times 5-12-13, so the other leg has length 15.

• **30°-60°-90°**

The sides of a 30°-60°-90° triangle are in a ratio of $1:\sqrt{3}:2$. You don't need to use the Pythagorean theorem.

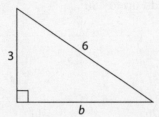

If the hypotenuse has length 6, then the shorter leg is half that, or 3, and the longer leg is equal to the short leg times $\sqrt{3}$, or $b = 3\sqrt{3}$.

• **45°-45°-90°**

The sides of a 45°-45°-90° triangle are in a ratio of $1:1:\sqrt{2}$.

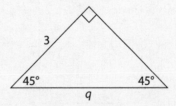

If one leg has length 3, then the other leg also has length 3, and the hypotenuse is equal to a leg times $\sqrt{2}$, or $q = 3\sqrt{2}$.

OTHER POLYGONS

86. SPECIAL QUADRILATERALS

Rectangle

A rectangle is a **four-sided figure with four right angles.** Opposite sides are equal. Diagonals are equal.

Quadrilateral *ABCD* above has three right angles. The fourth angle therefore also measures 90°, and *ABCD* is a rectangle. The **perimeter** of a rectangle is equal to the sum of the lengths of the four sides, which is equivalent to 2(length + width) or 2(length) + 2(width).

Parallelogram

A parallelogram has **two pairs of parallel sides.** Opposite sides are equal. Opposite angles are equal. Consecutive angles add up to 180°.

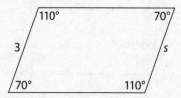

In the figure above, *s* is the length of the side opposite the 3, so $s = 3$.

Square

A square is a **rectangle with four equal sides.**

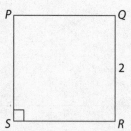

If *PQRS* is a square, all sides are the same length as *QR*. The **perimeter** of a square is equal to four times the length of one side.

Trapezoid

A **trapezoid** is a quadrilateral with one pair of parallel sides and one pair of nonparallel sides.

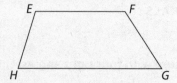

In the preceding quadrilateral, sides *EF* and *GF* are parallel, while sides *EH* and *FG* are not parallel. *EFGH* is therefore a trapezoid.

87. AREAS OF SPECIAL QUADRILATERALS

Area of Rectangle = Length × Width

The area of a 7-by-3 rectangle is $7 \times 3 = 21$.

Area of Parallelogram = Base × Height

The area of a parallelogram with a height of 4 and a base of 6 is $4 \times 6 = 24$.

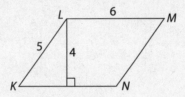

Area of Square = (Side)²

The area of a square with sides of length 5 is $5^2 = 25$.

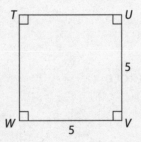

Area of Trapezoid = $\left(\dfrac{\text{base}_1 + \text{base}_2}{2} \right) \times \text{height}$

Think of it as the average of the bases (the two parallel sides) times the height (the length of the perpendicular altitude).

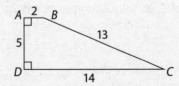

In trapezoid *ABCD*, you can use side *AD* for the height. The average of the bases is $\frac{2 + 14}{2} = 8$, so the area is 5×8, or 40.

88. INTERIOR ANGLES OF A POLYGON

The sum of the measures of the interior angles of a polygon is $(n - 2) \times 180$ degrees, where *n* is the number of sides.

Sum of the Angles = $(n - 2) \times 180$ degrees

The eight angles of an octagon, for example, add up to $(8 - 2) \times 180° = 1,080°$.

To find **one angle of a regular polygon,** divide the sum of the angles by the number of angles (which is the same as the number of sides). The formula, therefore, is:

$$\text{Interior Angle} = \frac{(n - 2) \times 180}{n}$$

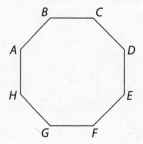

Each angle of the regular octagon above measures $\frac{1,080}{8} = 135$ degrees.

CIRCLES

89. CIRCUMFERENCE OF A CIRCLE

Circumference of a Circle = $2\pi r$

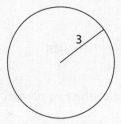

Here, the radius has a length of 3, so the circumference is $2\pi(3) = 6\pi$.

90. LENGTH OF AN ARC

An **arc** is a piece of the circumference of a circle. If n is the measure of the arc's central angle, then the formula for finding the length of the arc is:

Length of Arc $= \dfrac{n}{360}(2\pi)$

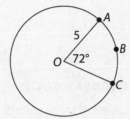

In the figure, the radius has length 5 and the measure of the central angle is 72°. The arc length is $\dfrac{72}{360}$, or $\dfrac{1}{5}$ of the circumference:

$$\left(\frac{72}{360}\right)2\pi\,(5) = \left(\frac{1}{5}\right)10\pi = 2\pi$$

91. AREA OF A CIRCLE

Area of a Circle $= \pi r^2$

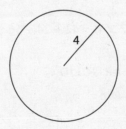

The area of the circle above is $\pi(4)^2 = 16\pi$.

92. AREA OF A SECTOR

A **sector** is a piece of the area of a circle. If n is the measure of the sector's central angle, then the formula is:

Area of a Sector $= \left(\dfrac{n}{360}\right)\left(\pi r^2\right)$

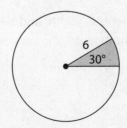

In the figure, the radius has length 6 and the measure of the sector's central angle is 30°. The sector represents $\frac{30}{360}$, or $\frac{1}{12}$ of the area of the circle:

$$\left(\frac{30}{360}\right)(\pi)\left(6^2\right) = \left(\frac{1}{12}\right)(36\pi) = 3\pi$$

SOLIDS

93. SURFACE AREA OF A RECTANGULAR SOLID

The surface of a rectangular solid consists of three pairs of identical faces. To find the surface area, find the area of each face and add them up. If the length is l, the width is w, and the height is h, the formula is:

Surface Area = 2*lw* + 2*wh* + 2*lh*

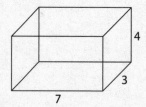

The surface area of the box above is:

$$(2 \times 7 \times 3) + (2 \times 3 \times 4) + (2 \times 7 \times 4) = 42 + 24 + 56 = 122$$

94. VOLUME OF A RECTANGULAR SOLID

Volume of a Rectangular Solid = *lwh*

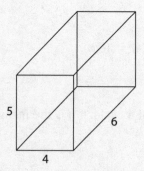

The volume of a 4-by-5-by-6 box is $4 \times 5 \times 6 = 120$.

A cube is a rectangular solid with length, width, and height all equal. If e is the length of an edge of the cube, the volume formula is:

Volume of a Cube = e^3

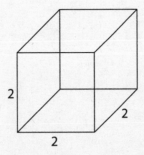

The volume of the cube above is $2^3 = 8$.

95. VOLUME OF OTHER SOLIDS

Volume of a Cylinder = $\pi r^2 h$

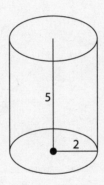

The volume of a cylinder where $r = 2$ and $h = 5$ is: volume $= \pi(2^2)(5) = 20\pi$.

Volume of a Cone $= \dfrac{1}{3}\pi r^2 h$

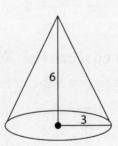

The volume of a cone where $r = 3$ and $h = 6$ is:

$$\text{Volume} = \frac{1}{3}\pi\left(3^2\right)(6) = 18\pi$$

Volume of a Sphere $= \dfrac{4}{3}\pi r^3$

The volume of a sphere where $r = 3$ is:

$$\text{Volume} = \frac{4}{3}\pi\left(3^3\right) = 36\pi$$

TRIGONOMETRY

96. SINE, COSINE, AND TANGENT OF ACUTE ANGLES

To find the sine, cosine, or tangent of an acute angle in a right triangle, use SOHCAHTOA, which is an abbreviation for the following definitions:

$$\text{Sine} = \frac{\text{Opposite}}{\text{Hypotenuse}}$$

$$\text{Cosine} = \frac{\text{Adjacent}}{\text{Hypotenuse}}$$

$$\text{Tangent} = \frac{\text{Opposite}}{\text{Adjacent}}$$

In the figure to the right:

$$\sin A = \frac{8}{17}$$

$$\cos A = \frac{15}{17}$$

$$\tan A = \frac{8}{15}$$

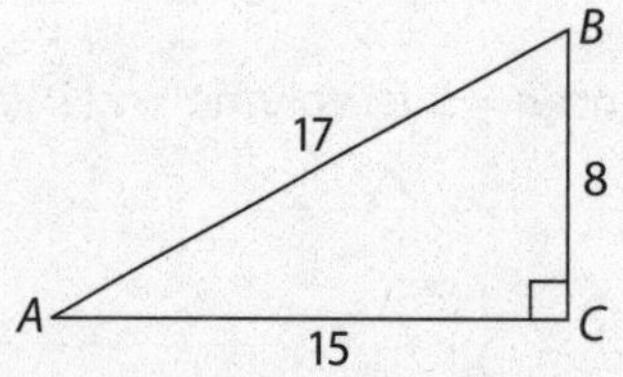

97. COTANGENT, SECANT, AND COSECANT OF ACUTE ANGLES

Think of the cotangent, secant, and cosecant as the reciprocals of the SOHCAHTOA functions:

$$\text{Cosecant} = \frac{1}{\text{Sine}} = \frac{\text{Hypotenuse}}{\text{Opposite}}$$

$$\text{Secant} = \frac{1}{\text{Cosine}} = \frac{\text{Hypotenuse}}{\text{Adjacent}}$$

$$\text{Cotangent} = \frac{1}{\text{Tangent}} = \frac{\text{Adjacent}}{\text{Opposite}}$$

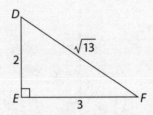

In the figure above:

$$\csc D = \frac{\sqrt{13}}{3}$$

$$\sec D = \frac{\sqrt{13}}{2}$$

$$\cot D = \frac{2}{3}$$

98. TRIGONOMETRIC FUNCTIONS OF OTHER ANGLES

To find a trigonometric function of an angle greater than 90°, sketch a circle of radius 1 and centered at the origin of the coordinate grid. Start from the point (1,0), and rotate the appropriate number of degrees counterclockwise.

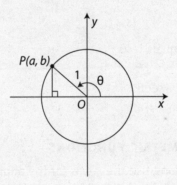

In the "unit circle" above, the basic trigonometric functions are defined in terms of the coordinates a and b:

$$\sin \theta = b$$

$$\cos \theta = a$$

$$\tan \theta = \frac{b}{a}$$

Example: $\sin 210° = ?$

Setup: Sketch a 210° angle in the coordinate plane:

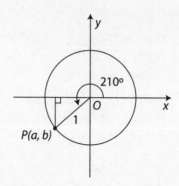

Because the triangle shown in the figure above is a 30°-60°-90° right triangle, you can determine that the coordinates of point P are $\left(-\dfrac{\sqrt{3}}{2}, -\dfrac{1}{2}\right)$.

The sine is therefore $-\dfrac{1}{2}$.

99. SIMPLIFYING TRIGONOMETRIC EXPRESSIONS

To simplify trigonometric expressions, use the inverse function definitions along with the fundamental trigonometric identity:

$$\sin^2 x + \cos^2 x = 1$$

Example: $\dfrac{\sin^2\theta + \cos^2\theta}{\cos\theta} = ?$

Setup: The numerator equals 1, so:

$$\frac{\sin^2\theta + \cos^2\theta}{\cos\theta} = \frac{1}{\cos\theta} = \sec\theta$$

100. GRAPHING TRIGONOMETRIC FUNCTIONS

To graph trigonometric functions, use the x-axis for the angle and the y-axis for the value of the trigonometric function. Use benchmark angles—0°, 30°, 45°, 60°, 90°, 120°, 135°, 150°, 180°, etc.—to plot key points.

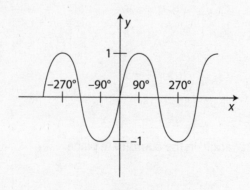

The figure above shows a portion of the graph of $y = \sin x$.

Practice Tests

ACT Practice Test One
ANSWER SHEET

MATHEMATICS TEST

1. Ⓐ Ⓑ Ⓒ Ⓓ Ⓔ	11. Ⓐ Ⓑ Ⓒ Ⓓ Ⓔ	21. Ⓐ Ⓑ Ⓒ Ⓓ Ⓔ	31. Ⓐ Ⓑ Ⓒ Ⓓ Ⓔ	41. Ⓐ Ⓑ Ⓒ Ⓓ Ⓔ	51. Ⓐ Ⓑ Ⓒ Ⓓ Ⓔ
2. Ⓕ Ⓖ Ⓗ Ⓙ Ⓚ	12. Ⓕ Ⓖ Ⓗ Ⓙ Ⓚ	22. Ⓕ Ⓖ Ⓗ Ⓙ Ⓚ	32. Ⓕ Ⓖ Ⓗ Ⓙ Ⓚ	42. Ⓕ Ⓖ Ⓗ Ⓙ Ⓚ	52. Ⓕ Ⓖ Ⓗ Ⓙ Ⓚ
3. Ⓐ Ⓑ Ⓒ Ⓓ Ⓔ	13. Ⓐ Ⓑ Ⓒ Ⓓ Ⓔ	23. Ⓐ Ⓑ Ⓒ Ⓓ Ⓔ	33. Ⓐ Ⓑ Ⓒ Ⓓ Ⓔ	43. Ⓐ Ⓑ Ⓒ Ⓓ Ⓔ	53. Ⓐ Ⓑ Ⓒ Ⓓ Ⓔ
4. Ⓕ Ⓖ Ⓗ Ⓙ Ⓚ	14. Ⓕ Ⓖ Ⓗ Ⓙ Ⓚ	24. Ⓕ Ⓖ Ⓗ Ⓙ Ⓚ	34. Ⓕ Ⓖ Ⓗ Ⓙ Ⓚ	44. Ⓕ Ⓖ Ⓗ Ⓙ Ⓚ	54. Ⓕ Ⓖ Ⓗ Ⓙ Ⓚ
5. Ⓐ Ⓑ Ⓒ Ⓓ Ⓔ	15. Ⓐ Ⓑ Ⓒ Ⓓ Ⓔ	25. Ⓐ Ⓑ Ⓒ Ⓓ Ⓔ	35. Ⓐ Ⓑ Ⓒ Ⓓ Ⓔ	45. Ⓐ Ⓑ Ⓒ Ⓓ Ⓔ	55. Ⓐ Ⓑ Ⓒ Ⓓ Ⓔ
6. Ⓕ Ⓖ Ⓗ Ⓙ Ⓚ	16. Ⓕ Ⓖ Ⓗ Ⓙ Ⓚ	26. Ⓕ Ⓖ Ⓗ Ⓙ Ⓚ	36. Ⓕ Ⓖ Ⓗ Ⓙ Ⓚ	46. Ⓕ Ⓖ Ⓗ Ⓙ Ⓚ	56. Ⓕ Ⓖ Ⓗ Ⓙ Ⓚ
7. Ⓐ Ⓑ Ⓒ Ⓓ Ⓔ	17. Ⓐ Ⓑ Ⓒ Ⓓ Ⓔ	27. Ⓐ Ⓑ Ⓒ Ⓓ Ⓔ	37. Ⓐ Ⓑ Ⓒ Ⓓ Ⓔ	47. Ⓐ Ⓑ Ⓒ Ⓓ Ⓔ	57. Ⓐ Ⓑ Ⓒ Ⓓ Ⓔ
8. Ⓕ Ⓖ Ⓗ Ⓙ Ⓚ	18. Ⓕ Ⓖ Ⓗ Ⓙ Ⓚ	28. Ⓕ Ⓖ Ⓗ Ⓙ Ⓚ	38. Ⓕ Ⓖ Ⓗ Ⓙ Ⓚ	48. Ⓕ Ⓖ Ⓗ Ⓙ Ⓚ	58. Ⓕ Ⓖ Ⓗ Ⓙ Ⓚ
9. Ⓐ Ⓑ Ⓒ Ⓓ Ⓔ	19. Ⓐ Ⓑ Ⓒ Ⓓ Ⓔ	29. Ⓐ Ⓑ Ⓒ Ⓓ Ⓔ	39. Ⓐ Ⓑ Ⓒ Ⓓ Ⓔ	49. Ⓐ Ⓑ Ⓒ Ⓓ Ⓔ	59. Ⓐ Ⓑ Ⓒ Ⓓ Ⓔ
10. Ⓕ Ⓖ Ⓗ Ⓙ Ⓚ	20. Ⓕ Ⓖ Ⓗ Ⓙ Ⓚ	30. Ⓕ Ⓖ Ⓗ Ⓙ Ⓚ	40. Ⓕ Ⓖ Ⓗ Ⓙ Ⓚ	50. Ⓕ Ⓖ Ⓗ Ⓙ Ⓚ	60. Ⓕ Ⓖ Ⓗ Ⓙ Ⓚ

SCIENCE TEST

1. Ⓐ Ⓑ Ⓒ Ⓓ	6. Ⓕ Ⓖ Ⓗ Ⓙ	11. Ⓐ Ⓑ Ⓒ Ⓓ	16. Ⓕ Ⓖ Ⓗ Ⓙ	21. Ⓐ Ⓑ Ⓒ Ⓓ	26. Ⓕ Ⓖ Ⓗ Ⓙ	31. Ⓐ Ⓑ Ⓒ Ⓓ	36. Ⓕ Ⓖ Ⓗ Ⓙ
2. Ⓕ Ⓖ Ⓗ Ⓙ	7. Ⓐ Ⓑ Ⓒ Ⓓ	12. Ⓕ Ⓖ Ⓗ Ⓙ	17. Ⓐ Ⓑ Ⓒ Ⓓ	22. Ⓕ Ⓖ Ⓗ Ⓙ	27. Ⓐ Ⓑ Ⓒ Ⓓ	32. Ⓕ Ⓖ Ⓗ Ⓙ	37. Ⓐ Ⓑ Ⓒ Ⓓ
3. Ⓐ Ⓑ Ⓒ Ⓓ	8. Ⓕ Ⓖ Ⓗ Ⓙ	13. Ⓐ Ⓑ Ⓒ Ⓓ	18. Ⓕ Ⓖ Ⓗ Ⓙ	23. Ⓐ Ⓑ Ⓒ Ⓓ	28. Ⓕ Ⓖ Ⓗ Ⓙ	33. Ⓐ Ⓑ Ⓒ Ⓓ	38. Ⓕ Ⓖ Ⓗ Ⓙ
4. Ⓕ Ⓖ Ⓗ Ⓙ	9. Ⓐ Ⓑ Ⓒ Ⓓ	14. Ⓕ Ⓖ Ⓗ Ⓙ	19. Ⓐ Ⓑ Ⓒ Ⓓ	24. Ⓕ Ⓖ Ⓗ Ⓙ	29. Ⓐ Ⓑ Ⓒ Ⓓ	34. Ⓕ Ⓖ Ⓗ Ⓙ	39. Ⓐ Ⓑ Ⓒ Ⓓ
5. Ⓐ Ⓑ Ⓒ Ⓓ	10. Ⓕ Ⓖ Ⓗ Ⓙ	15. Ⓐ Ⓑ Ⓒ Ⓓ	20. Ⓕ Ⓖ Ⓗ Ⓙ	25. Ⓐ Ⓑ Ⓒ Ⓓ	30. Ⓕ Ⓖ Ⓗ Ⓙ	35. Ⓐ Ⓑ Ⓒ Ⓓ	40. Ⓕ Ⓖ Ⓗ Ⓙ

MATHEMATICS TEST

60 Minutes—60 Questions

Directions: Solve each of the following problems, select the correct answer, and then fill in the corresponding oval on your Answer Grid.

Don't linger over problems that are too time-consuming. Do as many as you can, then come back to the others in the time permitted.

You may use a calculator on this test. Some questions, however, may be easier to answer without the use of a calculator.

Note: Unless the question says otherwise, assume all of the following:

1. Illustrative figures are *not* necessarily drawn to scale.

2. All geometric figures lie in a plane.

3. The term *line* indicates a straight line.

4. The term *average* indicates arithmetic mean.

---

1. A *rod* is a unit of length equivalent to 5.5 yards. If a field is 127 yards long, then how many rods long is the field, to the nearest tenth?

   A. 231.9
   B. 69.9
   C. 43.3
   D. 23.1
   E. 4.3

2. Because of increased rents in the area, a pizzeria needs to raise the cost of its $20.00 extra large pizza by 22%. What will the new cost be?

   F. $20.22
   G. $22.20
   H. $24.00
   J. $24.40
   K. $42.00

3. Increases in membership for 5 different organizations are indicated in the table below.

Organization	A	B	C	D	E
Increase in Membership	120	210	0	210	180

   What is the average increase in membership for the 5 organizations?

   A. 127.5
   B. 144
   C. 170
   D. 180
   E. 240

4. Train A travels 50 miles per hour for 3 hours; Train B travels 70 miles per hour for $2\frac{1}{2}$ hours. What is the *difference* between the number of miles traveled by Train A and the number of miles traveled by Train B?

   F. 0
   G. 25
   H. 150
   J. 175
   K. 325

GO ON TO THE NEXT PAGE

5. Which of the following is a value of b for which $(b - 3)(b + 4) = 0$?

   A. 3
   B. 4
   C. 7
   D. 10
   E. 12

6. In the parallelogram $RSTU$ shown below, $\overline{ST}$ is 8 feet long. If the parallelogram's perimeter is 42 feet, how many feet long is $\overline{UT}$?

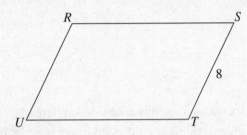

   F. 34
   G. 26
   H. 21
   J. 13
   K. $15\frac{1}{4}$

7. If the measure of each interior angle of a regular polygon is 60°, how many sides does the polygon have?

   A. 3
   B. 4
   C. 6
   D. 10
   E. 12

8. For all nonzero a, b, and c values, $\dfrac{12a^5bc^7}{-3ab^5c^2} = ?$

   F. $\dfrac{-4c^5}{a^4b^4}$
   G. $\dfrac{-4a^4c^5}{b^4}$
   H. $\dfrac{-4ac}{b}$
   J. $-4a^6b^6c^9$
   K. $-4a^4b^4c^5$

9. In the figure below, P and Q lie on the sides of $\triangle WXY$, and $\overline{PQ}$ is parallel to $\overline{WY}$. What is the measure of $\angle QPX$?

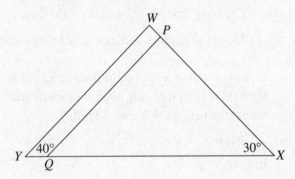

   A. 110°
   B. 120°
   C. 130°
   D. 140°
   E. 150°

10. $|-4| \cdot |2| = ?$

   F. −8
   G. −6
   H. −2
   J. 6
   K. 8

GO ON TO THE NEXT PAGE

11. A company conducted a taste test of its new soft drink. Of the 1250 participants, 800 liked the soft drink, 150 didn't like it, and the rest were undecided. What percent of the participants were undecided about the new soft drink?

    A. 24%

    B. 46%

    C. 64%

    D. 76%

    E. 300%

12. Two whole numbers have a greatest common factor of 15 and a least common multiple of 225. Which of the following pairs of numbers will satisfy this condition?

    F. 9 and 25

    G. 15 and 27

    H. 25 and 45

    J. 30 and 45

    K. 45 and 75

13. If $x = 2$ and $y = -3$, then $x^5y + xy^5 = ?$

    A. −60

    B. −192

    C. −390

    D. −582

    E. −972

14. How many units long is one side of a square with perimeter $16 - 24h$ units?

    F. $16 - 24h$

    G. $16 - 6h$

    H. $8h$

    J. $4 - 24h$

    K. $4 - 6h$

15. If $(x - k)^2 = x^2 - 26x + k^2$ for all real numbers x, then $k = ?$

    A. 13

    B. 26

    C. 52

    D. 104

    E. 208

16. Helena bought her daughter a game system and two game cartridges for her birthday, all on sale. The game system, regularly $180, was 10% off, and the game cartridges, regularly $40 each, were 20% off. What was the total price of the 3 items Helena bought? (Note: Assume there is no sales tax.)

    F. $186

    G. $194

    H. $221

    J. $226

    K. $250

17. Which of the following expressions gives the slope of the line connecting the points $(5,9)$ and $(-3,-12)$?

    A. $\dfrac{9 + (-12)}{-5 - (-3)}$

    B. $\dfrac{9 + (-12)}{-3 + 5}$

    C. $\dfrac{9 - (-12)}{5 - (-3)}$

    D. $\dfrac{9 - (-12)}{-3 - 5}$

    E. $\dfrac{9 - (-12)}{-5 + 3}$

GO ON TO THE NEXT PAGE ▷

18. In the standard (x,y) coordinate plane, how many times does the graph of $y = (x + 1)(x + 2)(x - 3)(x + 4)(x + 5)$ intersect the x-axis?

    F. 15
    G. 9
    H. 5
    J. 4
    K. 1

19. Which of the following is an equivalent, simplified version of $\dfrac{4 + 8x}{12x}$?

    A. $\dfrac{2x + 1}{3x}$

    B. $\dfrac{1 + 8x}{3x}$

    C. 1

    D. $\dfrac{7}{3}$

    E. $\dfrac{8}{3}$

20. Four friends about to share an airport shuttle for $21.50 for each ticket discover that they can purchase a book of 5 tickets for $95.00. How much would each of the 4 save if they can get a fifth person to join them and they divide the cost of the book of 5 tickets equally among all 5 people?

    F. $ 2.25
    G. $ 2.50
    H. $ 3.13
    J. $ 9.00
    K. $12.50

21. What is the sum of the polynomials $-2x^2y^2 + x^2y$ and $3x^2y^2 + 2xy^2$?

    A. $-6x^4y^4 + 2x^3y^3$
    B. $-2x^2y^2 + x^2y + 2xy^2$
    C. $x^2y^2 + x^2y + 2xy^2$
    D. $x^2y^2 + x^2y$
    E. $x^2y^2 + 3x^2y$

22. A 12 foot flagpole casts a 7 foot shadow when the angle of elevation of the sun is θ (see figure below). What is $\tan(\theta)$?

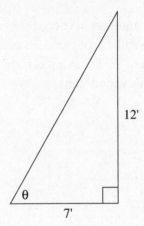

    F. $\dfrac{7}{12}$
    G. 1
    H. $\dfrac{12}{7}$
    J. 19
    K. 84

23. Yousuf was x years old 15 years ago. How old will he be 7 years from now?

    A. $x + 7$
    B. $(x - 15) + 7$
    C. $(x + 15) - 7$
    D. $(x - 15) - 7$
    E. $(x + 15) + 7$

GO ON TO THE NEXT PAGE ▷

24. Which of the following is a factor of $x^2 - 4x - 12$?

   F. $(x + 1)$

   G. $(x - 2)$

   H. $(x + 2)$

   J. $(x - 3)$

   K. $(x - 4)$

25. What is the length, in inches, of the hypotenuse of a right triangle with legs measuring 8 inches and 15 inches?

   A. 7

   B. 17

   C. 23

   D. $\sqrt{23}$

   E. $\sqrt{161}$

26. Which of the following expressions is a simplified form of $(-2x^5)^3$?

   F. $-6x^8$

   G. $8x^8$

   H. $-2x^{15}$

   J. $-6x^{15}$

   K. $-8x^{15}$

27. The *relative atomic mass* of an element is the ratio of the mass of the element to the mass of an equal amount of carbon. If 1 cubic centimeter of carbon has a mass of 12 grams, what is the relative atomic mass of an element that has a mass of 30 grams per cubic centimeter?

   A. 1

   B. 1.2

   C. 2.5

   D. 3

   E. 30

28. If $2x + 3 = -5$, what is the value of $x^2 - 7x$?

   F. -44

   G. -12

   H. -4

   J. 12

   K. 44

29. Which of the following is a graph of the solution set for $2(5 + x) < 2$?

   A.

   B.

   C.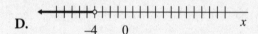

   D.

   E.

30. Which of the following equations has m varying directly as the cube of b and inversely as the square of c?

   F. $\dfrac{m^3}{c^2} = b$

   G. $\dfrac{b^3}{c^2} = m$

   H. $\dfrac{c^3}{b^2} = m$

   J. $\dfrac{\sqrt[3]{b}}{\sqrt{c}} = m$

   K. $\dfrac{b^3}{m^2} = c$

GO ON TO THE NEXT PAGE

31. Points $V(-2,-7)$ and $W(4,5)$ determine line segment $\overline{VW}$ in the standard (x,y) coordinate plane. If the midpoint of $\overline{VW}$ is $(1,p)$, what is the value of p?

    A. -2
    B. -1
    C. 1
    D. 2
    E. 6

32. If the graphs of $y = 3x$ and $y = mx + 6$ are parallel in the standard (x,y) coordinate plane, then $m = ?$

    F. -6
    G. $\dfrac{1}{3}$
    H. 2
    J. 3
    K. 6

33. When 3 times x is increased by 5, the result is less than 11. Which of the following is a graph of the real numbers x for which the previous statement is true?

    A.
    B.
    C.
    D.
    E.

34. It costs 54 cents to buy x pencils and 92 cents to buy y erasers. Which of the following is an expression for the cost, in cents, of 7 pencils and 3 erasers?

    F. $\dfrac{54}{7+x} + \dfrac{92}{3+y}$

    G. $3\left(\dfrac{54}{x}\right) + 7\left(\dfrac{92}{y}\right)$

    H. $7\left(\dfrac{x}{54}\right) + 3\left(\dfrac{y}{92}\right)$

    J. $7\left(\dfrac{54}{x}\right) + 3\left(\dfrac{92}{y}\right)$

    K. $7\left(\dfrac{92}{x}\right) + 3\left(\dfrac{54}{x}\right)$

35. When graphed in the standard (x,y) coordinate plane, 3 points from among $(-9,-7)$, $(-5,-3)$, $(-2,-1)$, $(1,-1)$ and $(10,-8)$ lie on the same side of the line $y - x = 0$. Which of the three points are they?

    A. $(-9,-7)$, $(-2,-1)$, $(-5,-3)$
    B. $(-9,-7)$, $(-2,-1)$, $(1,-1)$
    C. $(-9,-7)$, $(-5,-3)$, $(10,-8)$
    D. $(-9,-7)$, $(1,-1)$, $(10,-8)$
    E. $(-5,-3)$, $(1,-1)$, $(10,-8)$

GO ON TO THE NEXT PAGE

36. What is the sine of angle *E* in right triangle *DEF* below?

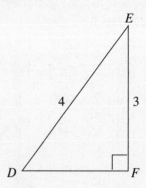

F. $\dfrac{\sqrt{7}}{3}$

G. $\dfrac{3}{4}$

H. $\dfrac{\sqrt{7}}{4}$

J. $\dfrac{3}{\sqrt{7}}$

K. $\dfrac{4}{\sqrt{7}}$

37. The graph of the solution set for the system of linear equations below is a single line in the (*x*,*y*) coordinate plane.

$$12x - 20y = 108$$

$$3x + ky = 27$$

What is the value of *k*?

A. −5

B. −3

C. $-\dfrac{1}{4}$

D. $\dfrac{3}{5}$

E. 4

38. A common rule of thumb is that each additional inch of height (*H*) will add 10 pounds to a person's weight (*W*). Doctors recommend finding your Body Mass Index (*BMI*) as a measure of health. *BMI* is computed as follows (*H* is in inches, and *W* is in pounds):

$$BMI = \dfrac{703W}{H^2}$$

If a 68 inch tall person typically weighs 150 pounds, which of the following is closest to the expected *BMI* of a 72 inch tall person?

F. 1

G. 2

H. 20

J. 26

K. 42

39. Dave's math tutor reminded him not to calculate $\left(\dfrac{x}{y}\right)^2$ as $\dfrac{x^2}{y}$. Dave thinks there are some numbers for which that calculation works. Eventually, he was able to show that $\left(\dfrac{x}{y}\right)^2$ equals $\dfrac{x^2}{y}$ if and only if:

(Note: Assume that y ≠ 0.)

A. *x* = 0

B. *x* = 1

C. *y* = 1

D. *x* = 0 and *y* = 1

E. *x* = 0 or *y* = 1

GO ON TO THE NEXT PAGE

40. In the figure below, $\overline{BD}$ is a perpendicular bisector of $\overline{AC}$ in equilateral triangle $\triangle ABC$. If $\overline{BD}$ is $4\sqrt{3}$ units long, how many units long is $\overline{BC}$?

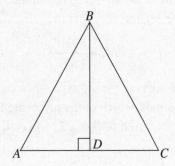

F. $2\sqrt{3}$

G. 4

H. 8

J. $8\sqrt{3}$

K. 16

41. What is the perimeter, in meters (m), of the figure below?

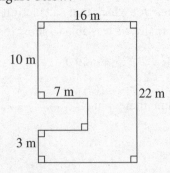

A. 58

B. 83

C. 90

D. 208

E. 352

42. Isosceles trapezoid $ABCD$ is inscribed in a circle with center O, as shown below. Which of the following is the most direct explanation of why $\triangle AOD$ is isosceles?

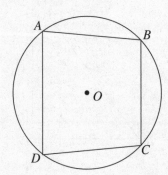

F. If two angles in a triangle are congruent, the sides opposite them are congruent.

G. Two sides are radii of the circle.

H. Side-angle-side congruence

J. Angle-side-angle congruence

K. Angle-angle-angle similarity

43. A circle with radius 4 meters is cut out of a circle with radius 12 meters, as shown in the figure below. Which of the following gives the area of the shaded figure, in square meters?

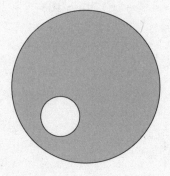

A. $\pi(12 - 2)^2$

B. $\pi 12^2 - 2^2$

C. $\pi 12^2 - 4^2$

D. $\pi(12 - 4^2)$

E. $\pi(12^2 - 4^2)$

44. A walkway, 31 by $32\frac{1}{2}$ feet, surrounds a pool that is $27\frac{1}{2}$ by 29 feet, as shown below.

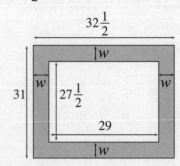

What is the width, w, of the walkway?

F. $1\frac{1}{4}$

G. 1

H. $1\frac{1}{2}$

J. $1\frac{3}{4}$

K. $3\frac{1}{2}$

45. The area of a rectangular floor is 323 square feet. The width of the floor is 21 feet less than twice the length. How many feet long is the floor?

A. 8.5

B. 11

C. 13.5

D. 17

E. 19

46. For the area of a circle to double, the new radius must be the old radius multiplied by:

F. $\frac{1}{2}$

G. $\sqrt{2}$

H. 2

J. π

K. 4

47. If $\log_x 64 = 3$, then $x = $?

A. 4

B. 8

C. $\frac{64}{3}$

D. $\frac{64}{\log 3}$

E. 64^3

48. If $A = \begin{bmatrix} 3 & -6 \\ 0 & 9 \end{bmatrix}$ and $B = \begin{bmatrix} -3 & 6 \\ 0 & -9 \end{bmatrix}$, then $A - B = $?

F. $\begin{bmatrix} 0 & 0 \\ 0 & 0 \end{bmatrix}$

G. $\begin{bmatrix} 1 & 0 \\ 0 & 1 \end{bmatrix}$

H. $\begin{bmatrix} 0 & -12 \\ 0 & 18 \end{bmatrix}$

J. $\begin{bmatrix} -6 & 0 \\ 0 & 0 \end{bmatrix}$

K. $\begin{bmatrix} 6 & -12 \\ 0 & 18 \end{bmatrix}$

49. If a and b are real numbers, and $a > 0$ and $b < a$, then which of the following inequalities must be true?

A. $b \leq 0$

B. $b \geq 0$

C. $b^2 \geq 0$

D. $b^2 \geq a^2$

E. $b^2 \leq a^2$

GO ON TO THE NEXT PAGE

50. The ratio of the lengths of the sides of a right triangle is $2:\sqrt{5}:3$. What is the cosine of the smallest angle in the triangle?

    F. $\dfrac{2}{3}$

    G. $\dfrac{\sqrt{5}}{3}$

    H. $\dfrac{2\sqrt{5}}{5}$

    J. $\dfrac{9}{10}$

    K. 2

51. What is the amplitude of the graph of the equation $y + 3 = 4\sin(5\theta)$?

    (Note: the amplitude is $\dfrac{1}{2}$ the difference between the maximum and the minimum values of y.)

    A. 3

    B. 4

    C. 5

    D. 7

    E. 10

52. Each of the following determines a unique plane in 3-dimensional Euclidian space EXCEPT:

    F. 1 line and 1 point NOT on the line.

    G. 3 distinct points NOT on the same line.

    H. 2 lines that intersect in exactly 1 point.

    J. 2 distinct parallel lines.

    K. 2 lines that are NOT parallel and do NOT intersect.

53. The measure of the vertex angle of an isosceles triangle is $(x - 10)°$. The base angles each measure $(3x + 18)°$. What is the measure in degrees of one of the base angles?

    A. 12

    B. 22

    C. $37\dfrac{1}{2}$

    D. $43\dfrac{1}{2}$

    E. 84

54. To make a set of potholders of various sizes to give as a gift, Margot needs the following amounts of fabric for each set:

Pieces of Fabric	Length (inches)
6	8
5	12
2	18

    If the fabric costs $1.95 per yard, which of the following would be the approximate cost of fabric for 5 sets of potholders?

    (Note: 1 yard = 36 inches)

    F. $ 8

    G. $ 24

    H. $ 39

    J. $ 58

    K. $117

GO ON TO THE NEXT PAGE

55. The formula for the surface area (*S*) of a rectangular solid with square bases (shown below) is $S = 4wh + 2w^2$, where *w* is the side length of the bases and *h* is the height of the solid. Doubling each of the dimensions (*w* and *h*) will increase the surface area to how many times its original size?

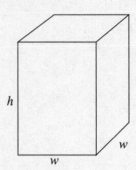

A. 2
B. 4
C. 6
D. 8
E. 24

56. The average of a set of four integers is 14. When a fifth number is included in the set, the average of the set increases to 16. What is the fifth number?

F. 16
G. 18
H. 21
J. 24
K. 26

57. Which of the following is the equation of the largest circle that can be inscribed in the ellipse with equation $\dfrac{(x-4)^2}{16} + \dfrac{y^2}{4} = 1$?

A. $(x - 4)^2 + y^2 = 64$
B. $(x - 4)^2 + y^2 = 16$
C. $(x - 4)^2 + y^2 = 4$
D. $x^2 + y^2 = 16$
E. $x^2 + y^2 = 4$

58. One of the graphs below is that of $y = x^3 + C$, where *C* is a constant. Which one?

F.

G.

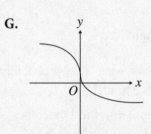

H.

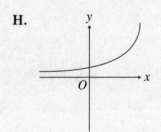

J.

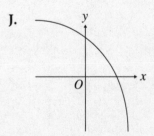

K.

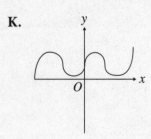

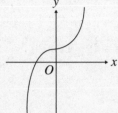

GO ON TO THE NEXT PAGE

59. How many points do the graphs of all three equations below have in common?

$$x = y + 8$$
$$-x = y - 8$$
$$6x = 2y + 4$$

A. 0
B. 1
C. 2
D. 3
E. Infinitely many

60. In 4 fair coin tosses, what is the probability of obtaining exactly 3 heads?

(Note: In a fair coin toss, the 2 outcomes, heads and tails, are equally likely.)

F. $\dfrac{1}{16}$

G. $\dfrac{1}{8}$

H. $\dfrac{3}{16}$

J. $\dfrac{1}{4}$

K. $\dfrac{1}{2}$

IF YOU FINISH BEFORE TIME IS CALLED, YOU MAY CHECK YOUR WORK ON THIS SECTION ONLY. DO NOT TURN TO ANY OTHER SECTION IN THE TEST. **STOP**

SCIENCE TEST

35 Minutes—40 Questions

Directions: There are several passages in this test. Each passage is followed by several questions. After reading a passage, choose the best answer to each question and fill in the corresponding oval on your Answer Grid. You may refer to the passages as often as necessary. You are NOT permitted to use a calculator on this test.

PASSAGE I

Soil, by volume, consists on the average of 45% minerals, 25% water, 25% air, and 5% organic matter (including both living and nonliving organisms). Time and topography shape the composition of soil and cause it to develop into layers known as *horizons*. The soil horizons are collectively known as the *soil profile*. The composition of soil varies in each horizon, as do the most common minerals (see Figure 1). Figure 1 also shows the depth of each horizon and the overall density of the soil.

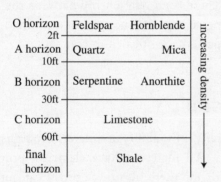

Figure 1

Table 1 lists the percents (%) of zinc and calcium in the minerals that compose soil.

Table 1

Mineral	Zinc content (%)	Calcium content (%)
Feldspar	35–40	0–10
Hornblende	30–35	10–20
Quartz	25–30	20–30
Mica	20–25	30–40
Serpentine	15–20	40–50
Anorthite	10–15	50–60
Limestone	5–10	60–70
Shale	0–5	70–80

Table 2 shows the percent of minerals that compose granite and sandstone, 2 rock types that are commonly found in soil.

Table 2

Mineral	Percent of mineral in:	
	Sandstone	Granite
Feldspar	30	54
Hornblende	2	0
Quartz	50	33
Mica	10	10
Serpentine	0	0
Anorthite	0	0
Limestone	5	0
Shale	0	0
Augite	3	3

GO ON TO THE NEXT PAGE ⇨

1. An analysis of an unknown mineral found in soil revealed its zinc content to be 32% and its calcium content to be 12%. Based on the data in Table 1, geologists would most likely classify this mineral as:

   A. hornblende.
   B. anorthite.
   C. serpentine.
   D. mica.

2. Geologists digging down to the A horizon would most likely find which of the following minerals?

   F. Limestone
   G. Shale
   H. Serpentine
   J. Mica

3. Based on the data presented in Figure 1 and Table 1, which of the following statements best describes the relationship between the zinc content of a mineral and the depth below surface level at which it is dominant? As zinc content increases:

   A. depth increases.
   B. depth decreases.
   C. depth first increases, then decreases.
   D. depth first decreases, then increases.

4. If geologists were to drill through to the top of the C horizon, which minerals would they most likely encounter?

   F. Quartz, mica, and limestone
   G. Feldspar, shale, and serpentine
   H. Feldspar, quartz, and anorthite
   J. Hornblende, limestone, and serpentine

5. If augite is most likely found at a depth between that of the other minerals found in granite, then augite would most likely be found at a depth of:

   A. 10 feet or less.
   B. between 10 feet and 30 feet.
   C. between 30 feet and 60 feet.
   D. greater than 60 feet.

6. How is the percentage of zinc content related to the percentage of calcium content in the minerals that make up soil?

   F. The percentage of zinc content increases as the percentage of calcium content increases.
   G. The percentage of zinc content increases as the percentage of calcium content decreases.
   H. Both the percentage of zinc content and the percentage of calcium content remain constant.
   J. There is no relationship between the percentage of zinc content and the percentage of calcium content.

PASSAGE II

Conductivity is the ability of a material to transmit electricity. All materials have electrical properties that divide them into three broad categories: conductors, insulators and semiconductors.

A conductor is a substance that allows an electric charge to travel from one object to another. An insulator is a substance that prevents an electric charge from traveling between objects. Substances with levels of conductivity between that of a conductor and that of an insulator are called semiconductors. A voltammeter is an instrument used to measure voltage (see Figure 1).

GO ON TO THE NEXT PAGE

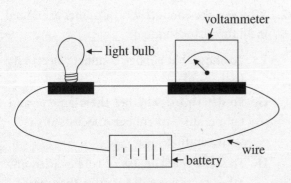

light bulb

voltammeter

wire

battery

Figure 1

Three studies were executed to determine the validity of the hypothesis that a wire's conductivity increases when either the diameter or the temperature of the wire decreases.

STUDY 1

Wires were made from 5 different materials. Each strand of wire had a diameter of exactly 4 millimeters (mm). The strand of wire connecting the battery, light bulb and voltammeter was 0.5 meters (m) long and was kept at a temperature of 50°C. Table 1 displays the voltage, in millivolts (mV), recorded by the voltammeter.

Table 1

Material	Voltammeter (mV)
Silicon carbide (SiC)	4.6
Copper (Cu)	9.4
Rubber	0.0
Zinc Telluride (ZnTe)	5.2
Steel	3.5

STUDY 2

The conditions in Study 1 were repeated, except the diameter of the wires was decreased to 2 mm. The length between the battery and the light bulb and the light bulb and the voltammeter was held constant at 0.5 m and the wires were kept at 50°C. Table 2 displays the findings.

Table 2

Material	Voltammeter (mV)
Silicon carbide (SiC)	6.5
Copper (Cu)	11.3
Rubber	0.0
Zinc Telluride (ZnTe)	7.1
Steel	8.6

STUDY 3

Study 2 was repeated at 30°C. Table 3 displays the findings.

Table 3

Material	Voltammeter (mV)
Silicon carbide (SiC)	7.3
Copper (Cu)	12.1
Rubber	0.0
Zinc Telluride (ZnTe)	8.9
Steel	6.6

7. Which of the following ranges represents the voltage of all five wires with diameters of 2 mm at 30°C?

   A. 6.5 mV to 11.3 mV

   B. 0.0 mV to 9.4 mV

   C. 0.0 mV to 11.3 mV

   D. 0.0 mV to 12.1 mV

8. The scientist hypothesized that decreasing the diameter of a wire increases its conductivity. The results from the studies for each of the following materials prove the scientist's hypothesis to be true, EXCEPT the results for:

   F. silicon carbide.

   G. rubber.

   H. copper.

   J. steel.

GO ON TO THE NEXT PAGE

9. According to the results of all 3 experiments, a wire made from ZnTe would have the highest conductivity with which of the following dimensions?

A. 1 mm diameter, 0.5 m length at 20°C

B. 4 mm diameter, 0.5 m length at 20°C

C. 4 mm diameter, 0.5 m length at 40°C

D. 10 mm diameter, 0.5 m length at 40°C

10. What would the voltammeter read if a scientist used two wires, one copper and one rubber, both with diameters of 2 mm, lengths of 0.5 m, and at 30°C, to conduct electricity to the light bulb?

F. 0.0 millivolts

G. 9.4 millivolts

H. 12.1 millivolts

J. 14.7 millivolts

11. How would the conductivity of the materials be affected if Study 3 was repeated and the temperature of the wires was increased to 100°C?

A. The conductivity would decrease with the exception of rubber.

B. The conductivity would remain unchanged.

C. The conductivity would increase only.

D. The conductivity would increase with the exception of rubber.

12. Why was the conductivity of rubber examined in all three studies?

F. To show that rubber conducts electricity well.

G. To determine whether the diameter and temperature of rubber affect its insulating abilities.

H. To show that the use of rubber with any other material will increase that material's conductivity.

J. To determine if the length of a rubber wire affects its insulating abilities.

13. Which of the following effects would be most appropriate for the scientists to test next to learn more about conductivity?

A. The changes in wire conductivity when diameter and temperature are modified

B. The effect of wire color on conductivity

C. The effect of wire temperature on conductivity

D. The effect of different wire lengths on conductivity

PASSAGE III

Engineers designing a roadway needed to test the composition of the soil that would form the roadbed. In order to determine whether their two sampling systems (System A and System B) give sufficiently accurate soil composition measurements, they first conducted a study to compare the two systems.

Soil samples were taken with varying levels of *humidity* (concentration of water). The concentrations of the compounds that form the majority of soil were measured. The results for the sampling systems were compared with data on file with the US Geological Survey (USGS), which compiles extremely

GO ON TO THE NEXT PAGE ⟶

accurate data. The engineers' and USGS' results are presented in the table below.

Table 1

Concentration (mg/L) of:	Level of Humidity				
	10%	25%	45%	65%	80%
Nitrogen (N)					
USGS	105.2	236	598	781	904
System A	111.6	342	716	953	1,283
System B	196.4	408	857	1,296	1,682
Potassium Oxide (K_2O)					
USGS	9.4	9.1	8.9	8.7	8.2
System A	9.4	9.0	8.7	8.5	8.0
System B	9.5	9.2	9.0	8.8	8.3
Calcium (Ca)					
USGS	39.8	24.7	11.4	5.0	44.8
System A	42.5	31.4	10.4	8.0	42.9
System B	37.1	23.2	11.6	11.1	45.1
Phosphorus Oxide (P_2O_5)					
USGS	69.0	71.2	74.8	78.9	122.3
System A	67.9	69.9	72.2	76.7	123.1
System B	74.0	75.6	78.7	82.1	126.3
Zinc (Zn)					
USGS	0.41	0.52	0.64	0.74	0.70
System A	0.67	0.80	0.88	0.97	0.93
System B	0.38	0.48	0.62	0.77	0.73

Note: Each system concentration measurement is the average of 5 measurements.

14. The hypothesis that increasing humidity increases the concentration (mg/L) of a compound is supported by all of the following EXCEPT:

F. nitrogen.

G. potassium oxide.

H. phosphorous oxide.

J. zinc.

15. At a humidity level of 25%, it could be concluded that System B least accurately measures the concentration of which of the following compounds, relative to the data on file with the USGS?

A. Nitrogen

B. Calcium

C. K_2O

D. P_2O_5

16. The engineers hypothesized that the concentration of potassium oxide (K_2O) decreases as the level of humidity increases. This hypothesis is supported by:

F. the data from the USGS only.

G. the System A measurements only.

H. the data from the USGS and the System B measurements only.

J. the data from the USGS, the System A measurements, and the System B measurements.

17. Do the results in the table support the conclusion that System B is more accurate than System A for measuring the concentration of zinc?

A. No, because the zinc measurements from System A are consistently higher than the zinc measurements from System B.

B. No, because the zinc measurements from System A are closer to the data provided by the USGS than the zinc measurements from System B.

C. Yes, because the zinc measurements from System B are consistently lower than the zinc measurements from System A.

D. Yes, because the zinc measurements from System B are closer to the data provided by the USGS than the zinc measurements from System A.

GO ON TO THE NEXT PAGE

18. The relationship between humidity level and calcium concentration, as measured by System B, is best represented by which of the following graphs?

F.

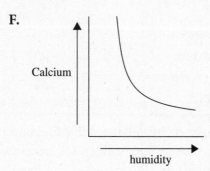

G.

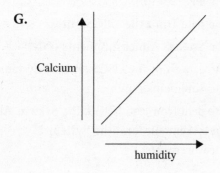

H.

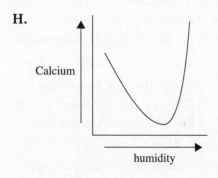

J.

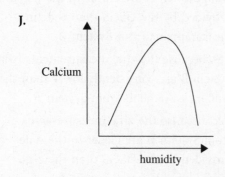

19. After conducting their comparisons, the engineers used System B to test a soil sample at the future road site. They measured the concentrations, in mg/L, of selected compounds in the sample and found that they were: potassium oxide (K_2O) = 9.1, calcium = 17.3, and zinc = 0.57. According to the data in the table, the engineers should predict that the level of humidity is approximately:

A. 16%.

B. 37%.

C. 49%.

D. 57%.

PASSAGE IV

An increasing number of individuals over 50 develop type II diabetes, which occurs when the body does not produce enough insulin or when the cells ignore the insulin and as a result, the body's blood sugar level rises dangerously. Although type II diabetes occurs in people of all ages and races, it is more common in adults. Several hypotheses have been proposed to explain the cause of type II diabetes.

DIETARY HYPOTHESIS

Most Americans consume too much sugar. Sugar from food is absorbed into the bloodstream and insulin is required for the body to be able to use that sugar. In a study of individuals 18–25 years old who consumed more than the recommended amount of sugar daily, and were thus considered at risk for developing type II diabetes, it was shown that the majority had significantly elevated levels of sugar in their blood but normal levels of insulin. When these individuals received small injections of insulin once a day, their blood sugar levels decreased to more normal levels. If abundant levels of sugar are supplied by the diet, sugar dissolving insulin injections should be given to avoid type II diabetes.

GO ON TO THE NEXT PAGE

GENETIC HYPOTHESIS

Genes, which primarily come from parents and grandparents, contribute to many medical problems that individuals will experience through life. Type II diabetes mainly depends on one's genes, but also on one's lifestyle. Diabetes occurs when the pancreas produces little or no insulin or when the insulin it produces does not work properly. As individuals grow older, the processes of their body do not run as efficiently as they did when the individuals were younger. Therefore, the same behaviors may be more detrimental to a person when he or she is older than when he or she was younger, especially for those over the age of 50. This is the main reason that type II diabetes is more common in adults.

Scientists compared the genetics and lifestyles of 4 groups of individuals over 50. The results are shown in the table below.

Table 1

Group	Attributes	% with type II diabetes
A	One parent with type II	55%
B	Healthy lifestyle, no parents with type II	20%
C	One parent with type II and healthy lifestyle	40%
D	Two parents with type II	70%

EXERCISE HYPOTHESIS

A lack of exercise results in high body fat content, and a high body fat content does not allow the body to work efficiently. Conversely, regular weight-bearing exercise can boost the body's efficiency. One study showed that 10 weeks of weight training lowered blood sugar in adults over 50. A second study on another group of adults over 50 showed that walking 2 miles a day for 12 weeks also lowered blood sugar levels.

20. The Dietary Hypothesis would be strengthened if it were proven that high blood sugar levels are indicative of:

   F. a low efficiency of insulin.
   G. a low-sugar diet.
   H. high levels of insulin produced by the body.
   J. sugar being stored elsewhere in the body.

21. The Genetic Hypothesis best explains why type II diabetes is more common in which of the following groups?

   A. Individuals under the age of 25 as opposed to individuals over the age of 25
   B. Individuals over the age of 25 as opposed to individuals under the age of 25
   C. Individuals over the age of 50 as opposed to individuals under the age of 50
   D. Individuals under the age of 50 as opposed to individuals over the age of 50

22. According to the Genetic Hypothesis, adults who have had their pancreas removed should exhibit:

   F. increased blood insulin levels.
   G. decreased blood sugar levels.
   H. increased blood sugar levels.
   J. decreased body fat content.

23. Supporters of the Dietary Hypothesis might criticize the experimental results in the Exercise Hypothesis for which of the following reasons?

   A. Not enough sugar was included in the diets of the test subjects in both groups.
   B. The sugar intake of the individuals in the two groups was not monitored.
   C. The genetics of each individual in both groups should have been determined.
   D. Type II diabetes is more common in children than adults.

GO ON TO THE NEXT PAGE

24. Assume that individuals with elevated blood sugar levels have a greater chance of developing type II diabetes. How would supporters of the Genetic Hypothesis explain the experimental results presented in the Dietary Hypothesis?

   F. The test subjects probably had high levels of sugar in their diets.

   G. The test subjects did not perform any weight-bearing exercise.

   H. The test subjects probably had no occurrence of type II diabetes in their genetic backgrounds.

   J. The test subjects were given too little insulin.

25. How might proponents of the Dietary Hypothesis explain the results of Group D in the Genetics Hypothesis experiment?

   A. Insulin supplements should not have been taken by this group.

   B. These individuals and both parents had an unhealthy diet that was high in sugar.

   C. More genetic background should be researched for all the groups.

   D. Not enough insulin was given to this group to affect the onset of type II diabetes.

26. The experiments cited in the Genetics Hypothesis and in the Exercise Hypothesis are similar in that each test subject:

   F. has at least one parent with type II diabetes.

   G. was given a shot of insulin.

   H. had their pancreas previously removed.

   J. was an adult over 50 years old.

PASSAGE V

Human blood is composed of approximately 45% *formed elements*, including blood cells, and 50% plasma. The formed elements of blood are further broken down into red blood cells, white blood cells, and platelets. The mass of a particular blood sample is determined by the ratio of formed elements to plasma, as the formed elements weigh approximately 1.10 grams per milliliter (g/mL), and plasma approximately 1.02 g/mL. This ratio varies according to an individual's diet, health, and genetic makeup.

The following experiments were performed by a phlebotomist to determine the composition and mass of blood samples from three different individuals, each of whom was required to fast overnight before the samples were taken.

EXPERIMENT 1

A 10 mL blood sample was taken from each of the three patients. The densities of the blood samples were measured using the *oscillator technique*, which determines fluid densities by measuring sound velocity transmission.

EXPERIMENT 2

Each 10 mL blood sample was spun for 20 minutes in a centrifuge to force the heavier formed elements to separate from the plasma. The plasma was then siphoned off and its mass recorded.

EXPERIMENT 3

The formed elements left over from Experiment 2 were analyzed using the procedure from Experiment 2, except this time they were spun at a slower speed for 45 minutes so that the red blood cells, white blood cells, and platelets could separate out. The mass of each element was then recorded. The results of the three experiments are shown below:

GO ON TO THE NEXT PAGE

Table 1

Patient	Plasma (g)	Red blood cells (g)	White blood cells (g)	Platelets (g)	Total density (g/mL)
A	4.54	2.75	1.09	1.32	1.056
B	4.54	2.70	1.08	1.35	1.054
C	4.64	2.65	1.08	1.34	1.050

27. The results of the experiments indicate that the blood sample with the lowest density is sample with the most:

 A. plasma.

 B. red blood cells.

 C. white blood cells.

 D. platelets.

28. Why did the phlebotomist likely require each patient to fast overnight before taking blood samples?

 F. It is more difficult to withdraw blood from patients who have not fasted.

 G. Fasting causes large, temporary changes in the composition of blood.

 H. Fasting ensures that blood samples are not affected by temporary changes caused by consuming different foods.

 J. Blood from patients who have not fasted will not separate when spun in a centrifuge.

29. Which of the following best explains why the amount of plasma, red blood cells, white blood cells, and platelets do not add up to 10.5 g?

 A. Some of the red blood cells might have remained in the plasma, yielding low red blood cell measurements.

 B. Some of the platelets might not have separated from the white blood cells, yielding high white blood cell counts.

 C. The centrifuge might have failed to fully separate the plasma from the formed elements.

 D. There are likely components other than plasma, red and white blood cells, and platelets in blood.

30. From the data presented in the experiment, it is possible to determine that, as total density increases, the mass of red blood cells:

 F. increases only.

 G. increases, then decreases.

 H. decreases only.

 J. decreases, then increases.

31. A 10 mL blood sample from a fourth individual contains 5 mL of plasma and 4 mL of formed elements. Approximately what is the mass of plasma and formed elements in this blood sample?

 A. 6.5 g

 B. 9.5 g

 C. 11.5 g

 D. 15.5 g

GO ON TO THE NEXT PAGE

32. The phlebotomist varied which of the following techniques from Experiment 2 to Experiment 3?

   F. The volume of blood taken from each patient

   G. The mass of blood taken from each patient

   H. The instrument used to separate the elements of the blood samples

   J. The amount of time the samples were left in the centrifuge

33. The patient with the greatest mass of red blood cells is:

   A. Patient 1.

   B. Patient 2.

   C. Patient 3.

   D. not possible to determine from the information given.

GO ON TO THE NEXT PAGE

PASSAGE VI

A student performed experiments to determine the relationship between the electrical conductivity of a metal rod and its length, mass density (mass per unit length of metal), and temperature.

EXPERIMENT 1

The student used several lengths of iron rods. The student weighed the rods and calculated their mass densities. The rods were then heated to the specified temperature by being held over a flame. To test the conductivity, pairs of rods were placed at opposite sides of a container containing an *electrolyte solution* (a solution containing positive ions with a positive electrical charge and negative ions with a negative electrical charge) and then connected to a battery. The movement of the ions in the solution was detected and displayed on the screen of an oscilloscope, where the conductivity could be measured. The results are presented in Table 1.

Table 1

Trial	Length (cm)	Mass density (g/cm)	Temperature (°C)	Conductivity (μΩ/cm)
1	16	100	20	240
2	16	100	80	120
3	16	400	20	60
4	16	400	80	30
5	8	100	20	120
6	8	100	80	60
7	8	400	20	30

EXPERIMENT 2

The student repeated the procedure in Experiment 1, this time using rods made from silver and tungsten. The results are presented in Table 2.

Table 2

Trial	Material	Length (cm)	Mass density (g/cm)	Temperature (°C)	Conductivity (μΩ/cm)
8	Silver	16	400	20	30
9	Silver	16	100	80	120
10	Silver	16	225	20	60
11	Silver	16	225	5	120
12	Tungsten	16	100	20	60
13	Tungsten	16	225	80	240

GO ON TO THE NEXT PAGE

34. Which of the following would most likely be the conductivity of a silver rod with a length 16 cm, a mass density of 225 g/cm, and a temperature of 15°C?

    F. 60 μΩ/cm
    G. 80 μΩ/cm
    H. 100 μΩ/cm
    J. 130 μΩ/cm

35. Between Trials 8 and 10, the student directly manipulated which of the following variables?

    A. The temperature of the metal rods
    B. The conductivity of the metal rods
    C. The material from which the metal rods were composed
    D. The mass density of the metal rods

36. Instead of immersing the rods in an electrolyte solution, the student could have measured conductivity by:

    F. shortening the rods to 8 cm.
    G. connecting the rods with a low-resistance wire.
    H. increasing the mass density of the rods to 200 g/cm.
    J. insulating the rods with rubber.

37. If the rods used in Trial 11 were heated to a temperature of 60°C, the conductivity would most likely be:

    A. less than 60 μΩ/cm.
    B. 80 μΩ/cm.
    C. 120 μΩ/cm.
    D. greater than 120 μΩ/cm.

38. Based on the results of both experiments, which of the following statements regarding the relationship between observed conductivity and the other physical variables is false?

    F. Increasing temperature increases observed conductivity.
    G. Increasing mass density decreases observed conductivity.
    H. Increasing length increases conductivity.
    J. The observed conductivity depends on the material.

39. A student is given several metal rods of identical length but unknown mass density. The rod with the lowest mass density could be found by:

    A. placing all the rods under identical temperature conditions and selecting the rod with the lowest conductivity.
    B. placing all the rods under identical temperature conditions and selecting the rod with the highest conductivity.
    C. varying the voltage until all the rods have the same conductivity and selecting the rod with the lowest temperature.
    D. selecting the rod capable of sustaining the highest temperature without melting.

40. An ordinary table lamp requires a conductivity level of less than 30 μΩ/cm. An electrician wants to use an iron rod with the mass density 400 g/cm at a temperature of 80°C in order to conduct electricity to the lamp. Approximately what length should the rod be for the electrician to connect the table lamp to the source of the electricity?

    F. 48 cm
    G. 24 cm
    H. 16 cm
    J. Less than 16 cm

IF YOU FINISH BEFORE TIME IS CALLED, YOU MAY CHECK YOUR WORK ON THIS SECTION ONLY. DO NOT TURN TO ANY OTHER SECTION IN THE TEST.   STOP

Practice Test One
ANSWER KEY

MATHEMATICS TEST

1. D	9. A	17. C	25. B	33. D	41. C	49. C	57. C
2. J	10. K	18. H	26. K	34. J	42. G	50. G	58. F
3. B	11. A	19. A	27. C	35. A	43. E	51. B	59. A
4. G	12. K	20. G	28. K	36. H	44. J	52. K	60. J
5. A	13. D	21. C	29. D	37. A	45. E	53. E	
6. J	14. K	22. H	30. G	38. J	46. G	54. H	
7. A	15. A	23. E	31. B	39. E	47. A	55. B	
8. G	16. J	24. H	32. J	40. H	48. K	56. J	

SCIENCE TEST

1. A	6. G	11. A	16. J	21. C	26. J	31. B	36. G
2. J	7. D	12. G	17. D	22. H	27. A	32. J	37. A
3. B	8. G	13. D	18. H	23. B	28. H	33. A	38. F
4. H	9. A	14. G	19. B	24. H	29. D	34. G	39. B
5. A	10. H	15. A	20. F	25. B	30. F	35. D	40. J

ANSWERS AND EXPLANATIONS

MATHEMATICS TEST

1. D
Category: Proportions and Probability
Difficulty: Low
Getting to the Answer: When you're converting units, writing out the units will help you determine if you've made a mistake.

127 yards $\cdot \dfrac{1 \text{ rod}}{5.5 \text{ yards}} \approx 23.1$ rods, (D)

2. J
Category: Proportions and Probability
Difficulty: Low
Getting to the Answer: Instead of finding the increase and adding it to the original cost, you can do the computation in one step by adding 100% to the percent increase. When the cost of the pizza is raised by 22%, the new cost will be 122% of the original cost.

$20 \cdot 1.22 = $24.40, (J)

3. B
Category: Proportions and Probability
Difficulty: Low
Getting to the Answer: Phrases like "average increase" may sound a little complicated, but there's nothing difficult going on here. The chart shows increases, so you just need to find the average to get the "average increase." Remember, the average of a set of terms is the sum of the terms divided by the number of terms.

$$\text{average} = \frac{120 + 210 + 0 + 210 + 180}{5} = \frac{720}{5} = 144,$$

which is (B).

4. G
Category: Proportions and Probability
Difficulty: Low
Getting to the Answer: The formula rate $\cdot$ time = distance will take you far on the ACT. Some people

find it easier to remember as rate $= \dfrac{\text{distance}}{\text{time}}$. It's the same equation no matter how you rearrange it. Train A travels 50 $\cdot$ 3 = 150 miles. Train B travels 70 $\cdot$ 2.5 = 175 miles. The difference is 175 − 150 = 25 miles, or (G).

If you got this one wrong, you probably didn't answer the right question; the distance each train traveled is a trap, waiting to catch you if you stop too soon.

5. A
Category: Variable Manipulation
Difficulty: Low
Getting to the Answer: When a factored product equals 0, one of the factors must be 0.

$b - 3 = 0$ or $b + 4 = 0$
$b = 3$ or $\quad b = -4$

Only one of these, 3, appears in the answers, as (A). Although B might be tempting if you looked too quickly, 4 is not the same as −4.

6. J
Category: Plane Geometry
Difficulty: Medium
Getting to the Answer: Many of the wrong answers are designed to catch you making a careless mistake. Just because you got one of the 5 answers doesn't mean you did the problem correctly, so work carefully! In a parallelogram, opposite sides have the same length, so $RU = ST = 8$ and $RS = UT$. The perimeter is the sum of the sides, so $RS + ST + UT + RU = 42$. Substitute in the known sides and solve for the length of UT:

$RS + 8 + UT + 8 = 42$
$2UT + 16 = 42$
$2UT = 26$
$UT = 13$, (J)

7. A

Category: Plane Geometry

Difficulty: Medium

Getting to the Answer: In a "regular polygon," all of the angles have the same measure. This is one you could eyeball. What figure with equal angles has 60° angles? If you said an equilateral triangle, you're absolutely right. If you're not sure off the top of your head, consider each possible answer choice. Choice (A): This is a triangle, and the interior angles of a triangle add up to 180 degrees. Therefore, if all three angles are the same, they each measure $\frac{180}{3}$ = 60 degrees— just right. Choice B: A regular four-sided polygon is a square, so all four angles are 90 degrees. Even if you didn't notice that at first, you should know that the sum of the interior angles of a quadrilateral is 360 degrees, so each angle in a regular quadrilateral must be $\frac{360}{4}$ = 90 degrees. This is clearly too large. Choice C: A six-sided figure's interior angles add up to 720 degrees. For a regular hexagon, each angle would be $\frac{720}{6}$ = 120 degrees. Notice the pattern here: As the number of sides goes up, so does the measure of each angle. It's clear that to get a small angle like 60°, you need a small number of sides, so there is no need to check D or E.

8. G

Category: Variable Manipulation

Difficulty: Low

Getting to the Answer: If you forget the rules of exponents, try writing out an example and canceling. For example, $\frac{a^2}{a^3} = \frac{a \cdot a}{a \cdot a \cdot a} = \frac{1}{a}$. When you're dividing, you subtract the exponents of powers with the same base.

$$\frac{12a^5bc^7}{-3ab^5c^2} = \frac{12}{-3} \cdot \frac{a^5}{a} \cdot \frac{b}{b^5} \cdot \frac{c^7}{c^2} = -4 \cdot a^4 \cdot \frac{1}{b^4} \cdot c^5$$

$$= \frac{-4a^4c^5}{b^4}, \text{ which is (G).}$$

9. A

Category: Plane Geometry

Difficulty: Medium

Getting to the Answer: Whenever you see parallel lines, look for corresponding angles and alternate interior angles. They're most obvious when you're just given two parallel lines and a transversal, so questions that include parallel lines as parts of shapes like triangles or parallelograms can be a little sneaky. Because is $\overline{PQ}$ parallel to $\overline{WY}$, by corresponding angles, $\angle YWX$ has the same measure as $\angle QPX$ and $\angle WYX$ has the same measure as $\angle POX$:

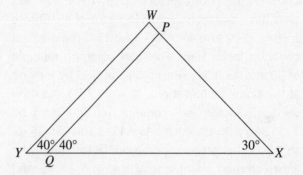

Using the fact that the angles of $\triangle PXQ$ sum to 180 degrees:

$40° + 30° + \angle QPX = 180°$

$\angle QPX = 110°$, which is (A).

10. K

Category: Operations

Difficulty: Low

Getting to the Answer: Absolute value is a special type of parentheses, so PEMDAS says to take the absolute values before multiplying. $|-4| \cdot |2| = 4 \cdot 2 = 8$, (K).

11. A

Category: Proportions and Probability

Difficulty: Low

Getting to the Answer: The math in this question isn't too challenging—just basic percents. The question is really testing whether you pay attention to detail and work carefully. 1250 – 800 – 150 = 300 participants were undecided. $\frac{300}{1250} = 0.24 = 24\%$, (A)

12. K

Category: Number Properties

Difficulty: Medium

Getting to the Answer: The greatest common factor is the largest factor that the two numbers share. The least common multiple is the smallest number that is a multiple of both numbers. Many questions about factors and multiples are made easier by considering the prime factorization of the numbers involved. Choice F: 15 is not a factor of 9 or 25, so these numbers can't have a greatest common factor of 15. Choice G: 15 is not a factor of 27, so these numbers can't have a greatest common factor of 15. Choice H: 15 is not a factor of 25, so these numbers can't have a greatest common factor of 15. Choice J: $30 = 15 \cdot 2$ and $45 = 15 \cdot 3$, so 15 is the greatest common factor here. The least common multiple of 30 and 45 is 90, which you can find by looking at the prime factorizations: $30 = 3 \cdot 2 \cdot 5$ and $45 = 5 \cdot 3 \cdot 3$, so the least common multiple must be $2 \cdot 3 \cdot 3 \cdot 5 = 90$. Choice (K): $45 = 15 \cdot 3$ and $75 = 15 \cdot 5$, so 15 is the greatest common factor here. The least common multiple is 225: $45 = 5 \cdot 3 \cdot 3$ and $75 = 5 \cdot 5 \cdot 3$, so the least common multiple is $5 \cdot 5 \cdot 3 \cdot 3 = 225$.

13. D

Category: Operations
Difficulty: Medium
Getting to the Answer: Remember that if you raise a negative number to an odd power, the result is negative. (If the power is even, the result is positive.) Any question that involves negative numbers will require extra attention. It's very easy to lose track of negative signs.

$x^5y + xy^5 = (2)^5(-3) + (2)(-3)^5 = (32)(-3) + (2)(-243) = -96 - 486 = -582$, which is (D).

14. K

Category: Plane Geometry
Difficulty: Medium
Getting to the Answer: The first step is to recall the definition of perimeter—the distance around the sides of a figure. For a square, the perimeter is four times the length of a side. $\frac{1}{4}(16 - 24h) = 4 - 6h$, which is (K). If you got this one wrong, you probably only took $\frac{1}{4}$ of one of the terms, not of the

entire expression for the perimeter. For example, J is $\frac{1}{4}(16) - 24h = 4 - 24h$.

15. A

Category: Variable Manipulation
Difficulty: Medium
Getting to the Answer: Memorizing the three classic quadratics will save you valuable time on questions like this. Remember, $(x - y)^2 = x^2 - 2xy + y^2$. Multiply out the quadratic, or better yet, write down the formula from memory. $(x - k)^2 = x^2 - 2kx + k^2 = x^2 - 26x + k^2$. Because the coefficient of x must be the same on both sides of the equation, $-2k = -26$. $k = 13$, which is (A).

16. J

Category: Proportions and Probability
Difficulty: Low
Getting to the Answer: Remember that 20% off means the sale price was 100% − 20% = 80% of the original price. The game system cost 0.9($180) = $162. Each game cartridge cost 0.8($40) = $32. The total was $162 + 2($32) = $226, (J). Be sure you have all the parts to avoid trap answers. Choice G, for instance, is the price of the system and *one* cartridge, so work carefully!

17. C

Category: Coordinate Geometry
Difficulty: Medium
Getting to the Answer: The slope of a line is $\frac{y_2 - y_1}{x_2 - x_1}$. Either point could be used first, but you must be sure that the first x- and y-coordinate come from the same point. If $(5,9) = (x_1,y_1)$ and $(-3,-12) = (x_2,y_2)$, then slope $= \frac{y_2 - y_1}{x_2 - x_1} = \frac{-12 - 9}{-3 - 5}$. Because this doesn't look like any of the answers, try the other ordering: If $(-3,-12) = (x_1,y_1)$ and $(5,9) = (x_2,y_2)$, then slope $= \frac{y_2 - y_1}{x_2 - x_1} = \frac{9 - (-12)}{5 - (-3)}$, which is (C).

18. H

Category: Coordinate Geometry
Difficulty: Medium

Getting to the Answer: You don't have to find the solutions, just the number of solutions. As long as you know the technique you would use to solve it, you can eyeball the answer in seconds. An equation crosses the x-axis when $y = 0$, so set the equation equal to 0: $(x + 1)(x + 2)(x - 3)(x + 4)(x + 5) = 0$ There are 5 factors, any of which could be 0 (for example, if $x + 1 = 0$, then $x = -1$), so there will be 5 roots, which is (H). Specifically, they are -1, -2, 3, -4, and -5, but you don't need to know that to answer the question. If you have a graphing calculator, you might be tempted to use it to graph this equation and see how many times the graph intersects the x-axis. Like solving for all the roots, this is more trouble than it's worth. The time taken up by typing the equation would be better used to think about the problem and realize that it actually asks for something very simple.

19. A
Category: Variable Manipulation
Difficulty: Medium
Getting to the Answer: To reduce, you must factor out the same number from the top and the bottom, then divide. You *cannot* only reduce the 4, neglecting the $8x$ (or vice versa).
$\dfrac{4 + 8x}{12x} = \dfrac{4(1 + 2x)}{4(3x)} = \dfrac{1 + 2x}{3x} = \dfrac{2x + 1}{3x}$, which is (A).

20. G
Category: Operations
Difficulty: Low
Getting to the Answer: Always read the problem carefully before you jump into your calculations. If 5 people buy 5 tickets for $95, each pays $\dfrac{\$95}{5} = \$19$. Because the individual rate is $21.50, this is a savings of $21.50 − $19 = $2.50 per person, which is (G). If you got this one wrong, you might have found the total savings for all five people, K, or split the savings just among the original four people, H.

21. C
Category: Variable Manipulation
Difficulty: Low
Getting to the Answer: Small mistakes will add up quickly—keep yourself focused! The key to this one is that x^2y and $2xy^2$ are <u>not</u> like terms. Like terms must have the same exponent on each variable. Combine the two first terms to get: $-2x^2y^2 + x^2y + 3x^2y^2 + 2xy^2 = x^2y^2 + x^2y + 2xy^2$, which is (C).

22. H
Category: Trigonometry
Difficulty: Low
Getting to the Answer: On sine and cosine questions, you can eliminate any answer that's not between −1 and 1, but remember that tangent can get very large or small. Use SOHCAHTOA to help you remember which trig function uses which sides of the triangle.
$\tan(\theta) = \dfrac{\text{opposite}}{\text{adjacent}} = \dfrac{12}{7}$, which is (H).

23. E
Category: Variable Manipulation
Difficulty: Low
Getting to the Answer: If you're not sure which operation is appropriate, try Picking Numbers. If Yousuf was 10 years old fifteen years ago, then he's 25 today. In another 7 years, he will be 32. Plug $x = 10$ into each answer choice to see which one equals 32:
A: $10 + 7 = 17$ No.
B: $(10 - 15) + 7 = 2$ No.
C: $(10 + 15) - 7 = 18$ No.
D: $(10 - 15) - 7 = -12$ No.
(E): $(10 + 15) + 7 = 32$ Yes!
The key to solving this problem algebraically is to realize that if Yousef was x years old 15 years ago, he is now $(x + 15)$ years old, not $(x - 15)$ years old. Seven years from now, he'll be another 7 years older, or $(x + 15) + 7$, (E).

24. H
Category: Variable Manipulation
Difficulty: Medium

Getting to the Answer: It's a good idea to check your factoring by multiplying it back out. You want two numbers that multiply to –12 and sum to –4; those numbers are –6 and +2: $x^2 - 4x - 12 = (x - 6)(x + 2)$. The latter factor is (H).

25. B
Category: Plane Geometry
Difficulty: Low
Getting to the Answer: When using the Pythagorean Theorem, $a^2 + b^2 = c^2$, remember that a and b are the legs and c is the hypotenuse. Wrong answer choices may come from plugging numbers into the wrong part of the formula.

$$8^2 + 15^2 = c^2$$
$$64 + 225 = c^2$$
$$289 = c^2$$
$$c = \sqrt{289} = 17, \text{(B)}$$

26. K
Category: Variable Manipulation
Difficulty: Medium
Getting to the Answer: As with every question involving negative numbers, be careful with the negative signs. Also pay attention to the location of the parentheses. Here you need to cube both the –2 and the x^5 because both are inside the parentheses.

$$(-2x^5)^3 = (-2)^3(x^5)^3 = -8x^{15}, \text{(K)}$$

27. C
Category: Proportions and Probability
Difficulty: High
Getting to the Answer: Don't worry when you see an unfamiliar term, like "relative atomic mass." On questions with terms the test makers don't expect you to be familiar with (including ones they just made up), they'll tell you everything that you need to know. 1 cubic centimeter of carbon masses 12 grams. 1 cubic centimeter of the unknown element is 30 grams. Because these are equal amounts (1 cubic centimeter of each), the relative atomic mass is simply the ratio of the mass of the unknown element to the mass of the carbon, which is $\frac{30}{12} = 2.5$, (C).

28. K
Category: Variable Manipulation
Difficulty: Medium
Getting to the Answer: Don't stop until you're sure you've answered the exact question asked. It's tempting to bubble in the value of x and move on, but that's not what this question is asking for.

$$2x + 3 = -5$$
$$2x = -8$$
$$x = -4$$
$$x^2 - 7x = (-4)^2 - 7(-4) = 16 + 28 = 44, \text{(K)}$$

Be wary of sign errors; it's easy to subtract 28 instead of adding.

29. D
Category: Coordinate Geometry
Difficulty: Medium
Getting to the Answer: Inequalities work exactly like equalities, except that the direction of the sign changes if you multiply or divide by a negative.

$$2(5 + x) < 2$$
$$5 + x < 1$$
$$x < -4$$

The numbers less than –4 are to the left, which means the correct graph is (D).

30. G
Category: Variable Manipulation
Difficulty: Medium
Getting to the Answer: When two things vary directly, one rises as the other rises and falls as the other falls. If x and y vary directly, they can be represented by the equation $y = kx$, where k is a constant. When two things vary inversely, one rises as the other falls and falls as the other rises. If x and y vary inversely, they can be represented by $y = \frac{k}{x}$. The question describes how m relates to the other variables, so the equation will be $m =$ something. Variables that it varies directly with will be in the numerator, and variables that it varies inversely with will be in the denominator. Therefore, $\frac{b^3}{c^2} = m$, which is (G). As the cube of b increases, so does m. As the square of c rises, m will fall.

31. B

Category: Coordinate Geometry

Difficulty: Low

Getting to the Answer: Occasionally you'll be given information in the problem that you don't need. Here, the x-coordinates are irrelevant because the variable is the midpoint of the y-coordinates. The midpoint formula states that the x- and y-coordinates of the midpoint are the averages of the x- and y-coordinates of the endpoints.

$$\frac{-7+5}{2} = p$$

$-1 = p$, so (B) is correct.

32. J

Category: Coordinate Geometry

Difficulty: Medium

Getting to the Answer: Understanding the slope-intercept equation of a line, $y = mx + b$, is essential on the ACT. In the equation $y = mx + b$, the slope is the coefficient of x. In the first equation, the slope is 3. In the second equation, the slope is m. To be parallel, the two equations must have the same slope, so m = 3, which is (J).

33. D

Category: Coordinate Geometry

Difficulty: Medium

Getting to the Answer: Problems like this that require two skills (translating English to algebra and graphing on a number line) can be skipped until you've done the more basic problems. "3 times x is increased by 5" translates to $3x + 5$, which is "less than 11":

$3x + 5 < 11$

$3x < 6$

$x < 2$

This is graphed with an open circle at 2 (because x cannot equal 2) and shaded to the left, where x is less than 2. Your graph should look like (D).

34. J

Category: Variable Manipulation

Difficulty: Medium

Getting to the Answer: If you decide to Pick Numbers, make sure they're easy to work with. For example, x = 6 would be a good number of pencils because 54 is divisible by 6. If x = 6, then each pencil costs $\frac{54}{6} = 9$ cents. If y = 4, then each eraser costs $\frac{92}{4} = 23$ cents. The cost of 7 pencils and 3 erasers is $9(7) + 23(3) = 63 + 69 = 132$ cents. Then plug x = 6 and y = 4 into each answer choice to see which also equals 132 cents. Alternatively, you could think through the algebra to avoid having to calculate all those complicated answer choices. If x pencils cost 54 cents, then each one costs $\left(\frac{54}{x}\right)$ cents. Similarly, each eraser will cost $\left(\frac{92}{y}\right)$ cents. To find the cost of 7 pencils, multiply 7 by the cost per pencil: $7\left(\frac{54}{x}\right)$. Similarly, the cost of 3 erasers is $3\left(\frac{92}{y}\right)$. Add these together to get (J).

35. A

Category: Coordinate Geometry

Difficulty: Medium

Getting to the Answer: You don't always need to make an exact plot. Make a rough sketch, and then, if you have to, you can go back and make it better. Rearrange $y - x = 0$ to get $y = x$. This is a line with slope 1 that goes through the origin. You could plot this line and the points, then find the three on the same side of the line. Another way to think about the line is that it's all the points where the x and y coordinates are equal. Above the line y will be greater than x (which is true of the first three points listed in the problem), and below the line x will be greater than y (which is true of the last two points). So the answer is the first 3 points, which is (A).

36. H

Category: Trigonometry

Difficulty: Medium

Getting to the Answer: Just because two of the sides are 3 and 4 doesn't mean it's a 3-4-5 triangle—remember that 3 and 4 are the legs, and 5 is the hypotenuse! The third side is $\sqrt{4^2 - 3^2} = \sqrt{7}$, so $\sin E = \dfrac{\text{opposite}}{\text{hypotenuse}} = \dfrac{\sqrt{7}}{4}$, which is (H).

37. A

Category: Coordinate Geometry
Difficulty: Medium
Getting to the Answer: If the solution set is a single line, then both equations describe the same line. When two equations are the same line, one is an exact multiple of the other. Look for a multiple. Because 3 • 4 = 12 and 27 • 4 = 108, you can get the first equation by multiplying the second by 4:

$4(3x + ky) = 4(27)$
$12x + 4ky = 108$

Therefore:

$4k = -20$
$k = -5$, (A)

38. J

Category: Variable Manipulation
Difficulty: High
Getting to the Answer: Sometimes on the ACT you will need to put together several pieces of information in the right way to get the answer. Because each extra inch adds approximately 10 pounds, a 72 inch person should weigh about 40 pounds more than a 68 inch person, for a total of 190 pounds. Now use the formula:

$BMI = \dfrac{703W}{H^2} = \dfrac{703(190)}{72^2} \approx 26$, which is (J).

39. E

Category: Variable Manipulation
Difficulty: High
Getting to the Answer: Sometimes questions will look harder than they actually are because they're written to intimidate you. Set the expressions equal to each other and see what you can deduce:

$\left(\dfrac{x}{y}\right)^2 = \dfrac{x^2}{y}$

$\dfrac{x^2}{y^2} = \dfrac{x^2}{y}$

$x^2 y = x^2 y^2$

At this point, you can divide by x^2y if x^2y is not equal to zero, leaving $1 = y$. If $x^2y = 0$, then either x or y equals 0. Because the question tells you to assume y does not equal 0, $x = 0$. There are two possible ways that $\left(\dfrac{x}{y}\right)^2$ can equal $\dfrac{x^2}{y}$: if $x = 0$ or if $y = 1$. If the algebra is confusing, Pick Numbers. Let $x = 0$ and $y = 2$. Then $\left(\dfrac{x}{y}\right)$ is $\left(\dfrac{0}{2}\right) = 0^2 = 0$ and $\left(\dfrac{x}{y}\right)$ is $\left(\dfrac{0}{2}\right)^2 = 0$.

These are equal. You can eliminate B, C, and D. To test whether A or (E) is true, try $x = 2$, $y = 1$:

$\left(\dfrac{x}{y}\right)^2 = \left(\dfrac{2}{1}\right)^2 = 2^2 = 4$

$\dfrac{x^2}{y} = \dfrac{2^2}{1} = \dfrac{4}{1} = 4$

These are equal, so (E) is correct.

40. H

Category: Plane Geometry
Difficulty: Medium
Getting to the Answer: Always be on the lookout for special triangles. What kind of triangles are formed when a perpendicular bisector is added to an equilateral triangle?

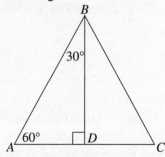

Because $\triangle ABC$ is equilateral, each of its angles is 60 degrees. Angle BDA is 90 degrees, because $\overline{BD}$ is perpendicular to $\overline{AC}$. Then each half of ABC is a 30-

60-90 triangle, with side ratios of $x:x\sqrt{3}:2x$. The side opposite the 60-degree angle is $x\sqrt{3} = 4\sqrt{3}$ units long, so $x = 4$. The hypotenuse, $\overline{BC}$, is $2x = 8$ units long, which is (H).

41. C
Category: Plane Geometry
Difficulty: Medium
Getting to the Answer: To find the perimeter of a complicated figure like this one, it can be easy to leave out some sides. Try marking each side as you add it so that you don't accidentally forget any or add any twice. First find the missing lengths:

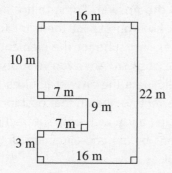

The small horizontal side must be the same length as the small horizontal side above it, because all the angles are right angles. The interior vertical side must be $22 - (10 + 3) = 9$ meters long, because the total vertical distance on each side of the figure must be the same. Now add the length of each side to find the perimeter: $16 + 22 + 16 + 10 + 7 + 9 + 7 + 3 = 90$, or (C).

42. G
Category: Plane Geometry
Difficulty: High
Getting to the Answer: The advantage of a multiple choice test is that you don't have to come up with the answer yourself; you just have to pick it out. However, remember that if you can predict your own answer and then compare, you're less likely to fall into a trap. Sides $\overline{AO}$ and $\overline{DO}$ will be radii of the circle, which means they are congruent. Choice (G) is the most direct explanation. All the other choices depend on this deduction to be proven true.

43. E
Category: Plane Geometry
Difficulty: Medium
Getting to the Answer: If your answer doesn't look quite like any of the answer choices, rearrange it so that it does. Don't just pick an answer that looks similar; make sure it actually means the same thing. Parentheses make a big difference, so you should carefully consider where they should be! The outer circle has area $\pi r^2 = \pi (12^2)$. The inner circle has area $\pi r^2 = \pi (4^2)$. The shaded area is $\pi (12^2) - \pi (4^2) = \pi (12^2 - 4^2)$, which is (E).

44. J
Category: Plane Geometry
Difficulty: High
Getting to the Answer: Be sure to examine the diagram carefully. Each dimension is $2w$ longer in the outer rectangle; there's w more above and w more below, and w more to the left and w more to the right. It will be very tempting to solve for $2w$ and stop, instead of finding the value the question asks for. You can use either dimension to find w. Using the length:

$$32\frac{1}{2} = 29 + 2w$$

$$3\frac{1}{2} = 2w$$

$$1\frac{3}{4} = w, \text{ so (J) is correct.}$$

Using the width:

$$27\frac{1}{2} + 2w = 31$$

$$2w = 3\frac{1}{2}$$

$$w = 1\frac{3}{4}, \text{ (J)}$$

45. E
Category: Variable Manipulation
Difficulty: Medium
Getting to the Answer: If you're trying to factor a quadratic and you're not getting anywhere, use one of the Kaplan strategies. Here, there are numbers in the answer choices, so Backsolving will work well. If you Backsolve, be sure you plug the answer choices

into the right part of the problem. Here, they represent possible lengths. Start with C: If the length of the floor is 13.5 feet, then the width is 2(13.5) – 21 = 27 – 21 = 6 feet. This would produce an area of 13.5(6) = 81 square feet—not big enough. Try D:

length = 17 feet

width = 2(17) – 21 = 34 – 21 = 13 feet

area = 17(13) = 221 square feet

This still isn't big enough, so (E) must be correct:

length = 19 feet

width = 2(19) – 21 = 38 – 21 = 17 feet

area = 19(17) = 323 square feet

Perfect!

You could also solve algebraically:

An equation for the width (W) in terms of the length (L) is $W = 2L - 21$.

$$\text{area} = L \cdot W = 323$$
$$L(2L - 21) = 323$$
$$2L^2 - 21L = 323$$
$$2L^2 - 21L - 323 = 0$$
$$(2L + 17)(L - 19) = 0$$

$2L + 17 = 0$ or $L - 19 = 0$

$2L = -17$ or $L = 19$

$L = \dfrac{-17}{2}$ or $L = 19$

Length must be positive, so the length is 19, (E).

46. G

Category: Plane Geometry
Difficulty: Medium
Getting to the Answer: Picking Numbers can make a theoretical problem much more concrete. Say the original radius was 1. Then the area of the circle would be $\pi r^2 = \pi(1^2) = \pi$. Twice this area would be 2π. Find the radius of a circle with area 2π:

$$\pi r^2 = 2\pi$$
$$r^2 = 2$$
$$r = \sqrt{2}$$

The new radius is $\sqrt{2}$ times the old radius, so (G) is correct.

47. A

Category: Operations
Difficulty: Medium
Getting to the Answer: On some questions you won't be able to depend on your calculator. You simply have to know the formula or rule. Rewrite this equation using an exponent: $x^3 = 64$. If you're not sure what to the third power will give you 64, you can always Backsolve. You'll find that $4^3 = 64$, which means (A) is correct.

48. K

Category: Patterns, Logic & Data
Difficulty: Medium
Getting to the Answer: Some things in math actually work like you expect them to. To subtract two matrices, just subtract the elements that are in the same position. After you've subtracted one position, eliminate the answer choices that don't have the correct number in that position. You may be able to get away with only subtracting one or two positions before you eliminate all the wrong answer choices.

$$\begin{bmatrix} 3 & -6 \\ 0 & 9 \end{bmatrix} - \begin{bmatrix} -3 & 6 \\ 0 & -9 \end{bmatrix} = \begin{bmatrix} (3)-(-3) & (-6)-(6) \\ (0)-(0) & (9)-(-9) \end{bmatrix} = \begin{bmatrix} 6 & -12 \\ 0 & 18 \end{bmatrix},$$

so (K) is correct.

49. C

Category: Number Properties
Difficulty: Medium
Getting to the Answer: When you're Picking Numbers for a question with few limits, don't forget to try both positives and negatives, integers and fractions. Try $a = 1$ and $b = -2$. This immediately eliminates B. Then $a^2 = 1$ and $b^2 = 4$, which eliminates E. Now try $a = 2$ and $b = 1$. This eliminates A. Then $a^2 = 4$ and $b^2 = 1$, which eliminates D. Choice (C) must be correct, because the square of a number cannot be negative. No matter what b is, b^2 must be greater than or equal to zero.

50. G
Category: Trigonometry
Difficulty: Medium
Getting to the Answer: On trigonometry problems, it usually helps to draw the triangle. Use SOHCAH-TOA to remember which trig function uses which sides of the triangle.

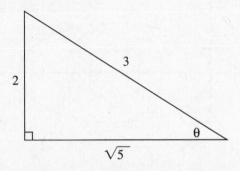

The smallest angle is the one opposite the shortest side. Here, it's marked θ. Based on the diagram,
$\cos\theta = \dfrac{\text{adjacent}}{\text{hypotenuse}} = \dfrac{\sqrt{5}}{3}$, which is (G).

51. B
Category: Trigonometry
Difficulty: High
Getting to the Answer: A graphing calculator can be a good backup, but understanding the math will always be faster. You could plug this equation into your graphing calculator, find the highest and lowest values, find their difference, and divide by 2. Alternatively, you could use algebra. Rewrite the equation as $y = 4\sin(5\theta) - 3$. The −3 at the end moves the entire graph down by 3; it doesn't affect the difference between the largest and smallest values. The 5 inside the sin affects how often the function repeats itself in the same space on the x-axis. It's the 4 in front that multiplies the y values and makes the extreme values of the function higher and lower. Sine usually goes from −1 to 1, so if you multiply all the values by 4, it will go from −4 to 4. The difference is 8, and half of the difference is 4. So the amplitude is 4, (B).

52. K
Category: Plane Geometry
Difficulty: Medium
Getting to the Answer: A line is like a pencil that goes forever in both directions, and a plane is like a piece of paper that goes forever in all directions. If you model each situation using pencils and paper, you'll see that you can only make one plane with the paper in every case but (K). To make (K), hold two pencils parallel, one above the other. Now rotate one. You cannot put a piece of paper flat against both of these lines, so they do not define a plane.

53. E
Category: Plane Geometry
Difficulty: Medium
Getting to the Answer: The wording of this problem is a giveaway. One of the angles is a "vertex angle" and the other two are "base angles." This will allow you to solve the problem even if you forgot what an isosceles triangle is! The angles sum to 180 degrees, so:

$$(x - 10) + (3x + 18) + (3x + 18) = 180$$
$$7x + 26 = 180$$
$$7x = 154$$
$$x = 22$$

Base angle = $3x + 18 = 3(22) + 18 = 84$, (E).
Notice that A and B are the answers to other questions; they are, respectively, the measure of the vertex angle and the value of x. Be sure to solve for the right thing.

54. H
Category: Proportions and Probability
Difficulty: Medium
Getting to the Answer: If you don't take the time to read carefully, you'll lose a lot of points on careless mistakes. You might want to leave a question you know will take a long time until after you've gotten points from easier questions. One set of potholders requires: $(6 \cdot 8) + (5 \cdot 12) + (2 \cdot 18) = 48 + 60 + 36 =$ 144 inches of fabric. This is 144 inches $\cdot \dfrac{1 \text{ yard}}{36 \text{ inches}}$ = 4 yards. Each yard costs \$1.95, so one set of

potholders costs 4($1.95) = $7.80. Margot is making five sets of potholders, so the total cost is 5($7.80) = $39, which is (H).

55. B

Category: Plane Geometry
Difficulty: High
Getting to the Answer: Picking Numbers is a great way to deal with abstract questions like this one. Say the original dimensions were $w = 2$, $h = 3$. The surface area would be $S = 4(2)(3) + 2(2)^2 = 24 + 8 = 32$. If you double each dimension, you'll have $w = 4$, $h = 6$, $S = 4(4)(6) + 2(4)^2 = 96 + 32 = 128$. The surface area has gone up by a factor of 4, (B), because $32 \cdot 4 = 128$.

56. J

Category: Proportions and Probability
Difficulty: Medium
Getting to the Answer: The average of a set of terms is the sum of the terms divided by the number of terms. Even if you're not sure what to do on an averages problem, plugging the given information into this formula can help you figure out where to go.

For the first 4 numbers: $\dfrac{\text{sum}}{4} = 14$

$$\text{sum} = 56$$

When you include the fifth number, x, the new sum will be $56 + x$. The new average is:

$$\frac{56+x}{5} = 16$$
$$56 + x = 80$$
$$x = 24$$

The fifth number is 24, which is (J). Don't forget to divide by 5 in the second equation, because there are now five numbers.

57. C

Category: Coordinate Geometry
Difficulty: High
Getting to the Answer: If you can tell at a glance that a problem is going to take several minutes, save it until you've done all the easier problems. In the equation of an ellipse, $\dfrac{(x-h)^2}{a^2} + \dfrac{(y-k)^2}{b^2} = 1$, the

center is at (h,k), the length of the horizontal axis is $2a$, and the length of the vertical axis is $2b$. This particular ellipse is:

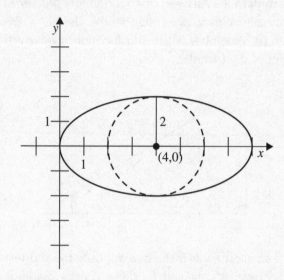

The largest circle possible is dotted on the diagram. Notice that it has the same center as the ellipse, and has radius 2 (the shortest dimension of the ellipse). In the equation of a circle, $(x-h)^2 + (y-k)^2 = r^2$, (h,k) is the center and r is the radius. Plug this information into the formula to find this circle's equation: $(x-4)^2 + y^2 = 4$, which is (C).

58. F

Category: Coordinate Geometry
Difficulty: Medium
Getting to the Answer: The highest power in the equation determines the general shape. The graph of $y = x^3$ will always have the general shape of the graph in F. Adding a constant C will move it up by C units (if C is positive—if C is negative the graph will move down). If you're not familiar with the shape of the graph $y = x^3$, you could either plug it into a graphing calculator or plot a few points. Don't worry too much about specific values. As soon as you realize that when x is negative, y will also be negative, you can eliminate all the graphs except (F).

59. A

Category: Coordinate Geometry
Difficulty: High

Getting to the Answer: Equations share points when the same values of x and y work for all the equations involved. Start with the first two equations. Rewrite the second so that both equations have the same term on one side, then set them equal to each other:

$$-x = y - 8$$
$$x = -y + 8$$
$$x = y + 8$$
$$-y + 8 = y + 8$$
$$-y = y$$

This is only true when y equals 0. Plug $y = 0$ into either equation and solve for x:

$$x = y + 8$$
$$x = 0 + 8$$
$$x = 8$$

The point (8,0) is shared between the first two equations. Does it work in the third?

$$6x = 2y + 4$$
$$6(8) = 2(0) + 4$$
$$48 = 4$$

That's not true, so this point is not shared between all three equations. There are no points that work in every equation, so (A) is correct.

60. J
Category: Proportions and Probability
Difficulty: High
Getting to the Answer: ACT probability questions are simple enough that you can write out all the possible outcomes if you need to. Remember that probability is the number of desired outcomes over the number of possible outcomes. The easiest way to think about this problem is to look at it backwards. If there are three heads, how many tails are there? In 4 coin tosses, if there are 3 heads there must be exactly one tail. That tail could be the first, second, third, or fourth coin, so there are four ways to get one tail and three heads. There are two possible positions for each coin, so the total number of possible arrangements of heads and tails is $2 \cdot 2 \cdot 2 \cdot 2 = 16$. Therefore, the probability that, in 4 tosses, there will be exactly 3 heads is $\frac{4}{16} = \frac{1}{4}$. (J) is correct.

If you're in doubt, write it out! All the possible arrangements of 4 coins are:

HHHH	**THHH**	HTTH	THTT
HHHT	HHTT	THTH	TTHT
HHTH	HTHT	TTHH	TTTH
HTHH	THHT	HTTT	TTTT

The ones with three heads are in bold type. There are 4 arrangements with exactly three heads, and 16 total possible arrangements. Again, the probability of getting exactly three heads is $\frac{4}{16} = \frac{1}{4}$, so (J) is correct.

SCIENCE TEST

PASSAGE I

1. A
Category: Figure Interpretation
Difficulty: Low
Getting to the Answer: The key to quickly answering many Figure Interpretation questions will be finding the data you need. When you examined the figures during Step 2 of the Kaplan Method for ACT Science, you should have noted that Table 1 lists the percents of calcium and zinc in a group of minerals. Look there for the answer. The question stem tells you that you're looking for a mineral composed of 32% zinc and 12% calcium. According to Table 1, hornblende is composed of 30 to 35 percent zinc and 10 to 20 percent calcium, so (A) is correct.

2. J
Category: Figure Interpretation
Difficulty: Low
Getting to the Answer: The key to answering this question is identifying the relevant figure. Where have you seen information about soil horizons? The answer is in Figure 1, which shows the two most common minerals in each horizon. A geologist digging down to the A horizon would encounter mostly quartz and mica. Quartz isn't listed as an answer choice, but mica is. Choice (J) is correct. Choice F is incorrect because limestone isn't commonly found until the

C horizon. Choice G is incorrect because shale isn't common until the final horizon. Choice H is incorrect because serpentine is commonly found in the B horizon.

3. B
Category: Patterns
Difficulty: Medium
Getting to the Answer: Be careful that you don't accidentally reverse the relationship in this otherwise straightforward question. First, look at Table 1. The minerals are arranged from highest zinc content to lowest zinc content. Next, use Figure 1 to check the depth at which each mineral is most commonly found. You can see that the minerals are arranged in Table 1 so that the shallowest are at the top of the column and the deepest are at the bottom. Be careful to note that the question stem gives the beginning of the relationship ("As zinc content increases:"). Zinc content increases if you read the table from the bottom up, which corresponds to the deepest minerals first. As zinc content *increases*, then, depth *decreases*, (B) .

4. H
Category: Figure Interpretation
Difficulty: Low
Getting to the Answer: You can answer this question either by looking through the answer choices for the minerals you would find or by crossing off choices that contain minerals you wouldn't find. The only minerals geologists wouldn't commonly find at a depth of 30 feet or less (the top of the C horizon) are limestone and shale. You can eliminate F, G, and J because these contain one of these minerals. Choice (H), then, is correct.

5. A
Category: Patterns
Difficulty: Medium
Getting to the Answer: The question stem doesn't name a figure or table for you to look at, so you'll have to ask yourself where best to find the answer. The mineral content of granite is the subject of

Table 2, so start there. Table 2 shows that granite is composed of feldspar, quartz, mica, and augite. If augite is found at a depth between the other minerals in granite, then it's found at a depth between that of feldspar, quartz, and mica. Now use Figure 1 to find the depths at which those three minerals are most commonly found. Feldspar is found in the O horizon, at a depth of 2 feet or less. Quartz and mica are found in the A horizon, at a depth of 10 feet or less. So you should expect to find augite at a depth of between 2 and 10 feet. Only (A) captures this range.

6. G
Category: Patterns
Difficulty: Low
Getting to the Answer: When asked to identify relationships, make sure you identify the correct table or figure to analyze. Zinc content percentage and calcium content percentage are found in Table 1. Zinc decreases as calcium increases. Choice (G) matches this nicely.

PASSAGE II

7. D
Category: Patterns
Difficulty: Medium
Getting to the Answer: Each of the tables includes rubber, which has a voltage of 0.0. Eliminate choice A. Use the information given in the question stem to determine the correct table to use. Study 2 and Study 3 both use 2 mm wire. Study 3 was conducted at 30°C, so use Table 3 to find the range. The highest value is 12.1 mV, so the range is 0.0 mV to 12.1 mV. Choice (D) matches this range.

8. G
Category: Figure Interpretation
Difficulty: Low
Getting to the Answer: Questions like this one require you to look at the results in the table; there is no need to even refer to the written paragraphs about the experiment. Trace the activity of each material in Tables 1

and 2. Silicon carbide conducts more voltage in Table 2, with a diameter of 2 mm, than in Table 1, with a diameter of 4 mm, so F proves the hypothesis and is incorrect. Copper's conductivity does the same thing, so H likewise supports the hypothesis. Steel's conductivity increases as its diameter decreases, so J again supports the hypothesis. Only rubber fails to support the hypothesis. It conducts 0 mV in Table 1 and Table 2, so decreasing the diameter does not increase the conductivity. Choice (G), therefore, fails to prove the hypothesis and is the correct answer.

9. A
Category: Patterns
Difficulty: Medium
Getting to the Answer: Look at one variable at a time when you're asked to determine the relationship between multiple variables. Start by looking at what happens to conductivity when the diameter of a Zinc Telluride (ZnTe) wire increases. Tables 1 and 2 show that when diameter increases, the amount of voltage conducted decreases. You can cross off B and D because they represent increased diameters at a constant temperature. Now check how temperature affects conductivity. Tables 2 and 3 show that when the temperature of a Zinc Telluride (ZnTe) wire increases, its conductivity decreases. Therefore, you're looking for the sample with the smallest diameter and lowest temperature. Choice (A) is correct.

10. H
Category: Figure Interpretation
Difficulty: Medium
Getting to the Answer: This question might seem confusing at first, but approach it one step at a time. You'll only need to refer to one table to determine the combined conductivity. Use Table 3 to find the voltage because the conditions in the question (2 mm diameter, 0.5 m long, and 30°C) match those in Study 3. Copper is listed in Table 3 as conducting 12.1 mV and rubber is listed as conducting 0.0 mV. No voltage will travel through the rubber wire, then, and only the voltage through the copper wire will be measured. The correct answer is 12.1 millivolts, (H).

11. A
Category: Patterns
Difficulty: Medium
Getting to the Answer: Start by determining which studies tested the effect of temperature on conductivity. Studies 2 and 3 held every variable constant except for temperature, so look there for the patterns in the data. Tables 2 and 3 show the amount of voltage conducted when temperature is decreased from 50°C to 30°C and the other variables are kept constant. You can see that when temperature is decreased, the conductivity of all the materials increases *except* rubber, which remains at 0.0 millivolts. If conductivity increases when temperature decreases, you can expect that conductivity will decrease when temperature increases. Choice (A) is correct. Choices B, C, and D all contradict the results shown in the tables.

12. G
Category: Scientific Reasoning
Difficulty: High
Getting to the Answer: Scientific Reasoning questions can usually be cracked either by focusing on the variables that were intentionally manipulated or by checking the opening paragraphs for information. You can see by looking at the tables that rubber conducts 0.0 millivolts of electricity throughout all three studies. Therefore, rubber is unable to conduct electricity. The second paragraph of the passage defines a material that is unable to conduct electricity as an insulator. Therefore, rubber must have been tested to see how temperature and diameter affect its insulating abilities. Choice (G) is correct. Choices F and H are incorrect because they contradict the results of the studies. Choice J could be immediately eliminated, as changes in wire length were never tested.

13. D
Category: Scientific Reasoning
Difficulty: Medium
Getting to the Answer: For this type of question, either look for an answer that further probes a variable that was shown to have some effect on the

outcome of the experiment but leaves more to be learned, or look for a choice that examines another logical variable. This question can best be answered by the process of elimination. You can eliminate A because the experiments in this passage already tested the effects of diameter and temperature. You can eliminate B because you're given no reason to suspect that color could have any effect on conductivity. You can eliminate C because, as in A, temperature was already tested in the studies. Choice (D) is the best answer. The studies showed that increasing the diameter of a wire affected its conductivity, so you can expect that changing the length would likewise provide additional information about conductivity.

PASSAGE III

14. G
Category: Patterns
Difficulty: Medium
Getting to the Answer: First look at nitrogen in Table 1. For all three systems, as the level of humidity (%) increases, the concentration of nitrogen also increases. Analyzing the other three elements confirms the same pattern for each of them except potassium oxide. As the level of humidity (%) increases, potassium oxide concentrations decrease in all three systems. This matches (G).

15. A
Category: Figure Interpretation
Difficulty: Low
Getting to the Answer: Don't complicate this type of question; the answer is right there in Table 1. The question is asking you to find the compound for which System B gives a concentration that is the farthest from the USGS concentration results. Make sure you're looking in the column that represents 25% humidity, and jot down the difference between System B's concentration results and the USGS's concentrations, so that you don't get confused. Choice (A) is correct. System B gives a nitrogen concentration of 408 mg/L at 25% humidity, and the USGS,

236 mg/L. That's a difference of more than 100 mg/L. Choice B is incorrect. System B gives a calcium concentration just 1.5 mg/L lower than the USGS. Choice C is incorrect. System B gives a K_2O concentration just 0.1 mg/L higher than the USGS. Choice D is likewise incorrect. System B gives a P_2O_5 concentration only 4.4 mg/L higher than the USGS.

16. J
Category: Patterns
Difficulty: Low
Getting to the Answer: When you're asked to identify how one variable changes with respect to another, check whether the variables always increase together, always decrease together, stay the same, or have no relationship at all. Look at the row for potassium oxide (K_2O) and determine if it continually decreases as humidity increases in the USGS data, System A, and System B measurements. From the table, you can see that potassium oxide concentration continually decreases from 9.4 to 8.2 as humidity increases from 10% to 80% in the USGS data, continually decreases from 9.4 to 8.0 in System A, and continually decreases from 9.5 to 8.3 in System B. Therefore, (J) is the correct answer.

17. D
Category: Patterns
Difficulty: Medium
Getting to the Answer: For yes and no questions, decide if the answer to the question is "yes" or "no" before reading each answer choice thoroughly. This way, you can eliminate two answer choices right away. The question asks you to determine which system gives measurements that are closer to the data from the USGS, not which system is higher than the other. Looking at the rows for zinc, you can see that the measurements from System B are closer to the data from the USGS than are the measurements from System A. Therefore, you know the answer to the question is yes, and you can eliminate A and B. Don't be tricked by C. System B does give

lower measurements than System A, but that alone doesn't mean it is more accurate. The measurements from System B are closer to the data from the USGS than are the measurements from System A, so the answer is (D).

18. H
Category: Patterns
Difficulty: Medium
Getting to the Answer: For questions that ask you to choose the best graph, identify the patterns in the data *before* you look at the answer choices. Look at the row in the table that represents calcium and System B. You can see that the numbers gradually decrease from 10% humidity to 65% humidity, then increase quickly from 65% to 85% humidity. The only graph that represents this curve is (H).

19. B
Category: Patterns
Difficulty: High
Getting to the Answer: When new data falls between two columns in the table, don't be afraid to use your pencil to draw in the new values. Using the table, determine which levels of humidity each of the new concentrations falls between for System B. A potassium oxide level of 9.1 falls between 25% and 45% humidity, a calcium level of 17.3 falls between 25% and 45% humidity, and a zinc level of 0.57 likewise falls between 25% and 45% humidity. Therefore, the level of humidity for this sample must be between 25% and 45%. Only (B) falls within this range.

PASSAGE IV

20. F
Category: Scientific Reasoning
Difficulty: Low
Getting to the Answer: Following the Kaplan Method for Conflicting Viewpoints passages, you should answer this question first, because it deals with the first hypothesis only. Start this question by making sure you understand what the Dietary Hypothesis says about blood sugar levels. The hypothesis says that individuals with high blood sugar levels and low insulin levels were able to lower their blood sugar levels by receiving insulin. To strengthen the hypothesis, look for a choice that supports the idea that high blood sugar is caused by or otherwise related to low insulin. Choice (F) does just that. Choice G is incorrect because it would weaken the Dietary Hypothesis, which states that high blood sugars are caused by a high-sugar diet. Choice H is incorrect. If it were proven that high blood sugars indicate high levels of insulin produced by the body, giving *more* insulin wouldn't lower blood sugar. Choice J is incorrect because the hypothesis doesn't address where sugar is stored.

21. C
Category: Scientific Reasoning
Difficulty: Low
Getting to the Answer: Review the Genetic Hypothesis to see what it says about age and type II diabetes. The last sentence of the first paragraph states, "Therefore, the same behaviors may be more detrimental to a person when he or she is older than when he or she was younger, especially for those over the age of 50. This is the main reason that type II diabetes is more common in adults." Choice (C), then, is the correct answer.

22. H
Category: Scientific Reasoning
Difficulty: Medium
Getting to the Answer: As always, use keywords from the question stem to point you to the part of the passage that contains the answer. In this case, those are "Genetic Hypothesis" and "pancreas." What does the Genetic Hypothesis say about the pancreas? The passage states that "Diabetes occurs when the pancreas produces little or no insulin at all or when the insulin it produces does not work properly." If diabetes is caused when the pancreas doesn't produce enough insulin, then removing the pancreas should lead to diabetes. None of the answer choices state this explicitly, but (H) gives the

major symptom of diabetes as stated in the introductory paragraph. Choices F and G are incorrect because they state the opposite of what you should expect. You can eliminate J because you're given no reason to suspect a link between the pancreas and body fat content.

23. B
Category: Scientific Reasoning
Difficulty: Medium
Getting to the Answer: Be sure you understand the crux of both hypotheses before you answer this question. The Dietary Hypothesis states, in effect, that eating too much sugar causes high blood sugar levels and that giving insulin can lower blood sugar. The Exercise Hypothesis states that adults who exercise regularly lower their blood sugar levels. You should expect, then, that supporters of the Dietary Hypothesis would want to know how much sugar was included in the diets of the exercisers over the course of the study. This matches (B). Choice A is incorrect because supporters of the Dietary Hypothesis wouldn't know how much sugar was in the diets of the exercisers. Therefore, they couldn't criticize the experiment for not including enough sugar. You can eliminate C because it discusses a detail from the Genetic Hypothesis and is therefore not a criticism supporters of the Dietary Hypothesis would likely levy. Choice D is incorrect according to the passage and, therefore, couldn't be a valid criticism.

24. H
Category: Scientific Reasoning
Difficulty: Medium
Getting to the Answer: Look for an answer choice that allows both the experimental results presented in the Dietary Hypothesis and those in the Genetic Hypothesis to be true. Supporters of the Genetic Hypothesis would look for an element of genetics in the results presented by the Dietary Hypothesis. If it were true that the test subjects had no diabetes in their genetic background, then the results from the Dietary Hypothesis experiment wouldn't violate the Genetic Hypothesis. Choice (H), then, is the cor-

rect answer. Choice F is incorrect because supporters of the Genetic Hypothesis aren't concerned with dietary sugar levels. Choice G is incorrect because it is a detail from the Exercise Hypothesis. Choice J is incorrect because supporters of the Genetic Hypothesis wouldn't be concerned with how much insulin the subjects were given; that's related to the Dietary Hypothesis.

25. B
Category: Scientific Reasoning
Difficulty: High
Getting to the Answer: Just as in the previous question, look for an answer choice that accounts for the results in the Genetics Hypothesis experiment but doesn't violate the Dietary Hypothesis. If you're unsure about the position of the Dietary Hypothesis, you'll have to review it before answering this question. Essentially, it states that eating too much sugar causes high blood sugar levels and adding insulin lowers blood sugar levels. Group D in the Genetics Hypothesis experiment showed that 70% of individuals who had two parents with type II diabetes also had type II diabetes. You can expect that supporters of the Dietary Hypothesis will look for a dietary reason for this link. Choice (B) is the correct answer because it blames the *diet* of the parents rather than genetics. Choices A and D are irrelevant because none of the groups took insulin supplements. Choice C is incorrect because Dietary Hypothesis supporters wouldn't look for a genetic explanation for the results.

26. J
Category: Scientific Reasoning
Difficulty: Low
Getting to the Answer: Move straight to the answer choices and eliminate those that aren't common to both experiments; it'll be quicker than trying to find the answer in the passage and then looking for a match in the choices. You're looking for what the subjects of both studies have in common, so eliminate those answer choices that apply to one or neither of the studies. Choice F is incorrect because parents were only discussed in the Genetics

Hypothesis study. Choice G refers only to the Dietary Hypothesis, so it is likewise incorrect. Choice H is incorrect because neither study mentions whether test subjects had their pancreases removed. Choice (J) is correct because both studies focused on adults over 50.

PASSAGE V

27. A
Category: Figure Interpretation
Difficulty: Low
Getting to the Answer: Don't read too much into easy questions. Here, you're simply asked to find which element is highest in the blood sample with the lowest density. Start by looking at Table 1. The lowest-density blood sample is 1.050 g/mL, that of Patient C. Looking at each column, you can see that Patient C has more platelets than Patient A but fewer than Patient B, fewer white blood cells than Patient A, the fewest red blood cells, but the most plasma. Choice (A), then, is correct.

28. H
Category: Scientific Reasoning
Difficulty: Medium
Getting to the Answer: Look for an answer choice that relates blood composition or density, the purpose of the experiment. You know that the phlebotomist is studying the blood composition of three different patients. His method, then, should include taking steps to make sure the samples weren't affected by temporary changes in composition. If what a patient eats can affect his blood composition, then it would make sense to require the patients to fast, (H). Choice F is incorrect because you're given no reason to suspect that taking blood is easier if a patient has fasted. Choice G is incorrect because if fasting could greatly change the composition of blood, then the phlebotomist would likely have made sure the patients ate before having blood withdrawn. The passage also states that diet can affect the composition of blood, so it would make sense that the phlebotomist tried to control

this factor by requiring the patients to fast. Choice J is incorrect because you're given no indication in the passage that anything can affect the ability of blood to separate.

29. D
Category: Scientific Reasoning
Difficulty: High
Getting to the Answer: Even for the most difficult questions, you know the answer *must* be true based on what's in the passage. You can always start by eliminating answer choices that aren't supported by the passage. If the resulting masses of the blood samples are less than their starting masses, you can assume that either some of the mass was lost during one or more of the experiments, or that there are more components to blood than plasma, red and white blood cells, and platelets. Choices A and C are incorrect because even if some of the formed elements remained in the plasma, they would have been weighed with the plasma, and their mass would still be included. Choice B is incorrect for the same reason. If platelets and white blood cells were mixed, the total amounts still wouldn't be affected. Additionally, the mass of the four elements was always lower than 10.5 g, so it's unlikely that any one element was overestimated. Choice (D) is correct. You're told in the opening paragraph that "Human blood is composed of approximately 45% *formed elements*, including blood cells, and 50% plasma." That leaves 5% of blood unaccounted for in the experiments.

30. F
Category: Patterns
Difficulty: Low
Getting to the Answer: Don't rush through the easier questions. Make sure you take the time to get them right and grab points. Go back to Table 1 and look at the columns for total density and red blood cell mass. (Circle each column if you tend to get distracted by the other information.) Reading the table from the bottom up, you can see that, as total density increases, the mass of the red blood cells also increases, (F).

31. B

Category: Patterns
Difficulty: Medium
Getting to the Answer: You can't use your calculator on the Science Test, but rest assured that any calculations required will be very simple. The beginning of the passage states that "formed elements weigh approximately 1.10 grams per milliliter (g/mL) and plasma approximately 1.02 g/mL." Here, you have 5 mL of plasma, so the total mass of plasma is (5 mL) (1.02 g/mL) = 5.1 g. You also have 4 mL of formed elements, so the total mass of formed elements is (4 mL)(1.10 g/mL) = 4.4 g. The total mass is then 5.1 g + 4.4 g = 9.5 g, (B).

32. J

Category: Scientific Reasoning
Difficulty: Low
Getting to the Answer: If you followed the Kaplan Method for ACT Science and circled the methods as you read the passage, this question offers some quick, easy points. The paragraph describing each experiment states that the phlebotomist placed the blood samples in a centrifuge for 20 minutes in Experiment 2 and at a slower speed for 45 minutes in Experiment 3. Only (J) captures any element of this difference. Choice F is incorrect because you're told at the beginning of the passage that 10 mL of blood were taken from each patient. Choice G is incorrect because, while the mass of the blood samples did vary from patient to patient, the masses weren't intentionally varied by the phlebotomist from Experiment 2 to Experiment 3. Choice H is incorrect because a centrifuge was used in both Experiment 2 and 3.

33. A

Category: Figure Interpretation
Difficulty: Medium
Getting to the Answer: Using Table 1, find the column labeled "Red blood cells (g)" and compare the data. Patient A has the greatest mass of red blood cells with 2.75 grams. Choice (A) matches this observation.

PASSAGE VI

34. G

Category: Patterns
Difficulty: Medium
Getting to the Answer: First, locate the correct table required to answer this question. Table 2 contains silver rods, so that is the best place to start. Looking at silver with a mass density of 225 yields two results. One result includes a temperature of 5°C and a conductivity of 120 μΩ/cm. The other result shows a temperature of 20°C and a conductivity of 60 μΩ/cm. This indicates that for every 5°C that the temperature is increased, the conductivity decreases by 20 μΩ/cm. Choice (G) matches this trend.

35. D

Category: Scientific Reasoning
Difficulty: Medium
Getting to the Answer: Remember, the answer is in the passage! Find where Trials 8 and 10 appear, and check to see which variable is different. The only differences in the two rows for Trial 8 and 10 are the mass density and the conductivity. Choice (D) is correct, because the student can manipulate the mass density, but conductivity is a measured property of the material that can't be *directly* manipulated by the student. Choice A is incorrect because the temperature of the rods in both trials was 20°C. Choice C is incorrect because both rods were made of silver.

36. G

Category: Scientific Reasoning
Difficulty: Low
Getting to the Answer: When you're asked to find another method, make sure the new method accomplishes the goals of the original. This question is just asking you to determine another way to measure electric current. You don't have to know anything about circuitry to recognize that only (G) even deals with the issue of measuring voltage. Choices F and H are incorrect because shortening the rods or increasing their mass density won't help the student measure conductivity. Choice J is incorrect because

insulating the rods will only make it more difficult to measure conductivity.

37. A
Category: Patterns
Difficulty: Medium
Getting to the Answer: To find the patterns in the data here, make sure you identify two trials where only temperature was varied. Look at the rows for Trials 10 and 11. Length and mass density are identical in those two trials, but temperature and conductivity are different. You can see that when length and mass density are kept constant, conductivity decreases as temperature increases. Trial 11 shows that the conductivity falls from 120 to 60 $\mu\Omega$/cm when the temperature rises from 5 to 20°C, so you can expect that at an even higher temperature, the conductivity will be even less than 60 $\mu\Omega$/cm. The only answer choice less than 60 $\mu\Omega$/cm is (A).

38. F
Category: Patterns
Difficulty: Medium
Getting to the Answer: Determine the relationship between each variable and the conductivity one at a time. Don't forget, you are looking for the statement that is *false*. Trials 10 and 11 show that when temperature increases, conductivity decreases. Choice (F) says the opposite of this, so it is false and the answer you're looking for. Trials 1 and 3 show that increasing mass density decreases conductivity, so G is true. Trials 1 and 5 show that increasing length increases conductivity, so H is true. Trials 3 and 8 show that when all other variables are held constant, different materials have different conductivities. Choice J is true.

39. B
Category: Scientific Reasoning
Difficulty: Medium
Getting to the Answer: In order to successfully manipulate just one variable, the rest of the variables must be kept constant. In the last question, you determined that as mass density increases

conductivity decreases. If you didn't remember this, you could check Trials 1 and 3. Therefore, the rod with the lowest mass density could be found by heating the rods to the same temperature and finding the rod with the highest conductivity, (B). Choice A is the opposite of what you're looking for. Choice C is incorrect because you're given no information regarding how voltage affects conductivity. Choice D is incorrect because the passage never relates mass density to melting point.

40. J
Category: Patterns
Difficulty: High
Getting to the Answer: This question involves several steps. Don't rush just because you're anxious to finish the test! Use Table 1 to answer this question, since it deals with an iron rod rather than a rod made of silver or tungsten. Trial 4 is the row with a mass density of 400 g/cm and a temperature of 80°C. The length of the rod in Trial 4 is 16 cm and the conductivity is 30 $\mu\Omega$/cm. There isn't a trial conducted with all of the same properties and a different length, so you will have to identify the pattern between length and conductivity by looking at two other rows. Use Trials 1 and 5. These rows have the same mass density and temperature but different lengths. These trials show that as length increases, conductivity increases. Therefore, the longer the rod, the more electricity it can conduct. Trial 4 has a conductivity of 30 $\mu\Omega$/cm, but you need conductivity less than that. To decrease the conductivity, then, you'll need a length shorter than the 16 cm used in Trial 4. Choice (J), then, is the correct answer.

ACT Practice Test Two
ANSWER SHEET

MATHEMATICS TEST

1. Ⓐ Ⓑ Ⓒ Ⓓ Ⓔ   11. Ⓐ Ⓑ Ⓒ Ⓓ Ⓔ   21. Ⓐ Ⓑ Ⓒ Ⓓ Ⓔ   31. Ⓐ Ⓑ Ⓒ Ⓓ Ⓔ   41. Ⓐ Ⓑ Ⓒ Ⓓ Ⓔ   51. Ⓐ Ⓑ Ⓒ Ⓓ Ⓔ
2. Ⓕ Ⓖ Ⓗ Ⓙ Ⓚ   12. Ⓕ Ⓖ Ⓗ Ⓙ Ⓚ   22. Ⓕ Ⓖ Ⓗ Ⓙ Ⓚ   32. Ⓕ Ⓖ Ⓗ Ⓙ Ⓚ   42. Ⓕ Ⓖ Ⓗ Ⓙ Ⓚ   52. Ⓕ Ⓖ Ⓗ Ⓙ Ⓚ
3. Ⓐ Ⓑ Ⓒ Ⓓ Ⓔ   13. Ⓐ Ⓑ Ⓒ Ⓓ Ⓔ   23. Ⓐ Ⓑ Ⓒ Ⓓ Ⓔ   33. Ⓐ Ⓑ Ⓒ Ⓓ Ⓔ   43. Ⓐ Ⓑ Ⓒ Ⓓ Ⓔ   53. Ⓐ Ⓑ Ⓒ Ⓓ Ⓔ
4. Ⓕ Ⓖ Ⓗ Ⓙ Ⓚ   14. Ⓕ Ⓖ Ⓗ Ⓙ Ⓚ   24. Ⓕ Ⓖ Ⓗ Ⓙ Ⓚ   34. Ⓕ Ⓖ Ⓗ Ⓙ Ⓚ   44. Ⓕ Ⓖ Ⓗ Ⓙ Ⓚ   54. Ⓕ Ⓖ Ⓗ Ⓙ Ⓚ
5. Ⓐ Ⓑ Ⓒ Ⓓ Ⓔ   15. Ⓐ Ⓑ Ⓒ Ⓓ Ⓔ   25. Ⓐ Ⓑ Ⓒ Ⓓ Ⓔ   35. Ⓐ Ⓑ Ⓒ Ⓓ Ⓔ   45. Ⓐ Ⓑ Ⓒ Ⓓ Ⓔ   55. Ⓐ Ⓑ Ⓒ Ⓓ Ⓔ
6. Ⓕ Ⓖ Ⓗ Ⓙ Ⓚ   16. Ⓕ Ⓖ Ⓗ Ⓙ Ⓚ   26. Ⓕ Ⓖ Ⓗ Ⓙ Ⓚ   36. Ⓕ Ⓖ Ⓗ Ⓙ Ⓚ   46. Ⓕ Ⓖ Ⓗ Ⓙ Ⓚ   56. Ⓕ Ⓖ Ⓗ Ⓙ Ⓚ
7. Ⓐ Ⓑ Ⓒ Ⓓ Ⓔ   17. Ⓐ Ⓑ Ⓒ Ⓓ Ⓔ   27. Ⓐ Ⓑ Ⓒ Ⓓ Ⓔ   37. Ⓐ Ⓑ Ⓒ Ⓓ Ⓔ   47. Ⓐ Ⓑ Ⓒ Ⓓ Ⓔ   57. Ⓐ Ⓑ Ⓒ Ⓓ Ⓔ
8. Ⓕ Ⓖ Ⓗ Ⓙ Ⓚ   18. Ⓕ Ⓖ Ⓗ Ⓙ Ⓚ   28. Ⓕ Ⓖ Ⓗ Ⓙ Ⓚ   38. Ⓕ Ⓖ Ⓗ Ⓙ Ⓚ   48. Ⓕ Ⓖ Ⓗ Ⓙ Ⓚ   58. Ⓕ Ⓖ Ⓗ Ⓙ Ⓚ
9. Ⓐ Ⓑ Ⓒ Ⓓ Ⓔ   19. Ⓐ Ⓑ Ⓒ Ⓓ Ⓔ   29. Ⓐ Ⓑ Ⓒ Ⓓ Ⓔ   39. Ⓐ Ⓑ Ⓒ Ⓓ Ⓔ   49. Ⓐ Ⓑ Ⓒ Ⓓ Ⓔ   59. Ⓐ Ⓑ Ⓒ Ⓓ Ⓔ
10. Ⓕ Ⓖ Ⓗ Ⓙ Ⓚ   20. Ⓕ Ⓖ Ⓗ Ⓙ Ⓚ   30. Ⓕ Ⓖ Ⓗ Ⓙ Ⓚ   40. Ⓕ Ⓖ Ⓗ Ⓙ Ⓚ   50. Ⓕ Ⓖ Ⓗ Ⓙ Ⓚ   60. Ⓕ Ⓖ Ⓗ Ⓙ Ⓚ

SCIENCE TEST

1. Ⓐ Ⓑ Ⓒ Ⓓ   6. Ⓕ Ⓖ Ⓗ Ⓙ   11. Ⓐ Ⓑ Ⓒ Ⓓ   16. Ⓕ Ⓖ Ⓗ Ⓙ   21. Ⓐ Ⓑ Ⓒ Ⓓ   26. Ⓕ Ⓖ Ⓗ Ⓙ   31. Ⓐ Ⓑ Ⓒ Ⓓ   36. Ⓕ Ⓖ Ⓗ Ⓙ
2. Ⓕ Ⓖ Ⓗ Ⓙ   7. Ⓐ Ⓑ Ⓒ Ⓓ   12. Ⓕ Ⓖ Ⓗ Ⓙ   17. Ⓐ Ⓑ Ⓒ Ⓓ   22. Ⓕ Ⓖ Ⓗ Ⓙ   27. Ⓐ Ⓑ Ⓒ Ⓓ   32. Ⓕ Ⓖ Ⓗ Ⓙ   37. Ⓐ Ⓑ Ⓒ Ⓓ
3. Ⓐ Ⓑ Ⓒ Ⓓ   8. Ⓕ Ⓖ Ⓗ Ⓙ   13. Ⓐ Ⓑ Ⓒ Ⓓ   18. Ⓕ Ⓖ Ⓗ Ⓙ   23. Ⓐ Ⓑ Ⓒ Ⓓ   28. Ⓕ Ⓖ Ⓗ Ⓙ   33. Ⓐ Ⓑ Ⓒ Ⓓ   38. Ⓕ Ⓖ Ⓗ Ⓙ
4. Ⓕ Ⓖ Ⓗ Ⓙ   9. Ⓐ Ⓑ Ⓒ Ⓓ   14. Ⓕ Ⓖ Ⓗ Ⓙ   19. Ⓐ Ⓑ Ⓒ Ⓓ   24. Ⓕ Ⓖ Ⓗ Ⓙ   29. Ⓐ Ⓑ Ⓒ Ⓓ   34. Ⓕ Ⓖ Ⓗ Ⓙ   39. Ⓐ Ⓑ Ⓒ Ⓓ
5. Ⓐ Ⓑ Ⓒ Ⓓ   10. Ⓕ Ⓖ Ⓗ Ⓙ   15. Ⓐ Ⓑ Ⓒ Ⓓ   20. Ⓕ Ⓖ Ⓗ Ⓙ   25. Ⓐ Ⓑ Ⓒ Ⓓ   30. Ⓕ Ⓖ Ⓗ Ⓙ   35. Ⓐ Ⓑ Ⓒ Ⓓ   40. Ⓕ Ⓖ Ⓗ Ⓙ

MATHEMATICS TEST

60 Minutes—60 Questions

Directions: Solve each of the following problems, select the correct answer, and then fill in the corresponding space on your answer sheet.

Don't linger over problems that are too time-consuming. Do as many as you can, then come back to the others in the time you have remaining.

The use of a calculator is permitted on this test. Though you are allowed to use your calculator to solve any questions you choose, some of the questions may be most easily answered without the use of a calculator.

Note: Unless otherwise noted, all of the following should be assumed.

1. Illustrative figures are *not* necessarily drawn to scale.
2. All geometric figures lie in a plane.
3. The term *line* indicates a straight line.
4. The term *average* indicates arithmetic mean.

1. In a class, 10 students are receiving honors credit. This number is exactly 20% of the total number of students in the class. How many students are in the class?

    A. 12
    B. 15
    C. 18
    D. 20
    E. 50

2. In the figure below, points A, B, and C are on a straight line. What is the measure of angle DBE?

    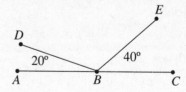

    F. 60°
    G. 80°
    H. 100°
    J. 120°
    K. 140°

3. What is the fifth term of the arithmetic sequence 7, 4, 1, … ?

    A. −5
    B. −2
    C. 1
    D. 4
    E. 14

4. What value of c solves the following proportion?

    $$\frac{20}{8} = \frac{c}{10}$$

    F. 4
    G. 16
    H. 18
    J. 22
    K. 25

GO ON TO THE NEXT PAGE ▷

5. If G, H, and K are distinct points on the same line, and $\overline{GK} \cong \overline{HK}$, then which of the following must be true?

   A. G is the midpoint of $\overline{HK}$
   B. H is the midpoint of $\overline{GK}$
   C. K is the midpoint of $\overline{GH}$
   D. G is the midpoint of $\overline{KH}$
   E. K is the midpoint of $\overline{KG}$

6. Four pieces of yarn, each 1.2 meters long, are cut from the end of a ball of yarn that is 50 meters long. How many meters of yarn are left ?

   F. 45.2
   G. 45.8
   H. 46.8
   J. 47.2
   K. 47.8

7. If $x = -2$, then $14 - 3(x + 3) = ?$

   A. -1
   B. 11
   C. 14
   D. 17
   E. 29

8. $-|-6| - (-6) = ?$

   F. -36
   G. -12
   H. 0
   J. 12
   K. 36

9. A car dealership expects an increase of 15% in its current annual sales of 3,200 cars. What will its new annual sales be?

   A. 3,215
   B. 3,248
   C. 3,680
   D. 4,700
   E. 4,800

10. If $x^4 = 90$ (and x is a real number), then x lies between which two consecutive integers?

    F. 2 and 3
    G. 3 and 4
    H. 4 and 5
    J. 5 and 6
    K. 6 and 7

11. If $47 - x = 188$, then $x = ?$

    A. -235
    B. -141
    C. 4
    D. 141
    E. 235

12. To complete a certain task, Group A requires 8 more hours than Group B, and Group B requires twice as long as Group C. If h is the number of hours required by Group C, how long does the task take Group A, in terms of h ?

    F. $10h$
    G. $16h$
    H. $10 + h$
    J. $2(8 + h)$
    K. $8 + 2h$

GO ON TO THE NEXT PAGE

13. In the standard (x,y) coordinate plane, three corners of a rectangle are $(2,-2)$, $(-5,-2)$, and $(2,-5)$. Where is the rectangle's fourth corner?

A. $(2,5)$

B. $(-2,5)$

C. $(-2,2)$

D. $(-2,-5)$

E. $(-5,-5)$

14. Which of the following is a simplified form of $5a - 5b + 3a$?

F. $5(a - b + 3)$

G. $(a - b)(5 + 3a)$

H. $a(8 - 5b)$

J. $8a - 5b$

K. $2a - 5b$

15. In the parallelogram below, what is the measure of angle *FEG* ?

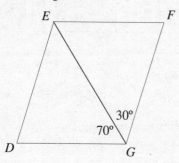

A. $30°$

B. $40°$

C. $50°$

D. $60°$

E. $70°$

16. What is the slope of any line parallel to the line $4x + 3y = 9$?

F. -4

G. $-\dfrac{4}{3}$

H. $\dfrac{4}{9}$

J. 4

K. 9

17. If $x > 0$ and $3x^2 - 7x - 20 = 0$, then $x = $?

A. $\dfrac{5}{3}$

B. 3

C. 4

D. 7

E. 20

18. The lengths of the sides of a triangle are 2, 5, and 8 centimeters. How many centimeters long is the shortest side of a similar triangle that has a perimeter of 30 centimeters?

F. 4

G. 7

H. 10

J. 15

K. 16

19. A shirt that normally sells for $24.60 is on sale for 15% off. How much does it cost during the sale, to the nearest dollar?

A. $ 4

B. $10

C. $20

D. $21

E. $29

GO ON TO THE NEXT PAGE ▷

20. Which of the following is a factored form of $3xy^4 + 3x^4y$?

    F. $3x^4y^4(y + x)$
    G. $3xy(y^3 + x^3)$
    H. $6xy(y^3 + x^3)$
    J. $3x^4y^4$
    K. $6x^5y^5$

21. If $x - 2y = 0$ and $3x + y = 7$, what is the value of x ?

    A. -1
    B. 0
    C. 1
    D. 2
    E. 3

22. There are three feet in a yard. If 2.5 yards of fabric cost $4.50, what is the cost per foot?

    F. $ 0.60
    G. $ 0.90
    H. $ 1.50
    J. $ 1.80
    K. $11.25

23. The figure below shows a square overlapping with a rectangle. One vertex of the rectangle is at the center of the square. What is the area of the shaded region, in square inches?

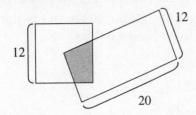

    A. 9
    B. 18
    C. 36
    D. 72
    E. 144

24. A salesperson earns $7h + 0.04s$ dollars, where h is the number of hours worked, and s is the total amount of her sales. What does she earn for working 15 hours with $120.50 in sales?

    F. $109.82
    G. $153.20
    H. $226.10
    J. $231.50
    K. $848.32

25. A floor has the dimensions shown below. How many square feet of tiles are needed to cover the entire floor?
    (Note: All angles are right angles)

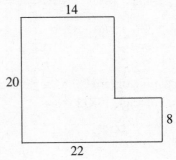

    A. 64
    B. 96
    C. 160
    D. 344
    E. 484

26. Which of the following is the graph of the solution set of $x - 2 < -4$?

    F.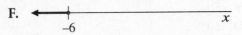
    G.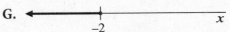
    H.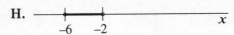
    J.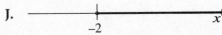
    K.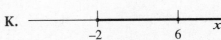

GO ON TO THE NEXT PAGE

27. Which of the following is less than $\frac{3}{5}$?

   A. $\frac{4}{6}$

   B. $\frac{8}{13}$

   C. $\frac{6}{10}$

   D. $\frac{7}{11}$

   E. $\frac{4}{7}$

28. What is the area, in square feet, of a right triangle with sides of length 7 feet, 24 feet, and 25 feet?

   F. 56

   G. 84

   H. $87\frac{1}{2}$

   J. 168

   K. 300

29. When the graduating class is arranged in rows of 6 people each, the last row is one person short. When it is arranged in rows of 7, the last row is still one person short. When arranged in rows of 8, the last row is *still* one person short. What is the least possible number of people in the graduating class?

   A. 23

   B. 41

   C. 71

   D. 167

   E. 335

30. A triangle has sides of length 3.5 inches and 6 inches. Which of the following CANNOT be the length of the third side, in inches?

   F. 2

   G. 3

   H. 4

   J. 5

   K. 6

31. For all $b > 0$, $\frac{4}{5} + \frac{1}{b} = ?$

   A. $\frac{4}{5b}$

   B. $\frac{5}{5b}$

   C. $\frac{4b+5}{5b}$

   D. $\frac{5}{5+b}$

   E. $\frac{4b+5}{5+b}$

32. In the right triangle below, how long is side $\overline{EF}$?

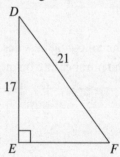

   F. $\sqrt{21^2 - 17^2}$

   G. $\sqrt{21^2 + 17^2}$

   H. $21^2 - 17^2$

   J. $21^2 + 17^2$

   K. $21 - 17$

GO ON TO THE NEXT PAGE

33. If the length of a square is increased by 2 inches and the width is increased by 3 inches, a rectangle is formed. If each side of the original square is b feet long, what is the area of the new rectangle, in square inches?

    A. $2b + 5$

    B. $4b + 10$

    C. $b^2 + 6$

    D. $b^2 + 5b + 5$

    E. $b^2 + 5b + 6$

34. If $\sin \beta = \dfrac{8}{17}$ and $\cos \beta = \dfrac{15}{17}$, then $\tan \beta = ?$

    F. $\dfrac{7}{17}$

    G. $\dfrac{8}{15}$

    H. $\dfrac{23}{17}$

    J. $\dfrac{15}{8}$

    K. $\dfrac{120}{17}$

35. Which of the following best describes the graph on the number line below?

    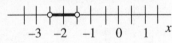

    A. $-|x| = -2$

    B. $-|x| < 0.5$

    C. $-3 < x < -1$

    D. $-1.5 < x < -2.5$

    E. $-1.5 > x > -2.5$

36. A basketball team made 1-point, 2-point, and 3-point baskets. 20% of their baskets were worth 1 point, 70% of their baskets were worth 2 points, and 10% of their baskets were worth 3 points. To the nearest tenth, what was the average point value of their baskets?

    F. 1.4

    G. 1.7

    H. 1.8

    J. 1.9

    K. 2.0

37. In the triangle below, if $\overline{CD}$ is 3 centimeters long, how many centimeters long is $\overline{CE}$?

    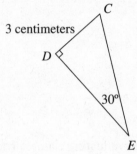

    A. 3

    B. $3\sqrt{2}$

    C. $3\sqrt{3}$

    D. 6

    E. 9

38. What is the largest possible product for two odd integers whose sum is 42 ?

    F. 117

    G. 185

    H. 259

    J. 377

    K. 441

GO ON TO THE NEXT PAGE ⇨

39. In the figure below, lines *l* and *m* are parallel, lines *n* and *p* are parallel, and the measures of two angles are as shown. What is the value of *x* ?

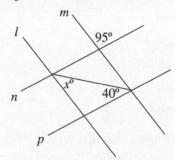

A. 40
B. 45
C. 50
D. 70
E. 85

40. In the (*x*,*y*) coordinate plane, what is the *y*-intercept of the line $12x - 3y = 12$?

F. –4
G. –3
H. 0
J. 4
K. 12

41. Among the points graphed on the number line below, which is closest to *e* ?
(Note: $e \approx 2.718281828$)

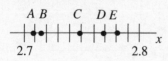

A. *A*
B. *B*
C. *C*
D. *D*
E. *E*

42. For what value of *a* would the following system of equations have no solution?

$$-x + 6y = 7$$
$$-5x + 10ay = 32$$

F. $\dfrac{5}{3}$
G. 3
H. 6
J. 30
K. 60

43. The expression $(360 - x)°$ is the degree measure of a nonzero obtuse angle if and only if:

A. $0 < x < 90$
B. $0 < x < 180$
C. $180 < x < 270$
D. $180 < x < 360$
E. $270 < x < 360$

44. If $p - q = -4$ and $p + q = -3$, then $p^2 - q^2 = ?$

F. 25
G. 12
H. 7
J. –7
K. –12

45. The sides of a triangle are 6, 8, and 10 meters long. What is the angle between the two shortest sides?

A. 30°
B. 45°
C. 60°
D. 90°
E. 135°

GO ON TO THE NEXT PAGE

46. In the standard (x,y) coordinate plane, if the x-coordinate of each point on a line is 9 more than three times the y-coordinate, the slope of the line is:

   F. –9

   G. –3

   H. $\dfrac{1}{3}$

   J. 3

   K. 9

47. A tree is growing at the edge of a cliff, as shown below. From the tree, the angle between the base of the cliff and the base of the house near it is 62°. If the distance between the base of the cliff and the base of the house is 500 feet, how many feet tall is the cliff?

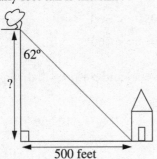

A. 500 cos 62°

B. 500 tan 62°

C. $\dfrac{500}{\sin 62°}$

D. $\dfrac{500}{\cos 62°}$

E. $\dfrac{500}{\tan 62°}$

48. Two numbers have a greatest common factor of 9 and a least common multiple of 54. Which of the following could be the pair of numbers ?

   F. 9 and 18

   G. 9 and 27

   H. 18 and 27

   J. 18 and 54

   K. 27 and 54

49. Five functions, each denoted $b(x)$ and each involving a real number constant $k > 1$, are listed below. If $a(x) = 5^x$, which of these 5 functions yields the greatest value of $a(b(x))$, for all $x > 2$?

   A. $b(x) = \dfrac{k}{x}$

   B. $b(x) = \dfrac{x}{k}$

   C. $b(x) = kx$

   D. $b(x) = x^k$

   E. $b(x) = \sqrt[k]{x}$

50. Line segments $\overline{WX}$, $\overline{XY}$, and $\overline{YZ}$, which represent the 3 dimensions of the rectangular box shown below, have lengths of 12 centimeters, 5 centimeters, and 13 centimeters, respectively. What is the cosine of $\angle ZWY$?

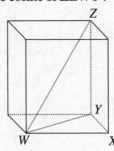

   F. $\dfrac{13\sqrt{2}}{12}$

   G. 1

   H. $\dfrac{12}{13}$

   J. $\dfrac{\sqrt{2}}{2}$

   K. $\dfrac{5}{13}$

GO ON TO THE NEXT PAGE

51. A certain circle has an area of 4π square centimeters. How many centimeters long is its radius?

    A. $\dfrac{1}{4}$

    B. 2

    C. 4

    D. 2π

    E. 4π

52. The equation of line l below is $y = mx + b$. Which of the following could be an equation for line q ?

    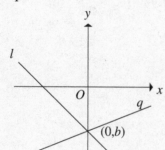

    F. $y = \dfrac{1}{2}mx$

    G. $y = \dfrac{1}{2}mx - b$

    H. $y = \dfrac{1}{2}mx + b$

    J. $y = -\dfrac{1}{2}mx - b$

    K. $y = -\dfrac{1}{2}mx + b$

53. The equation $x^2 - 6x + k = 0$ has exactly one solution for x. What is the value of k ?

    A. 0

    B. 3

    C. 6

    D. 9

    E. 12

54. In the standard (x,y) coordinate plane, what is the slope of the line through the origin and $\left(\dfrac{1}{3}, \dfrac{3}{4}\right)$?

    F. $\dfrac{1}{4}$

    G. $\dfrac{1}{3}$

    H. $\dfrac{5}{12}$

    J. $\dfrac{3}{4}$

    K. $\dfrac{9}{4}$

55. If R, S, and T are real numbers and $RST = 2$, which of the following *must* be true?

    A. $RT = \dfrac{2}{S}$

    B. R, S, and T are all positive

    C. Either $R = 2$, $S = 2$ or $T = 2$

    D. Either $R = 0$, $S = 0$ or $T = 0$

    E. Either $R > 2$, $S > 2$ or $T > 2$

56. A square has sides of length $(w + 5)$ units. Which of the following is the remaining area of the square, in square units, if a rectangle with sides of length $(w + 2)$ and $(w - 3)$ is removed from the interior of the square?

    F. 31

    G. $9w + 19$

    H. $11w + 31$

    J. $w^2 + 10w + 25$

    K. $2w^2 + 9w + 19$

GO ON TO THE NEXT PAGE

57. What is the smallest positive value for θ where sin 2θ reaches its minimum value?

A. $\dfrac{\pi}{4}$

B. $\dfrac{\pi}{2}$

C. $\dfrac{3\pi}{4}$

D. π

E. $\dfrac{3\pi}{2}$

58. In the standard (x,y) coordinate plane, if the distance between the points $(r,6)$ and $(10,r)$ is 4 coordinate units, which of the following could be the value of r?

F. 3

G. 4

H. 7

J. 8

K. 10

59. Calleigh puts 5 nickels into an empty hat. She wants to add enough pennies so that the probability of drawing a nickel at random from the hat is $\dfrac{1}{6}$. How many pennies should she put in?

A. 1

B. 5

C. 10

D. 25

E. 30

60. How many different integer values of x satisfy the inequality $\dfrac{1}{5} < \dfrac{3}{x} < \dfrac{1}{3}$?

F. 1

G. 2

H. 3

J. 4

K. 5

SCIENCE TEST

35 Minutes—40 Questions

Directions: There are several passages in this test. Each passage is followed by several questions. After reading a passage, choose the best answer to each question and fill in the corresponding oval on your Answer Grid. You may refer to the passages as often as necessary. You are NOT permitted to use a calculator on this test.

PASSAGE I

A panel of engineers designed and built a pressurized structure to be used for shelter by geologists during extended research missions near the South Pole. The design consisted of 4 rooms, each with its own separate heating and air pressure control systems. During testing, the engineers found the daily average air temperature, in degrees Celsius (°C), and daily average air pressure, in millimeters of mercury (mm Hg), in each room. The data for the first 5 days of their study are given in Table 1 and Table 2.

Table 1

Day	Daily average air temperature (°C)			
	Room 1	Room 2	Room 3	Room 4
1	19.64	19.08	18.67	18.03
2	20.15	19.20	18.46	18.11
3	20.81	19.19	18.62	18.32
4	21.06	19.51	19.08	18.91
5	21.14	19.48	18.60	18.58

Table 2

Day	Daily average air pressure (mm Hg)			
	Room 1	Room 2	Room 3	Room 4
1	748.2	759.6	760.0	745.2
2	752.6	762.0	758.7	750.3
3	753.3	760.2	756.5	760.4
4	760.1	750.8	755.4	756.8
5	758.7	757.9	754.0	759.5

1. The lowest daily average air pressure recorded during the first 5 days of the study was:

    A. 762.0 mm Hg.
    B. 745.2 mm Hg.
    C. 21.14 mm Hg.
    D. 18.03 mm Hg.

2. According to Table 2, daily average air pressures were recorded to the nearest:

    F. 0.01 mm Hg.
    G. 0.1 mm Hg.
    H. 1 mm Hg.
    J. 10 mm Hg.

GO ON TO THE NEXT PAGE ➡

3. Which of the following graphs best represents a plot of the daily average air temperature versus the daily average air pressure for Room 4?

A.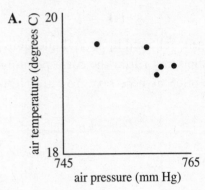

B.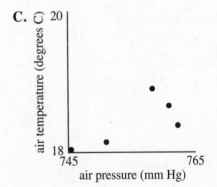

C.

D.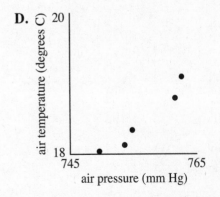

4. Which of the following most accurately describes the changes in the daily average air pressure in Room 3 during days 1–5?

F. The daily average air pressure increased from days 1 to 4 and decreased from days 4 to 5.

G. The daily average air pressure decreased from days 1 to 2, increased from days 2 to 4, and decreased again from days 4 to 5.

H. The daily average air pressure increased only.

J. The daily average air pressure decreased only.

5. Suppose the *heat absorption modulus* of a room is defined as the quantity of heat absorbed by the contents of the room divided by the quantity of heat provided to the entire room. Based on the data, would one be justified in concluding that the heat absorption modulus of Room 1 was higher than the heat absorption modulus of any of the other rooms?

A. Yes, because the quantity of heat provided to Room 1 was greater than the quantity of heat provided to any of the other rooms.

B. Yes, because the quantity of heat not absorbed by the contents of Room 1 was greater than the quantity of heat not absorbed by the contents of any of the other rooms.

C. No, because the quantity of heat absorbed by the contents of Room 1 was less than the quantity of heat absorbed by the contents of any of the other room.

D. No, because the information provided is insufficient to determine heat absorption modulus.

GO ON TO THE NEXT PAGE ⟹

6. If the geologists were to use equipment that malfunctions in warm environments, which room would be most likely to cause the equipment to malfunction?

F. Room 1

G. Room 2

H. Room 3

J. Room 4

PASSAGE II

Humans can experience toxic symptoms when concentrations of mercury (Hg) in the blood exceed 200 parts per billion (ppb). Frequent consumption of foods high in Hg content contributes to high Hg levels in the blood. On average, higher Hg concentrations are observed in people whose diets consist of more extreme amounts of certain types of seafood. A research group proposed that sea creatures that live in colder waters acquire greater amounts of Hg than those that reside in warmer waters. The researchers performed the following experiments to examine this hypothesis.

EXPERIMENT 1

Samples of several species of consumable sea life caught in the cold waters of the northern Atlantic Ocean were chemically prepared and analyzed using a cold vapor atomic fluorescence spectrometer (CVAFS), a device that indicates the relative concentrations of various elements and compounds found within a biological sample. Comparisons of the spectra taken from the seafood samples with those taken from samples of known Hg levels were made to determine the exact concentrations in ppb. Identical volumes of tissue from eight different specimens for each of four different species were tested, and the results

are shown in Table 1, including the average concentrations found for each species.

Table 1

Specimen	Hg concentration in cold-water species (ppb):			
	Cod	Crab	Swordfish	Shark
1	160	138	871	859
2	123	143	905	820
3	139	152	902	839
4	116	177	881	851
5	130	133	875	818
6	134	148	880	836
7	151	147	910	847
8	109	168	894	825
Average	133	151	890	837

EXPERIMENT 2

Four species caught in the warmer waters of the Gulf of Mexico were examined using the procedure from Experiment 1. The results are shown in Table 2.

Table 2

Specimen	Hg concentration in warm-water species (ppb):			
	Catfish	Crab	Swordfish	Shark
1	98	113	851	812
2	110	122	856	795
3	102	143	845	821
4	105	128	861	803
5	94	115	849	798
6	112	136	852	809
7	100	129	863	815
8	117	116	837	776
Average	105	125	852	804

GO ON TO THE NEXT PAGE →

7. According to Table 1 and Table 2, which species shows the greatest difference in average mercury concentration between cold and warm climates?

   A. Crab

   B. Swordfish

   C. Shark

   D. The greatest difference in the average mercury concentration cannot be determined from the information provided.

8. Given that shark and swordfish are both large predatory animals, and catfish and crab are smaller non-predatory animals, do the results of Experiment 2 support the hypothesis that the tissue of larger predatory fish exhibits higher levels of Hg than does the tissue of smaller species?

   F. Yes; the lowest concentration of Hg was found in swordfish.

   G. Yes; both swordfish and shark had Hg concentrations that were higher than those found in either catfish or crab.

   H. No; the lowest concentration of Hg was in catfish.

   J. No; both catfish and crab had concentrations of Hg that were higher than those found in either swordfish or shark.

9. A researcher, when using the CVAFS, was concerned that lead (Pb) in the tissue samples might be interfering with the detection of Hg. Which of the following procedures would best help the researcher explore this trouble?

   A. Flooding the sample with a large concentration of Pb before using the CVAFS

   B. Using the CVAFS to examine a non-biological sample

   C. Collecting tissue from additional species

   D. Testing a sample with known concentrations of Hg and Pb

10. Based on the results of the experiments and the data in the table below, sharks caught in which of the following locations would most likely possess the largest concentrations of Hg in February?

Location	Average water temperature (°F) for February
Northern Atlantic Ocean	33
Gulf of Mexico	70
Northern Pacific Ocean	46
Tampa Bay	72

   F. Northern Atlantic Ocean

   G. Northern Pacific Ocean

   H. Gulf of Mexico

   J. Tampa Bay

11. Which of the following factors was intentionally varied in Experiment 2?

   A. The volume of tissue tested

   B. The method by which the marine organisms were caught

   C. The species of marine organism tested

   D. The method of sample analysis

GO ON TO THE NEXT PAGE ⟹

12. Which of the following specimens would most likely have the highest concentration of Hg?

    F. A crab caught in cold water
    G. A swordfish caught in cold water
    H. A catfish caught in warm water
    J. A swordfish caught in warm water

13. How might the results of the experiments be affected if the chemical preparation described in Experiment 1 introduced Hg-free contaminants into the sample, resulting in a larger volume of tested material? The measured concentrations of Hg would be:

    A. the same as the actual concentrations for both cold-water and warm-water specimens.
    B. higher than the actual concentrations for both cold-water and warm-water specimens.
    C. lower than the actual concentrations for cold-water specimens, but higher than the actual concentrations for warm-water specimens.
    D. lower than the actual concentrations for both cold-water and warm-water specimens.

PASSAGE III

A student performed three exercises with a battery and four different light bulbs.

EXERCISE 1

The student connected the battery to a fixed outlet designed to accept any of the four bulbs. She then placed four identical light sensors at different distances from the outlet. Each sensor was designed so that a green indicator illuminated upon the sensor's detection of incident light, while a red indicator remained illuminated when no light was detected. The student darkened the room and recorded the state of each sensor while each bulb was lit. The results are shown in Table 1.

Table 1

Sensor distance (cm)	Sensor indicator color			
	Bulb 1	Bulb 2	Bulb 3	Bulb 4
50	green	green	green	green
100	red	green	green	green
150	red	red	green	green
200	red	red	red	green

EXERCISE 2

The battery produced an *electromotive force* of 12 volts. The student was given a device called an ammeter, which is used to measure the *current* passing through an electric circuit. She completed the circuit by connecting the battery, the ammeter, and each of the four light bulbs, one at a time. She measured the associated current in amperes (A) for each bulb and calculated the *impedance* (Z) in ohms (Ω) for each from the following formula:

$$Z = \text{electromotive force} \div \text{current.}$$

The results are shown in Table 2.

GO ON TO THE NEXT PAGE ▷

Table 2

Light bulb	Current (A)	Z (Ω)
1	0.2	60
2	0.3	40
3	0.4	30
4	0.6	20

EXERCISE 3

The *power rating* (P) of each light bulb was printed near its base. P gives the time rate of energy consumption of the bulb and is related to the *brightness* (B) of light at a given distance from the bulb. B is calculated in watts per meter squared (W/m^2) from the following formula:

$$B = \frac{P}{4\pi r^2}$$

where r is the distance in meters (m) from the bulb, and P is measured in watts (W).

The student calculated B for each bulb at a distance of 1 m. The results are shown in Table 3.

Table 3

Light bulb	P (W)	B (W/m2)
1	2.4	0.19
2	3.6	0.29
3	4.8	0.38
4	7.2	0.57

14. If the student had tested a fifth light bulb during Exercise 2 and measured the current passing through it to be 1.2 A, the Z associated with this bulb would have been:

    F.   1 Ω.

    G.   10 Ω.

    H.   14.4 Ω.

    J.   100 Ω.

15. Based on the results of Exercise 2, a circuit including the combination of which of the following batteries and light bulbs would result in the highest current in the circuit? (Assume Z is a constant for a given light bulb.)

    A.   A 10 V battery and Bulb 1

    B.   An 8 V battery and Bulb 2

    C.   A 6 V battery and Bulb 3

    D.   A 5 V battery and Bulb 4

16. With Bulb 3 in place in the circuit in Exercise 1, how many of the sensors were unable to detect any incident light?

    F.   1

    G.   2

    H.   3

    J.   4

17. Which of the following equations correctly calculates B (in W/m²) at a distance of 2 m from Bulb 2?

    A. $B = \dfrac{2}{4\pi(3.6)^2}$

    B. $B = \dfrac{2}{4\pi(2.4)^2}$

    C. $B = \dfrac{3.6}{4\pi(2)^2}$

    D. $B = \dfrac{2.4}{4\pi(2)^2}$

18. Another student used the approach given in Exercise 3 to calculate B at a distance of 1 m from a fifth light bulb. He determined that, for this fifth bulb, $B = 0.95$ W/m². Accordingly, P for this bulb was most likely closest to which of the following values?

    F.   0.1 W

    G.   6 W

    H.   12 W

    J.   18 W

GO ON TO THE NEXT PAGE ⟩

19. Exercise 1 and Exercise 2 differed in that in Exercise 1:

    A. 4 light sensors were used.

    B. 4 different light bulbs were used.

    C. the electromotive force of the battery was varied.

    D. the current was highest for Bulb 1.

20. According to Table 3, which light bulb with a brightness greater than 0.30 W/m² has the lowest power consumption?

    F. Bulb 1

    G. Bulb 2

    H. Bulb 3

    J. Bulb 4

PASSAGE IV

The electrons in a solid occupy *energy states* determined by the type and spatial distribution of the atoms in the solid. The probability that a given energy state will be occupied by an electron is given by the *Fermi-Dirac distribution function*, which depends on the material and the temperature of the solid. Fermi-Dirac distribution functions for the same solid at 3 different temperatures are shown in the figure below.

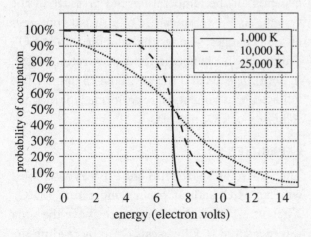

Figure 1

(Note: 1 electron volt (eV) = 1.66 × 10⁻¹⁹ joules (J); eV and J are both units of energy. At energies above 15 eV, the probability of occupation at each temperature continues to decrease.)

21. The information in the given figure supports which of the following statements about energy states?

    A. Cooler materials have a larger range of energy states than hotter materials.

    B. Materials have the same range of energy states regardless of temperature.

    C. Cooler materials are more capable of occupying higher energy states than hotter materials.

    D. Hotter materials are more capable of occupying higher energy states than cooler materials.

22. The steepness of the slope of each distribution function at the point where its value equals 50% is inversely proportional to the average *kinetic energy* of the atoms in the solid. Which of the following correctly ranks the 3 functions, from *least* to *greatest*, according to the average kinetic energy of the atoms in the solid?

    F. 25,000 K; 10,000 K; 1,000 K

    G. 25,000 K; 1,000 K; 10,000 K

    H. 10,000 K; 1,000 K; 25,000 K

    J. 1,000 K; 10,000 K; 25,000 K

23. Based on the figure, at a temperature of 1,000 K, the probability of a state at an energy of 20 eV being occupied by an electron will most likely be:

    A. less than 5%.

    B. between 5% and 50%.

    C. between 50% and 90%.

    D. greater than 90%.

GO ON TO THE NEXT PAGE

24. Based on the figure, which of the following sets of Fermi-Dirac distribution functions best represents an unknown solid at temperatures of 2,000 K, 20,000 K, and 50,000 K?

F.

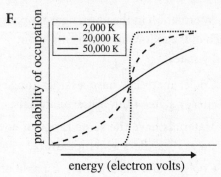

G.

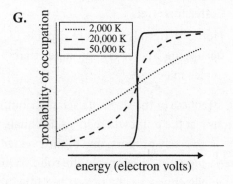

H.

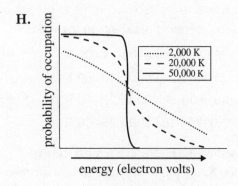

J.

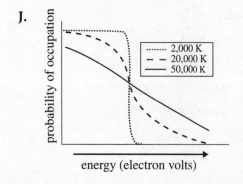

25. Based on the figure, the probability of a 5 eV energy state being occupied by an electron will equal 80% when the temperature of the solid is closest to:

A. 500 K.

B. 5,000 K.

C. 20,000 K.

D. 30,000 K.

26. The *de Broglie wavelength* of an electron energy state decreases as the energy of the state increases. Based on this information, over all energies in the figure, as the de Broglie wavelength of an electron energy state decreases, the probability of that state being occupied by an electron:

F. increases only.

G. decreases only.

H. increases, then decreases.

J. decreases, then increases.

PASSAGE V

A soda beverage is typically a solution of water, various liquid colorings and flavorings, and CO_2 gas. *Solubility* is defined as the ability of a substance to dissolve, and the solubility of CO_2 in a soda depends on the temperature and pressure of the system. As the temperature of a sealed container of soda changes, so does the solubility of the CO_2. This results in changes in the concentration of CO_2 in both the soda and the air in the container. The following experiments were performed to study the solubility of CO_2 in sodas.

EXPERIMENT 1

The apparatus shown in Figure 1 was assembled with an H_2O bath at room temperature (25°C). After 10 minutes, the air pressure above the soda was measured in kilopascals (kPa) by reading the

GO ON TO THE NEXT PAGE ⟶

value directly from the pressure gauge. Additional trials were performed at different temperatures and with other sodas in the container. The results are shown in Table 1.

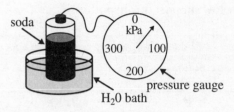

Figure 1

Table 1

Soda	Pressure (kPa) at:		
	0°C	**25°C**	**50°C**
A	230	237	256
B	214	234	253
C	249	272	294
D	223	243	282
E	209	228	247

EXPERIMENT 2

An apparatus similar to those used by companies that produce soda was constructed so that measured amounts of compressed CO_2 gas could be injected into each soda until the solution reached its maximum concentration of CO_2. The apparatus consisted of an air-sealed flask containing only soda and no air. Starting with sodas from which all of the CO_2 had been carefully removed, CO_2 was injected and the maximum CO_2 concentration for each soda was recorded. From the maximum concentrations, the solubility of CO_2 in each soda was calculated for three different temperatures at equal pressures. Solubility was recorded in centimolars per atmosphere (cM/atm), and the results are shown in Table 2.

Table 2

Soda	CO_2 solubility (cM/atm) at:		
	0°C	**25°C**	**50°C**
A	3.59	3.51	3.45
B	3.58	3.50	3.37
C	3.67	3.57	3.48
D	3.62	3.53	3.41
E	3.54	3.46	3.29

27. Which of the following bar graphs best expresses the pressures of the container contents from Experiment 1 at 25°C?

A.

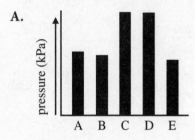

B.

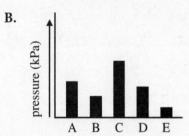

C.

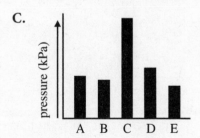

D.

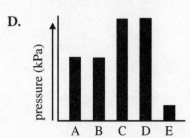

GO ON TO THE NEXT PAGE

28. Which of the following figures best depicts the change in position of the needle on the pressure gauge while attached to the container holding Soda C in Experiment 1?

needle position at 0°C	needle position at 50°C

F.

G.

H.

J.

29. A student hypothesized that, at a given pressure and temperature, the higher the sugar content of a soda, the higher the solubility of CO_2 in that soda. Do the results of Experiment 2 and all of the information in the table below support this hypothesis?

Soda	Sugar content (grams per 12 ounces)
A	23
B	32
C	38
D	40
E	34

A. Yes; Soda A has the lowest sugar content and the lowest CO_2 solubility.

B. Yes; Soda D has a higher sugar content and CO_2 solubility than Soda C.

C. No; the higher a soda's sugar content, the lower the soda's CO_2 solubility.

D. No; there is no clear relationship in these data between sugar content and CO_2 solubility.

30. According to the results of Experiment 2, as the temperature of the soda increases, the CO_2 solubility of the soda:

F. increases only.

G. decreases only.

H. increases, then decreases.

J. decreases, then increases.

GO ON TO THE NEXT PAGE

31. Which of the following figures best illustrates the apparatus used in Experiment 2?

A.

B. soda

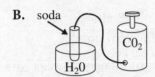

C.

D.

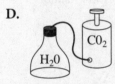

32. Which of the following statements best explains why, in Experiment 1, the experimenter waited 10 minutes before recording the pressure of the air above the soda? The experimenter waited to allow:

F. all of the CO_2 to be removed from the container.

G. time for the soda in the container to evaporate.

H. the contents of the container to adjust to the temperature of the H_2O bath.

J. time for the pressure gauge to stabilize.

33. Table 2 shows that the soda's CO_2 solubility:

A. increases as temperature increases.

B. decreases as temperature increases.

C. sometimes increases and sometimes decreases as temperature increases.

D. is not affected by changes in temperature.

PASSAGE VI

Straight-chain conformational isomers are carbon compounds that differ only by rotation about one or more single carbon bonds. Essentially, these isomers represent the same compound in a slightly different position. One example of such an isomer is butane (C_4H_{10}), in which two methyl (CH_3) groups are each bonded to the main carbon chain. The straight-chain conformational isomers of butane are classified into 4 categories.

1. In the *anti* conformation, the bonds connecting the methyl groups to the main carbon chain are rotated 180° with respect to each other.

2. In the *gauche* conformation, the bonds connecting the methyl groups to the main carbon chain are rotated 60° with respect to each other.

3. In the *eclipsed* conformation, the bonds connecting the methyl groups to the main carbon chain are rotated 120° with respect to each other.

4. In the *totally eclipsed* conformation, the bonds connecting the methyl groups to the main carbon chain are parallel to each other.

The anti conformation is the lowest energy and most stable state of the butane molecule, since it allows for the methyl groups to maintain maximum separation from each other. The methyl groups are much closer to each other in the gauche conformation, but this still represents a relative minimum or *meta-stable* state, due to the relative orientations of the other hydrogen atoms in the molecule. Molecules in the anti or gauche conformations tend to maintain their shape. The eclipsed conformation represents a relative maximum energy state, while the totally eclipsed conformation is the highest energy state of all of butane's straight-chain conformational isomers.

Two organic chemistry students discuss straight-chain conformational isomers.

GO ON TO THE NEXT PAGE

STUDENT 1

The *active shape* (the chemically functional shape) of a butane molecule is always identical to the molecule's lowest-energy shape. Any other shape would be unstable. Because the lowest-energy shape of a straight-chain conformational isomer of butane is the anti conformation, its active shape is always the anti conformation.

STUDENT 2

The active shape of a butane molecule is dependent upon the energy state of the shape. However, a butane molecule's shape may also depend on temperature and its initial isomeric state. Specifically, in order to convert from the gauche conformation to the anti conformation, the molecule must pass through either the eclipsed or totally eclipsed conformation. If the molecule is not given enough energy to reach either of these states, its active shape will be the gauche conformation.

34. According to the passage, molecules in conformation states with relatively low energy tend to:

    F. convert to the totally eclipsed conformation.

    G. convert to the eclipsed conformation.

    H. maintain their shape.

    J. chemically react.

35. The information in the passage indicates that when a compound changes from one straight-chain conformational isomer to another, it still retains its original:

    A. energy state.

    B. shape.

    C. number of single carbon bonds.

    D. temperature.

36. Student 2's views differ from Student 1's views in that only Student 2 believes that a butane molecule's active shape is partially determined by its:

    F. initial isomeric state.

    G. energy state.

    H. hydrogen bonding angles.

    J. proximity of methyl groups.

37. A student rolls a ball along the curved path shown below. Given that points closer to the ground represent states of lower energy, the ball coming to rest at the position shown corresponds to a butane molecule settling into which conformational isomer state?

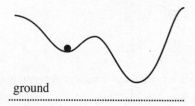

    A. Anti

    B. Gauche

    C. Eclipsed

    D. Totally eclipsed

38. Suppose butane molecules are cooled so that each molecule is allowed to reach its active shape. Which of the following statements is most consistent with the information presented in the passage?

    F. If Student 1 is correct, all of the molecules will be in the anti conformation.

    G. If Student 1 is correct, all of the molecules will have shapes different from their lowest-energy shapes.

    H. If Student 2 is correct, all of the molecules will be in the anti conformation.

    J. If Student 2 is correct, all of the molecules will have shapes different than their lowest-energy shapes.

GO ON TO THE NEXT PAGE ⟩

39. Which of the following diagrams showing the relationship between a given butane molecule's shape and its relative energy is consistent with Student 2's assertions about the energy of butane molecules, but is NOT consistent with Student 1's assertions about the energy of butane molecules?

A.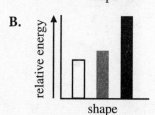

B.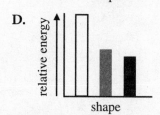

C.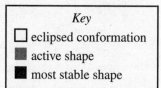

D.

Key
☐ eclipsed conformation
▨ active shape
■ most stable shape

40. Student 2 says that a butane molecule may settle into a moderately high-energy conformation. Which of the following findings, if true, could be used to *counter* this argument?

F. Once a molecule has settled into a given conformation, all of its single carbon bonds are stable.

G. Enough energy is available in the environment to overcome local energy barriers, driving the molecule into its lowest-energy conformation.

H. During molecule formation, the hydrogen bonds are formed before the carbon bonds.

J. Molecules that change their isomeric conformation tend to lose their chemical functions.

Practice Test Two
ANSWER KEY

MATHEMATICS TEST

1. E	9. C	17. C	25. D	33. E	41. B	49. D	57. C
2. J	10. G	18. F	26. G	34. G	42. G	50. J	58. K
3. A	11. B	19. D	27. E	35. E	43. C	51. B	59. D
4. K	12. K	20. G	28. G	36. J	44. G	52. K	60. K
5. C	13. E	21. D	29. D	37. D	45. D	53. D	
6. F	14. J	22. F	30. F	38. K	46. H	54. K	
7. B	15. E	23. C	31. C	39. B	47. E	55. A	
8. H	16. G	24. F	32. F	40. F	48. H	56. H	

SCIENCE TEST

1. B	6. F	11. C	16. F	21. D	26. G	31. A	36. F
2. G	7. B	12. G	17. C	22. J	27. C	32. H	37. B
3. C	8. G	13. D	18. H	23. A	28. G	33. B	38. F
4. J	9. D	14. G	19. A	24. J	29. D	34. H	39. D
5. D	10. F	15. D	20. H	25. C	30. G	35. C	40. G

ANSWERS AND EXPLANATIONS

MATHEMATICS TEST

1. E
Category: Proportions and Probability
Difficulty: Low
Getting to the Answer: This is a great question for Backsolving, because you know 20% of the answer should turn out to be 10. Alternatively, you could use your knowledge that Percent $= \dfrac{\text{part}}{\text{whole}} \cdot 100\%$ to set up an equation. Let x be the number of students in the class. Then 20% of x is 10:

$0.2x = 10$
$x = 50$, (E)

2. J
Category: Plane Geometry
Difficulty: Low
Getting to the Answer: Questions like this are simply testing whether you remember how many degrees are in a line. Remember that the arc between two points on a line is a half circle, or 180°.

$20° + 40° + x° = 180°$
$ x° = 120°$, (J)

3. A
Category: Patterns, Logic, & Data
Difficulty: Low
Getting to the Answer: Even if you forget what "arithmetic" means, you should still be able to recognize the pattern. Each term is 3 less than the previous term.

Fourth term $= 1 - 3 = -2$
Fifth term $= (-2) - 3 = -5$, (A)

Be sure not to stop too soon—the fourth term is a tempting, but wrong, answer choice.

4. K
Category: Proportions and Probability
Difficulty: Low
Getting to the Answer: Whenever you see a proportion (two fractions set equal to each other), you can cross-multiply to solve.

$$\frac{20}{8} = \frac{c}{10}$$
$20(10) = 8c$
$ c = 25$, (K)

5. C
Category: Coordinate Geometry
Difficulty: Low
Getting to the Answer: On problems like this without a diagram, drawing one is an excellent idea. $\overline{GK} \cong \overline{HK}$ means that the line segments $\overline{GK}$ and $\overline{HK}$ are congruent, or equal in length. For these two segments to have equal lengths, the points must be arranged like this:

K is the midpoint of $\overline{GH}$, so (C) is correct.

6. F
Category: Operations
Difficulty: Low
Getting to the Answer: On early problems, you can sometimes let your calculator do most of the work for you. Yarn cut off $= 4(1.2) = 4.8$ yards. Yarn remaining $= 50 - 4.8 = 45.2$ yards, (F).

7. B
Category: Variable Manipulation
Difficulty: Low
Getting to the Answer: Be careful with positives and negatives, and remember to follow the order of operations. You can count on some of the wrong answer choices resulting from careless mistakes.

$14 - 3[(-2) + 3]$
$= 14 - 3(1)$
$= 11$, (B)

8. H
Category: Operations
Difficulty: Low
Getting to the Answer: Use the order of operations. In PEMDAS, absolute value counts as parentheses, so you must evaluate it first.

$-|-6| - (-6)$
$= -(6) - (-6)$
$= -6 + 6$
$= 0$, (H)

9. C
Category: Proportions and Probability
Difficulty: Low
Getting to the Answer: Remembering how to convert between percents, fractions, and decimals will be a key skill on the ACT.

15% of 3,200 = 0.15(3,200) = 480
3,200 + 480 = 3,680, (C)

10. G
Category: Operations
Difficulty: Low
Getting to the Answer: You can use your calculator to quickly find the fourth root of 90, or you can compare the fourth power of the integers in the answer choices.

$2^4 = 16$
$3^4 = 81$
$4^4 = 256$
If $x^4 = 90$, then x must be between 3 and 4, because 90 is between 3^4 and 4^4, so (G) is correct.

11. B
Category: Variable Manipulation
Difficulty: Low
Getting to the Answer: You can use your calculator for the arithmetic here. It will take only a few seconds and will help you avoid mistakes.

$47 - x = 188$
$47 = 188 + x$
$-141 = x$, (B)

12. K
Category: Variable Manipulation
Difficulty: Low
Getting to the Answer: Don't jump at the first answer that seems right; on the other hand, don't think that the most complicated answer has to be right, either!
Time for Group C = h
Time for Group B = $2h$
Time for Group A = $8 + 2h$, (K)

13. E
Category: Coordinate Geometry
Difficulty: Low
Getting to the Answer: Draw a quick sketch. Notice that the answers are fairly different, so you just need a general idea in order to get the correct answer.

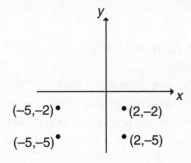

As you can see in the diagram above, the fourth coordinate must be (E), (−5,−5).

14. J
Category: Variable Manipulation
Difficulty: Low
Getting to the Answer: Before you try anything too fancy, check for like terms. Combine the like variables $5a$ and $3a$ to find that $5a - 5b + 3a = 8a - 5b$, (J).

15. E
Category: Plane Geometry
Difficulty: Medium
Getting to the Answer: Even if you forget all the properties of a parallelogram, you can figure out problems like this by using the fact that opposite sides are parallel. Redraw the diagram, and it's clear that *FEG* and *EGD* are alternate interior angles.

Therefore, the measure of *FEG* is also 70°, so (E) is correct.

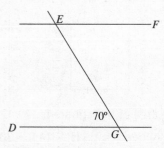

16. G
Category: Coordinate Geometry
Difficulty: Medium
Getting to the Answer: The easiest way to find the slope of a line is to write the equation in slope-intercept form, $y = mx + b$.
$$4x + 3y = 9$$
$$3y = -4x + 9$$
$$y = -\frac{4}{3}x + 3$$
The slope, *m*, is the coefficient of *x*, or $-\frac{4}{3}$, which matches (G).

17. C
Category: Variable Manipulation
Difficulty: Medium
Getting to the Answer: When you need to factor a quadratic equation, make sure one side is equal to zero before you begin. Then you know that one factor or the other must be equal to zero. Because 3 is a prime number, the two binomial factors must look like $(3x \pm \_\_)(x \pm \_\_)$. One of the last two numbers must be positive and the other negative, because they multiply to a negative number (−20). At this point you can use trial and error with the factors of −20 to find $(3x + 5)(x - 4) = 0$. If the product is equal to zero, then one of the factors must be equal to zero, so $3x + 5 = 0$ or $x - 4 = 0$. The left equation gives you a negative value for *x*, which contradicts the question stem, while the right equation gives you $x = 4$, which matches (C).

18. F
Category: Plane Geometry
Difficulty: Medium
Getting to the Answer: In similar shapes, each side is scaled up or down by the same factor, and the perimeter is scaled up or down by that same factor. The perimeter of the original triangle is $2 + 5 + 8 = 15$. Because the similar triangle has a perimeter twice as long, each side must also be twice as long. The smallest side is $2(2) = 4$, (F).

19. D
Category: Proportions and Probability
Difficulty: Low
Getting to the Answer: You can often save a step in percentage problems if you figure out what percentage is left. If the shirt is 15% off, then the sale price is $100\% - 15\% = 85\%$ of the original price. $\$24.60(0.85) = \$20.91 \approx \$21$, (D).

20. G
Category: Variable Manipulation
Difficulty: Medium
Getting to the Answer: One way to tackle problems like this is to multiply out the factored answers to see if they match the expression in the problem. Each term has a 3, an *x*, and a *y*, so you can take out a greatest common factor of $3xy$. What's left from each factor when you divide this out? The first term has no more coefficient or *x*, and it has 3 *y*s left, so the term turns into y^3. Similarly, the second term becomes x^3. The factored form is (G), $3xy(y^3 + x^3)$, which you can check by distributing.

21. D
Category: Variable Manipulation
Difficulty: Medium
Getting to the Answer: If the question asks you for *x*, that means you don't care about *y*—get rid of it! You can eliminate *y* using one of two methods.
Substitution:
$$3x + y = 7$$
$$y = 7 - 3x$$
$$x - 2y = x - 2(7 - 3x) = 0$$
$$x - 14 + 6x = 0$$
$$7x = 14$$
$$x = 2, \text{(D)}$$

Combination:

Multiply the second equation by 2: $2(3x + y) = 2(7)$

$$6x + 2y = 14$$

Add to the first equation:

$$
\begin{array}{r}
6x + 2y = 14 \\
+ \, x - 2y = 0 \\
\hline
7x = 14 \\
x = 2
\end{array}
$$

22. F

Category: Operations

Difficulty: Low

Getting to the Answer: Writing out the units on conversion problems will help you avoid mistakes. First find the cost per yard: $\dfrac{4.50 \text{ dollars}}{2.5 \text{ yards}} = 1.80 \dfrac{\text{dollars}}{\text{yards}}$.

Then find the cost per foot (notice that yards cancel):

$1.80 \dfrac{\text{dollars}}{\text{yards}} \cdot \dfrac{1 \text{ yard}}{3 \text{ feet}} = 0.60 \dfrac{\text{dollars}}{\text{foot}}$, (F)

23. C

Category: Plane Geometry

Difficulty: High

Getting to the Answer: Sometimes the answer choices give you a hint about how to solve the problem. Here, they are simple enough that you know you don't have to do any fancy calculations. In fact, they're different enough that you might even be able to eyeball the answer. Draw in lines that go from the center of the square to the edge at right angles.

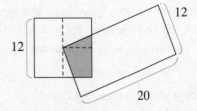

The gray triangle is the same size as the white triangle. (The portion of the upper 90° angle formed by the gray triangle is the same as the portion of the rectangle's 90° angle formed by the gray triangle. Therefore, the portion of the lower 90° angle formed by the white triangle is also the same.) If you move the gray triangle to where the white triangle is, the shaded area is exactly $\dfrac{1}{4}$ of the square.

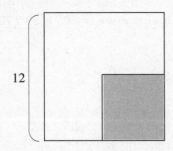

Because the area of the square is $12 \cdot 12 = 144$, the area of the shaded region is $\dfrac{1}{4}(144) = 36$, (C).

24. F

Category: Operations

Difficulty: Low

Getting to the Answer: Questions like this may seem complicated at first, but all you need to do is plug the given numbers into the formula.

$h = 15$ and $s = 120.50$

$7h + 0.04s$

$= 7(15) + 0.04(120.50)$

$= 105 + 4.82$

$= 109.82$, (F)

25. D

Category: Plane Geometry

Difficulty: Low

Getting to the Answer: Whenever you're faced with an odd shape, try to divide it into two or more shapes that are familiar. This shape can be divided into two rectangles. The width of the smaller one is the difference between the width of the entire shape and the width of the larger rectangle, or $22 - 14 = 8$. The area of any rectangle is length times width.

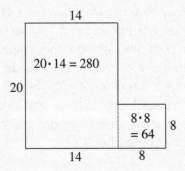

The total area of this shape is $280 + 64 = 344$, (D).

26. G

Category: Variable Manipulation

Difficulty: Low

Getting to the Answer: Don't forget that inequalities work exactly the same as equalities, except that the direction of the sign changes when you multiply or divide by a negative number. $x - 2 < -4$ Add 2 to both sides: $x < -2$ Everything less than -2 is everything to the left of -2, exactly what (G) shows.

27. E

Category: Operations

Difficulty: Low

Getting to the Answer: On the ACT Mathematics test, you have an average of one minute per question. If a problem can be solved quickly using your calculator, take advantage of it. For example, you can use your calculator to compare fractions by finding the decimal equivalents.

$$\frac{3}{5} = 0.6$$

A: $\frac{4}{6} = 0.\overline{66}$

B: $\frac{8}{13} = 061538\ldots$

C: $\frac{6}{10} = 0.6$

D: $\frac{7}{11} = 0.\overline{63}$

(E): $\frac{4}{7} = 0.57142\ldots$

Only (E) is smaller than 0.6.

28. G

Category: Plane Geometry

Difficulty: Medium

Getting to the Answer: In a right triangle, the hypotenuse is always the longest side, and the other two sides are the legs.

The legs are 7 and 24. Area $= \frac{1}{2}bh$

$$= \left(\frac{1}{2}\right)(7)(24)$$

$$= 84, \text{(G)}$$

29. D

Category: Number Properties

Difficulty: Medium

Getting to the Answer: You need to figure out how to turn this word problem into math. If the last row is one person short of 6, how does the number of students relate to multiples of 6? This is just another way of saying that the number of students is one less than a multiple of 6, one less than a multiple of 7, and one less than a multiple of 8. The least common multiple of 6, 7, and 8 is 168. One less is 167, (D). You can find the least common multiple by finding the prime factors of each number:

$6 = 2 \cdot 3$

$7 = 7$

$8 = 2 \cdot 2 \cdot 2$

To have all these factors in one number, you need one 3, one 7, and three 2s, or $3 \cdot 7 \cdot 2 \cdot 2 \cdot 2 = 168$. E produces the right arrangement, but it is not the smallest possible number. Choice A is 2 *more* then a multiple of 7, which means the last row will have 2 people (or be 5 people short). Choice B is 1 *more* than a multiple of 8, which means the last row will be 7 people short. Choice C is 1 *more* than a multiple of 7, which means the last row will be 6 people short.

30. F

Category: Plane Geometry

Difficulty: Low

Getting to the Answer: There will usually be one question that tests your knowledge of the Triangle Inequality Theorem. If you're not sure how to proceed, try drawing a sketch. The Triangle Inequality Theorem states that any side of a triangle is less than the sum of and more than the difference between the other two, so the third side must be at least $6 - 3.5 = 2.5$ inches. Choice (F) is too small.

31. C

Category: Variable Manipulation

Difficulty: Medium

Getting to the Answer: Fractions are always added in the same way, whether they include variables or not. You must write both fractions in terms of a common denominator, then add the numerators.

$$\frac{4}{5} = \frac{4b}{5b}$$

$$\frac{1}{b} = \frac{5}{5b}$$

$$\frac{4b}{5b} + \frac{5}{5b} = \frac{4b+5}{5b}, \text{(C)}$$

32. F
Category: Plane Geometry
Difficulty: Medium
Getting to the Answer: In the Pythagorean Theorem, $a^2 + b^2 = c^2$, c represents the hypotenuse (the longest side).

$$a^2 + b^2 = c^2$$
$$17^2 + b^2 = 21^2$$
$$b^2 = 21^2 - 17^2$$
$$b = \sqrt{21^2 - 17^2}, \text{(F)}$$

33. E
Category: Variable Manipulation
Difficulty: Medium
Getting to the Answer: When you multiply binomials, don't forget to use FOIL. If you only multiply the first terms together and the last terms together, you're missing two terms.

The new length is $b + 2$, and the new width is $b + 3$.
Area $= l \cdot w$
$$= (b + 2)(b + 3)$$
$$= b^2 + 3b + 2b + 6$$
$$= b^2 + 5b + 6, \text{(E)}$$

34. G
Category: Trigonometry
Difficulty: Low
Getting to the Answer: If you have sine and cosine, find tangent using the formula $\tan x = \dfrac{\sin x}{\cos x}$.

$$\tan x = \frac{\sin x}{\cos x} = \frac{\dfrac{8}{17}}{\dfrac{15}{17}} = \frac{8}{17} \cdot \frac{17}{15} = \frac{8}{15}, \text{(G)}$$

35. E
Category: Variable Manipulation
Difficulty: Medium
Getting to the Answer: If you're not sure which answer is correct, try plugging in points from the number line. The shaded region includes all the values between –1.5 and –2.5. Because –1.5 is larger than –2.5, the inequality should be $-1.5 > x > -2.5$. Plugging in $x = -2 \cdot 2.5$ eliminates A, B, and D. Choice C includes numbers that aren't in the shaded region, such as –2.9, so (E) is correct.

36. J
Category: Operations
Difficulty: Medium
Getting to the Answer: To make the question a little easier to follow, Pick Numbers for the total number of baskets. Imagine that they made a total of 100 baskets. If the team made 100 baskets, then they made 20 1-point baskets, 70 2-point baskets, and 10 3-point baskets. The average point value of all the baskets is the total number of points divided by the total number of baskets:

$$\frac{20(1) + 70(2) + 10(3)}{100}$$
$$= \frac{20 + 140 + 30}{100}$$
$$= \frac{190}{100}$$
$$= 1.9, \text{(J)}$$

37. D
Category: Plane Geometry
Difficulty: Medium
Getting to the Answer: If you can use 45-45-90 triangles or 30-60-90 triangles instead of trigonometry, do it. (Similarly, look for the Pythagorean Triplets before you use the Pythagorean Theorem.) The sides in a 30-60-90 triangle are in the ratio $x : x\sqrt{3} : 2x$. You have been given the side opposite the 30 degree angle, which is x. You want the side opposite the 90 degree angle, which is $2x$. Because $x = 3$, $CE = 2x = 6$, (D).

38. K
Category: Number Properties
Difficulty: Medium
Getting to the Answer: On this type of problem, you want the numbers to be either as close together or as far apart as possible. Try a few possibilities; you should see a pattern.

$3 \cdot 39 = 117$
$5 \cdot 37 = 185$
$7 \cdot 35 = 245$

The products are increasing, so it looks like you want the two numbers to be as close together as possible. Because half of 42 is 21, try $21 \cdot 21 = 441$. Because (K) is the largest possible answer choice, you can be sure it's correct.

39. B
Category: Plane Geometry
Difficulty: Medium
Getting to the Answer: Parallel lines provide lots of information—look for congruent and supplementary angles.

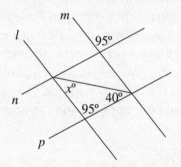

Because both sets of lines are parallel, the missing angle in the triangle corresponds to the angle marked 95°. The three interior angles of the triangle sum to 180°.

$40° + 95° + x° = 180°$
$x° = 45°$, (B)

40. F
Category: Coordinate Geometry
Difficulty: Medium
Getting to the Answer: Don't jump right in to using your graphing calculator; often some simple algebra is the best route to the correct answer. The easiest way to find the y-intercept (the value of y when the graph crosses the y-axis) is to plug in $x = 0$:

$12(0) - 3y = 12$
$-3y = 12$
$y = -4$, (F)

41. B
Category: Operations
Difficulty: Medium
Getting to the Answer: If the labels are missing on a number line, you can find the length of each interval by finding the difference in the endpoints (how much the total interval is) and dividing by the number of subintervals. The unmarked interval goes from 2.7 to 2.8, so it must be 0.1 units long. It's divided into 10 equally spaced subintervals, each which must be $\frac{0.1}{10} = 0.01$ units long.

E is approximately 2.718, so you want a point between 2.71 and 2.72. Choice (B) is the closest.

42. G
Category: Coordinate Geometry
Difficulty: High
Getting to the Answer: If a system has no solution, there are no values of x and y that make both equations true. If a system has infinitely many solutions, then every set of x and y that works in one equation will also work in the other. These two linear equations will have no solutions (points of intersection) if they are parallel. Write both equations in slope-intercept form, then set the slopes equal and solve for a.

$-x + 6y = 7$
$6y = x + 7$
$y = \frac{1}{6}x + \frac{7}{6}$

$-5x + 10ay = 32$
$10ay = 5x + 32$
$y = \frac{5}{10a}x + \frac{32}{10a}$

$$\frac{1}{6} = \frac{5}{10a}$$
$$10a = 30$$
$$a = 3, \text{(G)}$$

43. C

Category: Plane Geometry
Difficulty: Medium
Getting to the Answer: "If and only if" means the answer always gives you an obtuse angle, and it gives you all obtuse angles. An angle is obtuse if its measure is between 90° and 180°, non-inclusive. That's everything wider than a right angle and smaller than a straight line. Try plugging in the end values of each range. A: $0 < x < 90 : 360 - 0 = 360$, $360 - 90 = 270$. This answer gives angles from 270° to 360°. B: $0 < x < 180 : 360 - 0 = 360$, $360 - 180 = 180$. This answer gives angles from 180° to 360°. (C): $180 < x < 270 : 360 - 180 = 180$, $360 - 270 = 90$. This answer gives angles from 90° to 180°—exactly what you're looking for. D: $180 < x < 360 : 360 - 180 = 180$, $360 - 360 = 0$. This answer gives angles from 0° to 180°. E: $270 < x < 360 : 360 - 270 = 90$, $360 - 360 = 0$. This answer gives angles from 0° to 90°. Only (C) gives the entire set of obtuse angles and nothing more.

44. G

Category: Variable Manipulation
Difficulty: Medium
Getting to the Answer: Always be on the lookout for the three classic quadratics—they'll save you a lot of time.
$$p^2 - q^2 = (p + q)(p - q)$$
$$= (-3)(-4)$$
$$= 12, \text{(G)}$$

45. D

Category: Plane Geometry
Difficulty: Low
Getting to the Answer: Remember, time is short on the ACT. Do they really expect you to apply something like the law of cosines here? If you recognized the Pythagorean Triplet (3:4:5) scaled up by 2, then you realized that this is a right triangle. The right angle is always between the two shortest sides

(and opposite the longest side, the hypotenuse), so the angle is 90 degrees, (D).

46. H

Category: Coordinate Geometry
Difficulty: Medium
Getting to the Answer: Be sure to read the question carefully. This one gives you an equation where x is in terms of y, not y in terms of x. First, translate the description into an equation: $x = 9 + 3y$. Then put it into slope-intercept form:
$$x = 9 + 3y$$
$$\frac{1}{3}x - 3 = y$$
The slope is the coefficient of x, which is $\frac{1}{3}$, (H).

47. E

Category: Trigonometry
Difficulty: High
Getting to the Answer: Use SOHCAHTOA. When you're presented with a trigonometry problem, your first step should be to identify the sides you're given and the side you're trying to find. Then figure out which trig function gives you a relationship between the side you know and the side you want to know. The distance between the cliff and the house is opposite the given angle. The height of the cliff is adjacent. The trig function that gives a relationship between the opposite and the adjacent sides is tangent.
$$\tan 62° = \frac{500}{\text{height}}$$
$$\text{height} \cdot \tan 62° = 500$$
$$\text{height} = \frac{500}{\tan 62°}, \text{(E)}$$

48. H

Category: Number Properties
Difficulty: Medium
Getting to the Answer: Prime factorizations can help you find the greatest common factor and least common multiple.
$$9 = 3 \cdot 3$$
$$18 = 2 \cdot 3 \cdot 3$$
$$27 = 3 \cdot 3 \cdot 3$$
$$54 = 2 \cdot 3 \cdot 3 \cdot 3$$

	GCF	LCM
9 and 18	3 • 3 = 9	2 • 3 • 3 = 18
9 and 27	3 • 3 = 9	3 • 3 • 3 = 27
18 and 27	3 • 3 = 9	2 • 3 • 3 • 3 = 54
18 and 54	2 • 3 • 3 = 18	2 • 3 • 3 • 3 = 54
27 and 54	3 • 3 • 3 = 27	2 • 3 • 3 • 3 = 54

Only (H) fits the description. All the other pairs have either a lower common multiple or a greater common factor.

49. D
Category: Patterns, Logic, & Data
Difficulty: High
Getting to the Answer: You can make abstract questions like this one easier to handle by Picking Numbers. Be sure the numbers you pick obey any restrictions in the question stem. Try $k = 2$ and $x = 4$. (These numbers will make the radical in E easy to calculate.)

A $b(4) = \dfrac{2}{4} = \dfrac{1}{2}$

B $b(4) = \dfrac{4}{2} = 2$

C $b(4) = 2(4) = 8$

D $b(4) = 4^2 = 16$

E $b(4) = \sqrt[2]{4} = 2$

The largest $b(x)$ is (D). When this is plugged into $a(b(x)) = 5^{b(x)}$, the largest value of $b(x)$ will produce the largest value of $a(b(x))$. For the values of k and x allowed in this question, (D) is always the largest, so it is correct. Another way to approach this problem is to use a graphing calculator. Pick a value for k, then graph each answer choice on your calculator at the same time to see which one has the largest value when x is greater than 2.

50. J
Category: Trigonometry
Difficulty: High
Getting to the Answer: If you're not sure where to get started, try working backwards. What are you looking for? The cosine of $\angle ZWY$. What do you need to find that? The lengths of the hypotenuse and the adjacent leg, WZ and WY. How can you find those

lengths? By using the given side lengths and your knowledge of right triangles. Now that you've figured out how to get from what you have to what you need, you can go ahead and get there. WXY is a right triangle. You know that WX is 12 and XY is 5, so WY must be 13. If you didn't spot the 5:12:13 triplet, you could have used the Pythagorean Theorem.) WYZ is also a right triangle. You know that WY and YZ are both 13, so WZ must be $13\sqrt{2}$. (Again, you could have used the Pythagorean Theorem if you didn't spot the 45–45–90 triangle.) The cosine of an angle is the adjacent leg over the hypotenuse, so the cosine of $\angle ZWY$ is WY over WZ, or $\dfrac{13}{13\sqrt{2}} = \dfrac{1}{\sqrt{2}} = \dfrac{\sqrt{2}}{2}$.

51. B
Category: Plane Geometry
Difficulty: Low
Getting to the Answer: On circle questions, it's important to distinguish between radius and diameter. Area $= \pi r^2 = 4\pi$

$$r^2 = 4$$
$$r = 2, \text{(B)}$$

52. K
Category: Coordinate Geometry
Difficulty: Medium
Getting to the Answer: Lines with positive slope rise to the right, and lines with negative slope rise to the left. The lines have the same y-intercept, so line q must also have "+ b" at the end of its equation. You may have been fooled by the fact that b is negative, but imagine that the y-intercept was at –2: The equation would be $y = mx + (-2)$. This eliminates F, G, and J. Line l has a negative slope, so $m < 0$, and line q has a positive slope. Because $\dfrac{1}{2}m < 0$, this cannot be the slope of line q, so its equation must be $y = -\dfrac{1}{2}mx + b$, (K).

53. D
Category: Variable Manipulation
Difficulty: Medium
Getting to the Answer: When you solve a quadratic equation by factoring, each factor will give you a solution. If the equation has only one solution, that means the factors are the same. To get a middle

term of −6x, you must have factors of (x − 3)(x − 3). Multiply it out to get (x − 3)(x − 3) = x^2 − 3x − 3x + 9 = x^2 − 6x + 9. This means that k = 9, (D).

54. K

Category: Coordinate Geometry
Difficulty: Low
Getting to the Answer: Most coordinate geometry questions rely on the slope. Make sure you remember that slope is $\frac{\text{rise}}{\text{run}}$.

$$\frac{y_2-y_1}{x_2-x_1} = \frac{\frac{3}{4}-0}{\frac{1}{3}-0}$$

$$= \frac{\frac{3}{4}}{\frac{1}{3}}$$

$$= \frac{3}{4} \cdot \frac{3}{1}$$

$$= \frac{9}{4}, (K)$$

55. A

Category: Number Properties
Difficulty: Medium
Getting to the Answer: Don't accidentally add your own assumptions. For example, the question never says that the variables are integers. (A) is true. It follows immediately from dividing both sides by S. Choices B, C, and E are disproved by R = −3, S = −2, and T = $\frac{1}{3}$. Choice D can never be true, because then the product would be 0.

56. H

Category: Plane Geometry
Difficulty: High
Getting to the Answer: Although this is a geometry question, the part that's most likely to trip you up is the variable manipulation. Work carefully, and remember that you can Pick Numbers if you run into trouble with w. The area of the square is $(w + 5)^2$ = w^2 + 10w + 25. The area of the rectangle is (w + 2)(w − 3) = w^2 − w − 6. The area of the square after the rectangle is removed is w^2 + 10w + 25 − (w^2 − w − 6) =

11w + 31, (H). If you had any trouble with the math, Picking Numbers can simplify things. Let w = 5. Then the square has sides of length 10 and an area of 100. The rectangle's sides are length 7 and length 2, so its area is 14. The area of the square after the rectangle is removed is 100 − 14 = 86. Plug w = 5 into each answer choice to find that only (H) equals 86.

57. C

Category: Trigonometry
Difficulty: Medium
Getting to the Answer: First figure out where sine reaches a minimum, then worry about where the 2θ comes in. The sine function first reaches its minimum of −1 at $\frac{3\pi}{2}$, so:

$$2\theta = \frac{3\pi}{2}$$

$$\theta = \frac{3\pi}{4}, (C)$$

58. K

Category: Coordinate Geometry
Difficulty: High
Getting to the Answer: Backsolving is a great option here if you're not sure how to set this problem up algebraically or if you're not confident about your variable manipulation skills. Plug the given coordinates into the distance formula ($\sqrt{(x_1-x_2)^2+(y_1-y_2)^2}$), set the distance equal to 4, and solve for r:

$$\sqrt{(r-10)^2+(6-r)^2} = 4$$
$$(r-10)^2 + (6-r)^2 = 16$$
$$(r^2-20r+100)+(36-12r+r^2) = 16$$
$$2r^2 - 32r + 136 = 16$$
$$2r^2 - 32r + 120 = 0$$
$$r^2 - 16r + 60 = 0$$
$$(r-10)(r-6) = 0$$
$$r = 10 \text{ or } r = 6$$

Only 10 is an answer choice, so (K) is correct. To Backsolve, you still need to know the distance formula to figure out whether F, G, H, and J are correct, but notice that (K) gives the points (10,6) and (10,10). Because the x-coordinate is the same, you can see that the distance between the points is 10 − 6 = 4 without using the formula.

59. D
Category: Proportions and Probability
Difficulty: Medium
Getting to the Answer: This is a great candidate for Backsolving. For each answer, add that number to 5 to find the total. Is the probability of getting a nickel $\frac{1}{6}$? You know that $\frac{1}{6}$ of the total should be 5 nickels. Let x be the total:

$$\frac{1}{6}x = 5$$
$$x = 30$$

If the total is 30, the number of pennies is 30 − 5 = 25, (D).

60. K
Category: Variable Manipulation
Difficulty: High
Getting to the Answer: Each question is worth the same amount, so don't spend too much time on any one. Problems that ask how many different values of something there are tend to be particularly lengthy, so it's a good idea to save them for the end of the test. One way to solve this is to make all of the numerators 3:

$$\frac{3}{15} < \frac{3}{x} < \frac{3}{9}$$

This inequality is true for all values of x between 9 and 15. That's 10, 11, 12, 13, and 14: five integer values of x, making (K) correct. You can also solve this algebraically. The fastest way is to first take the reciprocals, which will reverse the inequalities (Because $\frac{1}{3} < \frac{1}{2}$ but 3 > 2):

$$5 > \frac{x}{3} > 3$$

Multiply everything by 3 to find 15 > x > 9. Again, the five integer values of x are 10, 11, 12, 13, and 14. There are five different integer values of x, so (K) is correct.

SCIENCE TEST

PASSAGE I

1. B
Category: Figure Interpretation
Difficulty: Low
Getting to the Answer: This question represents a case where it's faster to go to the answer choices first. Searching for the lowest value of the 20 in Table 2 is time consuming. Instead, start with the lowest value among the answer choices. Choice D, like C, is far lower than the values in Table 2—in fact, both are values from Table 1. Choice (B) is the lowest value in Table 2. It was recorded in Room 4 on Day 1. Choice A is the highest value in Table 2.

2. G
Category: Figure Interpretation
Difficulty: Low
Getting to the Answer: Don't answer low-difficulty questions like this one from memory. It's worth taking an extra ten seconds to look back at the table to be sure you get the right answer. All of the values in Table 2 are recorded to the nearest 0.1 mm Hg. Choice (G) is correct. Choice F is a trap for those who mistakenly refer to Table 1, in which values are recorded to the nearest 0.01°C.

3. C
Category: Patterns
Difficulty: Medium
Getting to the Answer: Pick one data point at a time and check it against the answer choices. Eliminate any that don't contain that data point. Start by checking the answer choices for the point from Day 1 (745.2 mm Hg, 18.03°C). Eliminate A, B, and D, as they don't contain this point. Only (C) correctly represents this data point, so this must be the correct answer.

4. J
Category: Patterns
Difficulty: Low
Getting to the Answer: For questions that ask you

to describe a data trend in words, be sure to make a prediction before checking the answer choices. Circle the Room 3 column in Table 2, so your eye doesn't accidentally wander. Reading down the column, you can see that the average daily air pressure decreases every day. Choice (J) matches this perfectly. Choice F describes the pressure changes for Room 1. Choice G describes the temperature changes (Table 1) for Room 3. Choice H is the opposite of the correct answer.

5. D

Category: Scientific Reasoning
Difficulty: Medium
Getting to the Answer: Pay close attention to the definitions of new terms introduced in the question stem. Make sure to account for all referenced quantities. Choice (D) is correct because you are given only the temperatures of each room. You know nothing about the exact quantities of heat provided to the room or the amount absorbed by the room's contents. Choices A, B, and C all suggest knowledge of these quantities.

6. F

Category: Figure Interpretation
Difficulty: Low
Getting to the Answer: In Table 1, the room with the highest daily average air temperate is Room 1, so (F) is correct.

PASSAGE II

7. B

Category: Figure Interpretation
Difficulty: High
Getting to the Answer: The last rows of Tables 1 and 2 provide the average Hg concentration for each species of sea creature. To find the difference, subtract the cold-water value (Table 1) from the warm-water value (Table 2). Swordfish, with a difference of 38 ppb, is the sea creature with the greatest difference, (B).

8. G

Category: Scientific Reasoning
Difficulty: Low
Getting to the Answer: Before looking at the choices, determine for yourself what factors would support the given hypothesis or cause it to be rejected. In this case, the hypothesis would be supported only if the results showed higher Hg concentrations in swordfish and shark than in catfish and crab. The results do indeed show higher Hg concentrations for swordfish and shark, so the hypothesis is confirmed as in (G). The "Yes" part of F is correct, but the reason given contradicts the data in Table 2, so it is incorrect. Choice H is also incorrect; the lowest Hg concentration was indeed in catfish, but this does not cause the hypothesisto be rejected. Choice J contradicts the data in Table 2.

9. D

Category: Scientific Reasoning
Difficulty: High
Getting to the Answer: Refer to the passage to understand the process in question on Scientific Method questions. In this case, refer to the description of CVAFS. The passage mentions that CVAFS "indicates the relative concentrations of various elements and compounds." A properly working CVAFS, then, should be able to correctly measure the relative concentrations of Hg and Pb. Using a sample of known concentrations of Hg and Pb, as in (D), would support or reject the accuracy of the CVAFS in detecting the presence of Pb. Nothing in the passage supports either A or C. Choice B is similarly unrelated to the passage.

10. F

Category: Patterns
Difficulty: Medium
Getting to the Answer: Try to find a correlation between any new information given in the question stem and the information in the passage. Because new data on water temperature is introduced, try to find a correlation between water temperature and Hg concentration. Compare the average Hg concentrations in Tables 1 and 2 to see that, for all three

common species (crab, swordfish, and shark), the cold-water specimens have higher Hg concentrations than do the warm-water specimens. Choice (F) is correct because the Northern Atlantic Ocean is the coldest location.

11. C
Category: Scientific Reasoning
Difficulty: Low
Getting to the Answer: The factors intentionally varied are the ones researchers purposefully changed in the course of the experiment. Look back at Table 2. It shows that researchers intentionally tested four different kinds of fish, which matches (C). Choice A is incorrect because the volume of tissue was held constant. Choice B is incorrect because all four species in Experiment 2 were extracted from water of the same temperature. Choice D is incorrect because CVAFS is the only method of analysis mentioned.

12. G
Category: Patterns
Difficulty: Medium
Getting to the Answer: Use Tables 1 and 2 to locate the Hg concentrations for each of the answer choices. A swordfish caught in cold water would have the highest Hg concentration. Choice (G) is correct.

13. D
Category: Scientific Reasoning
Difficulty: Medium
Getting to the Answer: Think about the experimental method used, and try to predict an answer before looking at the choices. Increasing the volume of tested material while maintaining the same volume of Hg in the sample would lead to a smaller fraction of Hg content, regardless of water temperature. Choice (D) is perfect. Choice C is incorrect because nothing in the passage indicates that the contamination would affect warm- and cold-water fish differently.

PASSAGE III

14. G
Category: Patterns
Difficulty: Medium
Getting to the Answer: When a question introduces new data, try to relate it to patterns in the data given. Occasionally, some basic math may be required. The equation in Exercise 2 tells you how to calculate the answer. Z = electromotive force ÷ current = (12 V) ÷ (1.2 A) = 10 Ω, (G). Even if you missed the equation, you can note from Table 2 that Z decreases as current increases and at least eliminate J. Also, if you noticed that Z and current are inversely proportional (either by looking at the equation or Table 2), you could see that *double* the current of Bulb 4 results in *half* the Z of Bulb 4.

15. D
Category: Patterns
Difficulty: High
Getting to the Answer: Challenging pattern analysis questions may require you to do some basic math. Here, noticing the pattern of decreasing Z with increasing current is not enough. You must actually apply the given equation. Because Z = electromotive force ÷ current, current = electromotive force ÷ Z. You have to apply this equation to each answer choice and calculate which gives the highest current. You must use the Z values for each light bulb from Table 2.

For A, current = (10 V) ÷ (60 Ω) = $\frac{1}{6}$ A.

For B, current = (8 V) ÷ (40 Ω) = $\frac{1}{5}$ A.

For C, current = (6 V) ÷ (30 Ω) = $\frac{1}{5}$ A.

For (D), current = (5 V) ÷ (20 Ω) = $\frac{1}{4}$ A. (D), then, gives the largest current.

16. F
Category: Figure Interpretation
Difficulty: Medium
Getting to the Answer: Sometimes, a question may require you to combine a detail from the pas-

sage with data in a figure. Read the question carefully for clues regarding which part of the passage and which table you'll need. According to Table 1, Bulb 3 produced one red indicator light and three green indicator lights. The text above Table 1 explains that green indicators illuminate when light is detected and red indicators illuminate when no light is detected. For Bulb 3, then, only one sensor did not detect any light, (F). Beware of the trap in H for those who confuse the meaning of red and green indicators.

17. C
Category: Figure Interpretation
Difficulty: Medium
Getting to the Answer: This question certainly looks more intimidating than it is. Work methodically and you'll get the answer in no time. First, find the equation for B in the passage. It's listed in *Exercise* 3 as $B = \dfrac{P}{4\pi r^2}$. The text explains that P is the power rating in watts and r is the distance in meters from the bulb. You're given r in the question stem (2 meters), but you'll need to find P. P is listed in Table 3 as 3.6 watts for Bulb 2. Plugging in $P = 3.6$ and $r = 2$ gives you (C). Choice A is the result if you accidentally swap the values of P and r. Choices B and D are the results if you mistakenly use the P for Bulb 1.

18. H
Category: Scientific Reasoning
Difficulty: High
Getting to the Answer: Because you are not permitted to use your calculator during the Science test, there must be a way to estimate seemingly involved calculations. Look for ways to make the math easier than it appears. Because $B = \dfrac{P}{4\pi r^2}$, $P = 4\pi r^2 B$. You're given in the question stem that the bulb is 1 m away, so you can plug $r = 1$ into the equation. You're also given that $B = 0.95$ W/m². Plugging r and B into the equation for P gives: $P = 4\pi(1 \text{ m})^2(0.95 \text{ W/m}^2)$. Because π is slightly larger than 3, and 0.95 is slightly smaller than 1, $P \approx 4(3)(1 \text{ m}^2)(1 \text{ W/m}^2) = 12$ W, (H). Even without calculating,

you can eliminate F and G by comparing B for the fifth bulb with all of the B values from Table 3. 0.95 is just less than double the B value for Bulb 4, which had a P value of 7.2 watts. Because B increases with P, you know your answer must be greater than 7.2.

19. A
Category: Scientific Reasoning
Difficulty: Low
Getting to the Answer: Look first at the tables for the similarities and differences between the two exercises. It's often easier to read information from those than from the text. No light sensors were used in Exercise 2, so (A) is correct. Choice B is incorrect because 4 different light bulbs were used in both exercises. Choice C is incorrect because a 12 V battery was used for both exercises. Choice D is incorrect because Table 2 shows that the current was lowest for Bulb 1.

20. H
Category: Figure Interpretation
Difficulty: Medium
Getting to the Answer: According to Table 3, Bulb 1 and Bulb 2 are not bright enough to consider. Bulb 3's P is greater than 0.30 W/m² and has a lower power rating than Bulb 4. Thus, (H) is the best answer given the criteria.

PASSAGE IV

21. D
Category: Patterns
Difficulty: High
Getting to the Answer: The figure shows how energy relates to the probability of occupation. The passage states that the figure shows the same solid at 3 different temperatures. The hottest solid (25,000 K) is able to reach beyond 14 electron volts, which is greater than both of the other temperature solids. Therefore (D) is the correct choice.

22. J
Category: Scientific Reasoning
Difficulty: High
Getting to the Answer: Don't be intimidated by presentation of unfamiliar science. All three curves happen to intersect at 50%, so look at their slopes at this point of intersection. "Inversely proportional" means that the average kinetic energy is lower for steeper slopes. The 1,000 K curve has the steepest slope and therefore the lowest average kinetic energy. The 10,000 K curve has the next steepest slope and therefore the next lowest kinetic energy, and the 25,000 K curve has the least steep slope and the highest kinetic energy. Choice (J) is correct.

23. A
Category: Patterns
Difficulty: Low
Getting to the Answer: Locate the correct data set and extrapolate beyond the given data by using your pencil to continue the curve. The 1,000 K curve appears to reach 0% at energies above approximately 8 eV. The note under the graph says that the curves all continue to decrease beyond 15 eV, so the value of the 1,000 K curve should still be 0% at an energy of 20 eV. Only (A) agrees with this.

24. J
Category: Patterns
Difficulty: Medium
Getting to the Answer: The correct answer will depict the same trends as the figure in the passage. Identify those trends before you check the answer choices. There are two trends in the figure that accompanies Passage IV: probability of occupation decreases as energy increases, and the curves get shallower as temperature increases. Look for both of these patterns to be represented in the correct answer. You can immediately eliminate F and G, because the curves in these choices increase in value with an increase in energy, behavior opposite to that of the curves in the figure. Choice (J) is perfect because these curves decrease with increasing energy

and because the lowest temperature (2,000 K) curve has the steepest slope, just as the figure shows.

25. C
Category: Figure Interpretation
Difficulty: High
Getting to the Answer: Sometimes a question will ask you to predict the placement of data points that are in between those actually shown on a graph. The answer choices for these questions will be spaced far enough apart so that a rough estimate will be enough. Locate the point corresponding to an 80% probability of occupation of a 5 eV energy state. It lies between the 10,000 K and 25,000 K curves. The temperature of the solid, then, must be between 10,000 K and 25,000 K. Choice (C) must be correct, because it is the only choice in this range.

26. G
Category: Patterns
Difficulty: High
Getting to the Answer: Questions that introduce technical terms can be very confusing. Remember, though, that you don't need to understand what they mean, just how they fit into the question. It doesn't matter whether you've ever heard of the de Broglie wavelength before (or whether you can pronounce it!). Everything you need to know is in the question stem. As the de Broglie wavelength decreases, the energy increases. What happens to probability as energy increases? The figure clearly shows that the values on all three curves decrease with increasing energy. Choice (G), then, is the correct answer. You can eliminate H and J because the figure doesn't show any curve changing from an increase to a decrease, or vice versa, so these can't be correct.

PASSAGE V

27. C
Category: Patterns
Difficulty: Medium
Getting to the Answer: On questions that ask you to generate a graph from a data set, identify a few points and check for the answer choice in which

they all appear. The bars on the graph need to represent the values in the 25°C column of Table 1. Soda C gives the largest value, and Soda D gives the second largest value. Eliminate A and D because neither shows this trend. Choice (C) shows Sodas A, B, and E all very close to each other, and B shows Soda A higher than Sodas B and E. Checking back with Table 1, you'll see that the pressures in Sodas A, B, and E are fairly equally spaced at 25°C, so (C) is correct.

28. G
Category: Patterns
Difficulty: Low
Getting to the Answer: Make sure you understand the meaning of new and unusual diagrams. Start by looking back at Table 1 and jotting down the pressures of Soda C at 0°C (249 kPa) and 50°C (294 kPa). Choice (G) shows the gauge at 249 kPa at 0°C and 294 kPa at 50°C, so this is the correct answer. Choice F shows the opposite relationship between 0°C and 50°C. Beware of H, which shows the values for Soda B.

29. D
Category: Scientific Reasoning
Difficulty: Medium
Getting to the Answer: Pick a few data points that stand out and look for correlations between the two relevant sets of data (in this case, sugar content and solubility). Soda D has the highest sugar content (40 grams per 12 ounces), but, according to Table 2, it has lower solubility than Soda C. This means higher sugar content doesn't always give higher solubility. The theory is not true, so eliminate A and B. Choice C is incorrect because, while Soda D has the highest sugar content, it doesn't have the lowest solubility (Soda E does). Only (D), then, is true.

30. G
Category: Patterns
Difficulty: Low
Getting to the Answer: Read across each row to identify the trend in the data. If you read across each row in Table 2, you'll see that the solubil-

ity values for each soda decrease with increasing temperature, with no exceptions. Choice (G), then, is the correct answer.

31. A
Category: Scientific Reasoning
Difficulty: Medium
Getting to the Answer: Refer to the passage for a description of the experiment in question, paying close attention to detail. Experiment 2 says, "The apparatus consisted of an air-sealed flask containing only soda and no air," and "CO_2 was injected." So you're looking for an apparatus built to allow CO_2 to be injected into a sealed container of soda. Only (A) shows this setup. Choice B is incorrect because it shows the soda immersed in an H_2O bath, which was only the case in Experiment 1. Choice C is incorrect because it shows an apparatus that injects H_2O into soda. Choice D is incorrect because it shows an apparatus that injects CO_2 into H_2O.

32. H
Category: Scientific Reasoning
Difficulty: Medium
Getting to the Answer: Sometimes a question requires you to infer things not directly stated in the passage. Draw only conclusions that MUST follow from the passage. The passage states that the H_2O bath is at room temperature, but soda temperatures are what matter. You can predict that 10 minutes allows the soda to reach the same temperature as the H_2O. Choice (H) correctly explains how the H_2O and soda reach the same temperature. Choices F and G are incorrect because CO_2 and soda evaporation aren't mentioned in Experiment 1. In J the chronology is backwards—you need the pressure gauge to stabilize during the reading of the pressure, not beforehand.

33. B
Category: Figure Interpretation
Difficulty: Low
Getting to the Answer: Table 2 shows that for Soda A, as temperature increases from 0 to 50 degrees Celsius, the value for CO_2 solubility decreases. Elimi-

nate A and D. By checking the other 4 sodas, you can see the same pattern. Choice (B) describes this trend.

PASSAGE VI

34. H
Category: Scientific Reasoning
Difficulty: Medium
Getting to the Answer: Reread where the passage discusses "low energy," paying close attention to the details. The passage states, "The anti conformation is the lowest energy and most stable state of the butane molecule," and "Molecules in the anti or gauche conformations tend to maintain their shape." Choice (H), then is a perfect match. Choices F and G are the opposite of what you're looking for—the low energy molecules *don't* change their shape. Choice J is incorrect because the passage never mentions chemical reactions.

35. C
Category: Scientific Reasoning
Difficulty: Medium
Getting to the Answer: Refer to the passage, paying close attention to detail. Eliminate choices that contradict the passage or ones that don't logically follow from statements made in the passage. The passage states that *"Straight-chain conformational isomers* are carbon compounds that differ only by rotation about one or more single carbon bonds," and "isomers represent the same compound in a slightly different position." The number of carbon bonds, then, must not vary between different isomers of the same compound. Choice (C) matches perfectly. Choices A, B, and D contradict the statements in the passage.

36. F
Category: Scientific Reasoning
Difficulty: Medium
Getting to the Answer: When the question stem says "only Student 2 believes," you know you're looking for an answer choice with which Student 1 would disagree. Student 1 believes that a molecule's active shape is *always* identical to its lowest-energy shape. Student 2 believes that "The active shape of a butane

molecule is dependent upon the energy state of the shape," but also that there are two other factors, namely temperature and initial isomeric state, that affect active shape. Temperature does not appear in the choices, so (F) is the only possibility.

37. B
Category: Scientific Reasoning
Difficulty: High
Getting to the Answer: Don't be intimidated by questions that ask you to apply the logic of the passage to a totally different situation. Try to find structural similarities between the passage and the new situation. The question stem tells you that the two dips in the path represent states of lower energy. The lowest dip is closer to the ground and represents the lowest energy state, and the smaller, higher dip is a higher energy state. The ball is in the dip that is higher off the ground, so check the passage to find the name of the second-lowest energy state. According to the information at the beginning of the passage, the anti conformation is the lowest energy state, and the gauche conformation is the next lowest ("The methyl groups are much closer to each other in the gauche conformation, but this still represents a relative minimum or *meta-stable* state"). Choice (B), then, is the correct answer.

38. F
Category: Scientific Reasoning
Difficulty: Medium
Getting to the Answer: Make sure you understand the fundamentals of both arguments presented in the passage before you tackle this question. Student 1 believes that a molecule's active shape and its lowest-energy shape are the same thing. Therefore, Student 1 believes that all molecules in their active shape are in the anti conformation. This matches (F). Student 2 believes that some of the molecules will settle into the gauche conformation, so you can eliminate H. Choice G is incorrect and the opposite of what Student 1 believes. Choice J is incorrect because Student 2 only believes that some of the molecules will have shapes different from that of their lowest-energy shapes.

39. D

Category: Scientific Reasoning

Difficulty: High

Getting to the Answer: Look for agreements between the graphs and the viewpoint of each student. Student 2 believes that the energy of a molecule's active shape may be slightly higher than that of its most stable shape, while Student 1 believes that a molecule's most stable shape and its active shape are always the same. Choice (D), then, depicts a situation in which the active shape has a higher energy than the most stable shape, so it is the correct answer. Choice C is incorrect because Student 1 believes that the active shape and most stable shape are always the same, and you're looking for a choice Student 1 would disagree with. Both students agree that the energy of the eclipsed conformation is higher than that of either the active or most stable shape, so you could immediately eliminate A and B.

40. G

Category: Scientific Reasoning

Difficulty: High

Getting to the Answer: Be sure to understand an argument thoroughly before attempting to counter it. Student 2 says that a butane molecule may settle into a moderately high-energy conformation. This, he says, is because "in order to convert from the gauche conformation to the anti conformation, the molecule must pass through either the eclipsed or totally eclipsed conformation. If the molecule is not given enough energy to reach either of these states, its active shape will be the gauche conformation." According to Student 2, then, without being given enough energy, the butane molecule can't always settle from the gauche (second-lowest energy) to the anti (lowest energy) state. If (G) was true, and the molecule could get enough energy from the environment to get into the lowest state, then this would counter his argument. Choices H and J are incorrect because hydrogen bonds, carbon bonds, and chemical functions aren't related to Student 2's argument. Choice F is not clearly relevant, and, if anything, supports the argument of Scientist 2.

ACT Practice Test Three
ANSWER SHEET

MATHEMATICS TEST

1. Ⓐ Ⓑ Ⓒ Ⓓ Ⓔ
2. Ⓐ Ⓑ Ⓒ Ⓓ Ⓔ
3. Ⓐ Ⓑ Ⓒ Ⓓ Ⓔ
4. Ⓕ Ⓖ Ⓗ Ⓙ Ⓚ
5. Ⓐ Ⓑ Ⓒ Ⓓ Ⓔ
6. Ⓕ Ⓖ Ⓗ Ⓙ Ⓚ
7. Ⓐ Ⓑ Ⓒ Ⓓ Ⓔ
8. Ⓕ Ⓖ Ⓗ Ⓙ Ⓚ
9. Ⓐ Ⓑ Ⓒ Ⓓ Ⓔ
10. Ⓕ Ⓖ Ⓗ Ⓙ Ⓚ

11. Ⓐ Ⓑ Ⓒ Ⓓ Ⓔ
12. Ⓕ Ⓖ Ⓗ Ⓙ Ⓚ
13. Ⓐ Ⓑ Ⓒ Ⓓ Ⓔ
14. Ⓕ Ⓖ Ⓗ Ⓙ Ⓚ
15. Ⓐ Ⓑ Ⓒ Ⓓ Ⓔ
16. Ⓕ Ⓖ Ⓗ Ⓙ Ⓚ
17. Ⓐ Ⓑ Ⓒ Ⓓ Ⓔ
18. Ⓕ Ⓖ Ⓗ Ⓙ Ⓚ
19. Ⓐ Ⓑ Ⓒ Ⓓ Ⓔ
20. Ⓕ Ⓖ Ⓗ Ⓙ Ⓚ

21. Ⓐ Ⓑ Ⓒ Ⓓ Ⓔ
22. Ⓕ Ⓖ Ⓗ Ⓙ Ⓚ
23. Ⓐ Ⓑ Ⓒ Ⓓ Ⓔ
24. Ⓕ Ⓖ Ⓗ Ⓙ Ⓚ
25. Ⓐ Ⓑ Ⓒ Ⓓ Ⓔ
26. Ⓕ Ⓖ Ⓗ Ⓙ Ⓚ
27. Ⓐ Ⓑ Ⓒ Ⓓ Ⓔ
28. Ⓕ Ⓖ Ⓗ Ⓙ Ⓚ
29. Ⓐ Ⓑ Ⓒ Ⓓ Ⓔ
30. Ⓕ Ⓖ Ⓗ Ⓙ Ⓚ

31. Ⓐ Ⓑ Ⓒ Ⓓ Ⓔ
32. Ⓕ Ⓖ Ⓗ Ⓙ Ⓚ
33. Ⓐ Ⓑ Ⓒ Ⓓ Ⓔ
34. Ⓕ Ⓖ Ⓗ Ⓙ Ⓚ
35. Ⓐ Ⓑ Ⓒ Ⓓ Ⓔ
36. Ⓕ Ⓖ Ⓗ Ⓙ Ⓚ
37. Ⓐ Ⓑ Ⓒ Ⓓ Ⓔ
38. Ⓕ Ⓖ Ⓗ Ⓙ Ⓚ
39. Ⓐ Ⓑ Ⓒ Ⓓ Ⓔ
40. Ⓕ Ⓖ Ⓗ Ⓙ Ⓚ

41. Ⓐ Ⓑ Ⓒ Ⓓ Ⓔ
42. Ⓕ Ⓖ Ⓗ Ⓙ Ⓚ
43. Ⓐ Ⓑ Ⓒ Ⓓ Ⓔ
44. Ⓕ Ⓖ Ⓗ Ⓙ Ⓚ
45. Ⓐ Ⓑ Ⓒ Ⓓ Ⓔ
46. Ⓕ Ⓖ Ⓗ Ⓙ Ⓚ
47. Ⓐ Ⓑ Ⓒ Ⓓ Ⓔ
48. Ⓕ Ⓖ Ⓗ Ⓙ Ⓚ
49. Ⓐ Ⓑ Ⓒ Ⓓ Ⓔ
50. Ⓕ Ⓖ Ⓗ Ⓙ Ⓚ

51. Ⓐ Ⓑ Ⓒ Ⓓ Ⓔ
52. Ⓕ Ⓖ Ⓗ Ⓙ Ⓚ
53. Ⓐ Ⓑ Ⓒ Ⓓ Ⓔ
54. Ⓕ Ⓖ Ⓗ Ⓙ Ⓚ
55. Ⓐ Ⓑ Ⓒ Ⓓ Ⓔ
56. Ⓕ Ⓖ Ⓗ Ⓙ Ⓚ
57. Ⓐ Ⓑ Ⓒ Ⓓ Ⓔ
58. Ⓕ Ⓖ Ⓗ Ⓙ Ⓚ
59. Ⓐ Ⓑ Ⓒ Ⓓ Ⓔ
60. Ⓕ Ⓖ Ⓗ Ⓙ Ⓚ

SCIENCE TEST

1. Ⓐ Ⓑ Ⓒ Ⓓ
2. Ⓕ Ⓖ Ⓗ Ⓙ
3. Ⓐ Ⓑ Ⓒ Ⓓ
4. Ⓕ Ⓖ Ⓗ Ⓙ
5. Ⓐ Ⓑ Ⓒ Ⓓ

6. Ⓕ Ⓖ Ⓗ Ⓙ
7. Ⓐ Ⓑ Ⓒ Ⓓ
8. Ⓕ Ⓖ Ⓗ Ⓙ
9. Ⓐ Ⓑ Ⓒ Ⓓ
10. Ⓕ Ⓖ Ⓗ Ⓙ

11. Ⓐ Ⓑ Ⓒ Ⓓ
12. Ⓕ Ⓖ Ⓗ Ⓙ
13. Ⓐ Ⓑ Ⓒ Ⓓ
14. Ⓕ Ⓖ Ⓗ Ⓙ
15. Ⓐ Ⓑ Ⓒ Ⓓ

16. Ⓕ Ⓖ Ⓗ Ⓙ
17. Ⓐ Ⓑ Ⓒ Ⓓ
18. Ⓕ Ⓖ Ⓗ Ⓙ
19. Ⓐ Ⓑ Ⓒ Ⓓ
20. Ⓕ Ⓖ Ⓗ Ⓙ

21. Ⓐ Ⓑ Ⓒ Ⓓ
22. Ⓕ Ⓖ Ⓗ Ⓙ
23. Ⓐ Ⓑ Ⓒ Ⓓ
24. Ⓕ Ⓖ Ⓗ Ⓙ
25. Ⓐ Ⓑ Ⓒ Ⓓ

26. Ⓕ Ⓖ Ⓗ Ⓙ
27. Ⓐ Ⓑ Ⓒ Ⓓ
28. Ⓕ Ⓖ Ⓗ Ⓙ
29. Ⓐ Ⓑ Ⓒ Ⓓ
30. Ⓕ Ⓖ Ⓗ Ⓙ

31. Ⓐ Ⓑ Ⓒ Ⓓ
32. Ⓕ Ⓖ Ⓗ Ⓙ
33. Ⓐ Ⓑ Ⓒ Ⓓ
34. Ⓕ Ⓖ Ⓗ Ⓙ
35. Ⓐ Ⓑ Ⓒ Ⓓ

36. Ⓕ Ⓖ Ⓗ Ⓙ
37. Ⓐ Ⓑ Ⓒ Ⓓ
38. Ⓕ Ⓖ Ⓗ Ⓙ
39. Ⓐ Ⓑ Ⓒ Ⓓ
40. Ⓕ Ⓖ Ⓗ Ⓙ

MATHEMATICS TEST

60 Minutes—60 Questions

Directions: Solve each of the following problems, select the correct answer, and then fill in the corresponding space on your answer sheet.

Don't linger over problems that are too time-consuming. Do as many as you can, then come back to the others in the time you have remaining.

The use of a calculator is permitted on this test. Though you are allowed to use your calculator to solve any questions you choose, some of the questions may be most easily answered without the use of a calculator.

Note: Unless otherwise noted, all of the following should be assumed.

1. Illustrative figures are *not* necessarily drawn to scale.
2. All geometric figures lie in a plane.
3. The term *line* indicates a straight line.
4. The term *average* indicates arithmetic mean.

1. What is the average of 230, 155, 320, 400, and 325?

   A. 205
   B. 286
   C. 300
   D. 430
   E. 490

2. Sarah has a wooden board that is 12 feet long. If she cuts three 28-inch pieces from the board, how much board will she have left?

   F. 14 inches
   G. 28 inches
   H. 36 inches
   J. 60 inches
   K. 72 inches

3. If $4x + 18 = 38$, then $x =$

   A. 3
   B. 4.5
   C. 5
   D. 12
   E. 20

4. John weighs 1.5 times as much as Ellen. If John weighs 165 lb, how much does Ellen weigh?

   F. 100 lb
   G. 110 lb
   H. 150 lb
   J. 165 lb
   K. 175 lb

5. What is the average of 237, 482, 375, and 210?

   A. 150
   B. 185
   C. 210
   D. 260
   E. 326

6. If $\sqrt[3]{x} = 4$, then $x =$

   F. 4
   G. 12
   H. 36
   J. 64
   K. 256

GO ON TO THE NEXT PAGE

7. If $x^2 + 14 = 63$, then $x =$

   A. 4.5
   B. 7
   C. 14
   D. 24.5
   E. 2.8

8. Which of the following is equivalent to $\sqrt{54}$?

   F. $2\sqrt{3}$
   G. $3\sqrt{6}$
   H. 15
   J. 9
   K. $9\sqrt{6}$

9. What whole number is closest to the solution of $\sqrt{90} \times \sqrt{32}$?

   A. 7
   B. 11
   C. 36
   D. 44
   E. 54

10. $5.2^3 + 6.8^2 =$

    F. 46.24
    G. 94.872
    H. 120.534
    J. 140.608
    K. 186.848

11. If x is a real number such that $x^3 = 512$, what is the value of x^2?

    A. 8
    B. 16
    C. 64
    D. 81
    E. 135

12. $3^3 \div 9 + (6^2 - 12) \div 4 =$

    F. 3
    G. 6.75
    H. 9
    J. 12
    K. 15

13. If bananas cost \$0.24 and oranges cost \$0.38, what is the total cost of x bananas and y oranges?

    A. $(x + y)(\$0.24 + \$0.38)$
    B. $\$0.24x + \$0.38y$
    C. $\$0.62(x + y)$
    D. $\dfrac{\$0.24}{x} + \dfrac{\$0.38}{y}$
    E. $\$0.38x + \$0.24y$

14. If $4x + 13 = 16$, what is the value of x?

    F. 0.25
    G. 0.50
    H. 0.55
    J. 0.70
    K. 0.75

15. What is 6% of 1,250?

    A. 75
    B. 208
    C. 300
    D. 500
    E. 750

GO ON TO THE NEXT PAGE ⇨

16. On her first three geometry tests, Sarah scored an 89, a 93, and an 84. If there are four tests total and Sarah needs at least a 90 average for the four, what is the lowest score she can receive on the final test?

    F. 86
    G. 90
    H. 92
    J. 94
    K. 96

17. What is the solution set of $3x - 11 \geq 22$?

    A. $x \geq -11$
    B. $x < -3$
    C. $x \geq 0$
    D. $x > 3$
    E. $x \geq 11$

18. The eighth grade girls' basketball team played a total of 13 games this season. If they scored a total of 364 points, what was their average score per game?

    F. 13
    G. 16
    H. 20
    J. 28
    K. 32

19. If $6x + 4 = 11x - 21$, what is the value of x?

    A. 2
    B. 3
    C. 4
    D. 5
    E. 6

20. The school band has a collection of 300 pieces of music. Of these, 10% are movie theme songs. Out of the rest of the pieces of music, 80 of the pieces are marches. How many of the band's pieces of music are neither marches nor movie theme songs?

    F. 190
    G. 198
    H. 210
    J. 220
    K. 270

21. A cooking class has 20 spaces available for each daily session. Data showed that 19 people attended the first session, 17 people attended the second session, and 15 people attended each of the remaining sessions. If the average number of attendees was exactly 16 per class session, how many total sessions of the cooking class were there?

    A. 3
    B. 4
    C. 6
    D. 11
    E. Cannot be determined from the given information

22. In the following figure, all of the small triangles are the same size. What percent of the entire figure is shaded?

    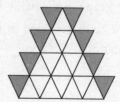

    F. 8
    G. 24
    H. $33\frac{1}{3}$
    J. 50
    K. $66\frac{2}{3}$

GO ON TO THE NEXT PAGE

23. A piece of letter-sized paper is $8\frac{2}{3}$ inches wide and 11 inches long. Suppose you want to cut strips of paper that are $\frac{5}{8}$ inch wide and 11 inches long. What is the maximum number of strips of paper you could make from 1 piece of letter-sized paper?

  A. 5
  B. 6
  C. 12
  D. 13
  E. 14

24. A scientist was studying a meadow and the birds that lived in the meadow. He kept a count of the birds that appeared in the meadow by tagging them so that the individual birds could be distinguished from one another. There are only three types of birds that live in the meadow: buntings, larks, and sparrows. He found that the ratio of buntings to total birds in the meadow was 35:176, while the ratio of larks to total birds was 5:11. If the scientist randomly chooses one individual bird that he had previously counted, which type of bird is he most likely to choose?

  F. Bunting
  G. Lark
  H. Sparrow
  J. All bird types are equally likely
  K. Cannot be determined from the given information

25. For all x, $(x + 4)(x - 4) + (2x + 2)(x - 2) = ?$

  A. $x^2 - 2x - 20$
  B. $3x^2 - 12$
  C. $3x^2 - 2x - 20$
  D. $3x^2 + 2x - 20$
  E. $3x^2 + 2x + 20$

26. Which of these is equivalent to $(4x - 1)(x + 5)$?

  F. $4x^2 + 8x$
  G. $4x^2 - 10x - 5$
  H. $4x^2 - 15x + 5$
  J. $4x^2 + 19x - 5$
  K. $4x^2 + 19x + 5$

27. In the following triangle, if $\cos \angle BAC = 0.6$ and the hypotenuse of the triangle is 15, what is the length of side BC?

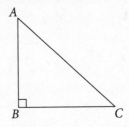

  A. 3
  B. 5
  C. 10
  D. 12
  E. 15

28. If a car drives 80 miles per hour for x hours and 60 miles per hour for y hours, what is the car's average speed, in miles, for the total distance traveled?

  F. $\dfrac{480}{xy}$

  G. $\dfrac{80}{x} + \dfrac{60}{y}$

  H. $\dfrac{80}{x} \times \dfrac{60}{y}$

  J. $\dfrac{80x + 60y}{x + y}$

  K. $\dfrac{x + y}{80x + 60y}$

GO ON TO THE NEXT PAGE

29. Four numbers are in a sequence with 8 as its first term and 36 as its last term. The first 3 numbers are in an arithmetic sequence with a common difference of –7. The last 3 numbers are in a geometric sequence. What is the common ratio of the last 3 terms of the sequence?

   A. –10

   B. –6

   C. 0

   D. 10

   E. 32

30. If $3^{3x+3} = 27^{\frac{2}{3}x - \frac{1}{3}}$, then $x = $?

   F. –4

   G. $-\dfrac{7}{4}$

   H. $-\dfrac{10}{7}$

   J. 2

   K. 4

31. A scientist had a container of liquid nitrogen that was at a temperature of –330°F. If the temperature of the room was 72°F, how much must the temperature of the liquid nitrogen change to become the room's temperature? (Note: "+" indicates a rise in temperature, and "–" indicates a drop in temperature.)

   A. –330°F

   B. –258°F

   C. +72°F

   D. +402°F

   E. +474°F

32. A playground is $(x + 7)$ units long and $(x + 3)$ units wide. If a square of side length x is sectioned off from the playground to make a sandpit, which of the following could be the remaining area of the playground?

   F. $x^2 + 10x + 21$

   G. $10x + 21$

   H. $2x + 10$

   J. 21

   K. $21x$

33. If u is an integer, then $(u - 3)^2 + 5$ must be:

   A. an even integer.

   B. an odd integer.

   C. a positive integer.

   D. a negative integer.

   E. Cannot be determined by the information given.

34. The point $(-3,-2)$ is the midpoint of the line segment in the standard (x,y) coordinate plane joining the point $(1,9)$ and (m,n). Which of the following is (m,n) ?

   F. $(-7,-13)$

   G. $(-1,7)$

   H. $(-2,5.5)$

   J. $(2,5.5)$

   K. $(5,20)$

35. If $f(x) = \dfrac{1}{3}x + 13$ and $g(x) = 3x^2 + 6x + 12$, what is the value of $f(g(x))$?

   A. $x^2 + 12x + 4$

   B. $\dfrac{x^2}{3} + 2x + 194$

   C. $x^2 + 2x + 17$

   D. $x^2 + 2x + 25$

   E. $x^2 + 2x + 54$

GO ON TO THE NEXT PAGE

36. What is the length of side *AC* in triangle *ABC* graphed on the following coordinate plane?

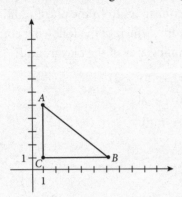

  F. 3
  G. 4
  H. 5
  J. 6
  K. 7

37. If $f(x) = 16x^2 - 20x$, what is the value of $f(3)$?

  A. −12
  B. 36
  C. 84
  D. 144
  E. 372

38. What is the equation of a line that is perpendicular to the line $y = \dfrac{2}{3}x + 5$ and contains the point $(4,-3)$?

  F. $y = \dfrac{2}{3}x + 4$

  G. $y = -\dfrac{2}{3}x + 3$

  H. $y = -\dfrac{3}{2}x + 3$

  J. $y = -\dfrac{3}{2}x - 9$

  K. $y = -\dfrac{3}{2}x + 9$

39. What is the slope of the line through the points $(-10,0)$ and $(0,-6)$?

  A. $-\dfrac{5}{3}$

  B. $-\dfrac{3}{5}$

  C. $\dfrac{3}{5}$

  D. $\dfrac{5}{3}$

  E. 0

40. The figure on the coordinate plane is a rectangle. What coordinate pair corresponds to point *B*?

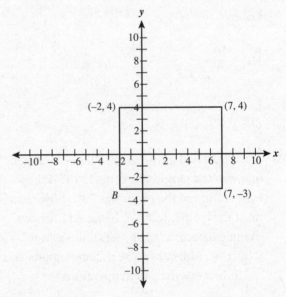

  F. $(-2,-3)$
  G. $(-2,4)$
  H. $(2,-3)$
  J. $(-3,4)$
  K. $(3,-2)$

GO ON TO THE NEXT PAGE

41. What is the length of a line segment with endpoints (3,−6) and (−2,6)?

    A. 1
    B. 5
    C. 10
    D. 13
    E. 15

42. What is the midpoint of the line segment in the following graph?

    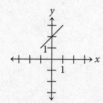

    F. (0,1)
    G. (0,2)
    H. (1,2)
    J. (1,1)
    K. (1,3)

43. What is the volume of a sphere with a diameter of 6?

    A.   3π
    B.   9π
    C.   27π
    D.   36π
    E.   288π

44. A rectangle has a side length of 8 and a perimeter of 24. What is the area of the rectangle?

    F. 16
    G. 24
    H. 32
    J. 64
    K. 96

45. Isosceles triangle *ABC* has an area of 48. If $\overline{AB}$ = 12, what is the perimeter of *ABC*?

    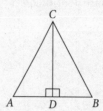

    A. 32
    B. 36
    C. 48
    D. 64
    E. 76

46. The rectangular backyard of a house is 130 feet by 70 feet. If the backyard is completely fenced in, what is the length, in feet, of the fence?

    F. 130
    G. 200
    H. 260
    J. 400
    K. 420

47. In the following figure, lines *m* and *l* are parallel and ∠*a* = 68°. What is the measure of ∠*f*?

    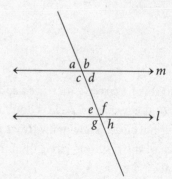

    A.   22°
    B.   68°
    C.   80°
    D.   112°
    E.   292°

GO ON TO THE NEXT PAGE ▷

48. Square *ABCD* is shown, with one side measuring 6 centimeters. What is the area of triangle *BCD*, in square centimeters?

    F. 3

    G. 6

    H. 12

    J. 18

    K. 36

49. The hypotenuse of right triangle *RST* is 16. If the measure of $\angle R = 30°$, what is the length of *RS*?

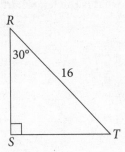

    A. 4

    B. 8

    C. $8\sqrt{3}$

    D. 12

    E. 16

50. The radius of a circle is increased so that the radius of the resulting circle is double that of the original circle. How many times larger is the area of the resulting circle than that of the original circle?

    F. 0.5

    G. 1

    H. 2

    J. π

    K. 4

51. What is the length of the diagonal of a square with sides of length 7?

    A. 7

    B. $7\sqrt{2}$

    C. 14

    D. 21

    E. 28

52. What is the perimeter of a regular hexagon with a side of 11?

    F. 33

    G. 44

    H. 66

    J. 72

    K. 78

53. A rectangle has a perimeter of 28, and its longer side is 2.5 times the length of its shorter side. What is the length of the diagonal of the rectangle, rounded to the nearest tenth?

    A. 4.0

    B. 10.0

    C. 10.8

    D. 12.4

    E. 14.2

GO ON TO THE NEXT PAGE ⇨

54. In the following figure, $\overline{MN}$ and $\overline{PQ}$ are parallel. Point A lies on MN, and points B and C lie on $\overline{PQ}$. If $AB = AC$ and $\angle MAB = 55°$, what is the measure of $\angle ACB$?

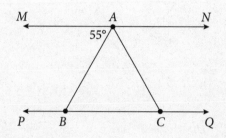

F. 35°

G. 55°

H. 65°

J. 80°

K. 125°

55. The following chord is 8 units long. If the chord is 3 units from the center of the circle, what is the area of the circle?

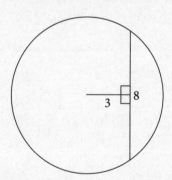

A. 9π

B. 16π

C. 18π

D. 25π

E. 28π

56. If isosceles triangle QRS has a base of length 16 and sides of length 17, what is the area of the triangle?

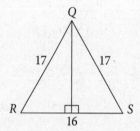

F. 50

G. 80

H. 110

J. 120

K. 180

57. In a high school senior class, the ratio of girls to boys is 5:3. If there are a total of 168 students in the senior class, how many girls are there?

A. 63

B. 100

C. 105

D. 147

E. 152

GO ON TO THE NEXT PAGE

58. What is the tangent of ∠EFD shown?

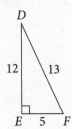

F. $\dfrac{5}{13}$

G. $\dfrac{5}{12}$

H. $\dfrac{12}{13}$

J. $\dfrac{13}{12}$

K. $\dfrac{12}{5}$

59. In the following triangle, what is the value of sin ∠QRS?

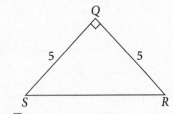

A. $\dfrac{\sqrt{2}}{6}$

B. $\dfrac{\sqrt{2}}{5}$

C. $\dfrac{\sqrt{2}}{2}$

D. $2\sqrt{2}$

E. $5\sqrt{2}$

60. In the following right triangle, JL = 17 and KL = 8. What is the value of sin ∠JLK?

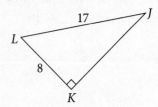

F. $\dfrac{8}{15}$

G. $\dfrac{8}{17}$

H. $\dfrac{8}{20}$

J. $\dfrac{3}{4}$

K. $\dfrac{15}{17}$

SCIENCE TEST

35 Minutes—40 Questions

Directions: There are several passages in this test. Each passage is followed by several questions. After reading a passage, choose the best answer to each question and fill in the corresponding oval on your answer document. You may refer to the passages as often as necessary. You are NOT permitted to use a calculator on this test.

PASSAGE I

A *binary star system* consists of two stars that are gravitationally bound to each other. If two stars that orbit each other are viewed along a line of sight that is not perpendicular to the orbital plane, they will alternately appear to eclipse each other. The orbit of *eclipsing binary* System Q is shown in Figure 1.

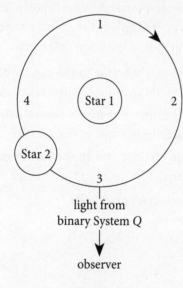

Figure 1

Notes: Diagram is not drawn to scale. Star 1 is brighter than Star 2.

Astronomers deduce that a given star is an eclipsing binary from its *light curve*—the plot of its surface brightness (observed from a fixed position) against time. The light curve of an eclipsing binary typically displays a deep primary minimum and a shallower secondary minimum. Figure 2 shows the light curve of System Q.

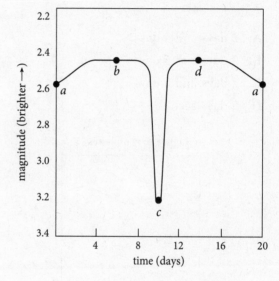

Figure 2

1. The point on the light curve labeled *c* corresponds to the position in Figure 1 labeled:

   A. 1.

   B. 2.

   C. 3.

   D. 4.

GO ON TO THE NEXT PAGE

2. The period of revolution for eclipsing binary
   Q is about:

   F.   4 days.
   G. 10 days.
   H. 12 days.
   J. 20 days.

3. The stars in eclipsing binary Q alternately
   eclipse each other for periods of approximately:

   A. 2 days and 4 days.
   B. 2 days and 5 days.
   C. 2 days and 8 days.
   D. 5 days each.

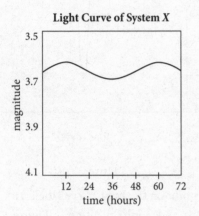

Light Curve of System X

4. The light curves for two eclipsing binaries,
   Systems X and Z, are shown at the left. Which
   of the following hypotheses would account for
   the deeper primary minimum of System Z?

   F.   There is a more extreme difference
        between the magnitudes of the two stars
        of System X than between those of the
        two stars of System Z.

   G.   There is a more extreme difference
        between the magnitudes of the two stars
        of System Z than between those of the
        two stars of System X.

   H.   System X has a longer period of revolu-
        tion than does System Z.

   J.   System Z has a longer period of
        revolution than does System X.

5. The greatest total brightness shown on the
   light curve of an eclipsing binary system
   corresponds to the point in the orbit when:

   A. the brighter star in the binary pair is
       directly in front of the darker star.

   B. the larger star in the binary pair is
       directly in front of the smaller star.

   C. the brighter star in the binary pair is
       directly in front of the smaller star.

   D. both stars are visible.

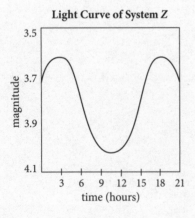

Light Curve of System Z

GO ON TO THE NEXT PAGE ▷

PASSAGE II

The utilization and replenishing of the Earth's carbon supply is a cyclic process involving all living matter. This cycle is shown here.

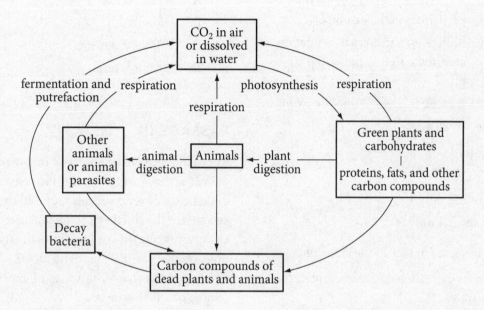

6. Carbon fixation involves the removal of CO_2 from the air and the incorporation of various compounds. According to the cycle shown, carbon fixation can form which compound(s)?

    I. Fat

    II. Starch

    III. Protein

F. II only

G. I and II only

H. II and III only

J. I, II, and III

7. What effect would a sudden drop in the amount of the Earth's decay bacteria have on the amount of carbon dioxide in the atmosphere?

A. The CO_2 level will drop to a life-threatening level since the bacteria is the sole source of CO_2.

B. The CO_2 level will rise because the bacteria usually consume CO_2.

C. The CO_2 level may decrease slightly, but there are other sources of CO_2.

D. The CO_2 level will increase slightly due to an imbalance in the carbon cycle.

GO ON TO THE NEXT PAGE

8. Which of the following statements are consistent with the carbon cycle as presented in the diagram?

    I. A non-plant-eating animal does not participate in the carbon cycle.

    II. Both plant and animal respiration contribute CO_2 to the earth's atmosphere.

    III. All CO_2 is released into the air by respiration.

  F. I only

  G. II only

  H. I and II only

  J. II and III only

9. A direct source of CO_2 in the atmosphere is:

  A. the fermentation of green-plant carbohydrates.

  B. the photosynthesis of tropical plants.

  C. the digestion of plant matter by animals.

  D. the respiration of animal parasites.

10. Which of the following best describes the relationship between animal respiration and photosynthesis?

  F. Respiration and photosynthesis serve the same function in the carbon cycle.

  G. Animal respiration provides vital gases for green plants.

  H. Animal respiration prohibits photosynthesis.

  J. There is no relationship between respiration and photosynthesis.

11. The elimination of which of the following would cause the earth's carbon cycle to grind to a complete halt?

  A. Green plants

  B. Animals

  C. Animal digestion

  D. Decay bacteria

PASSAGE III

Microbiologists have observed that certain species of bacteria are magnetotactic, i.e., sensitive to magnetic fields. Several species found in the bottom of swamps in the Northern Hemisphere tend to orient themselves toward magnetic north (the northern pole of the earth's magnetic field). Researchers conducted the following series of experiments on magnetotactic bacteria.

STUDY 1

A drop of water filled with magnetotactic bacteria was observed under high magnification. The direction of the first 500 bacterial migrations across the field of view was observed for each of five trials and the tally for each trial recorded in Table 1. Trial 1 was conducted under standard laboratory conditions. In Trial 2, the microscope was shielded from all external light and electric fields. In Trials 3 and 4, the microscope was rotated clockwise 90° and 180°, respectively. For Trial 5, the microscope was moved to another laboratory at the same latitude.

GO ON TO THE NEXT PAGE ⟹

Table 1

Trial #	Direction			
	North	**East**	**South**	**West**
1	474	7	13	6
2	481	3	11	5
3	479	4	12	5
4	465	9	19	7
5	480	3	11	6

STUDY 2

The north pole of a permanent magnet was positioned near the microscope slide. The magnet was at the 12:00 position for Trial 1 and was moved 90° clockwise for each of three successive trials. All other conditions were as in Trial 1 of Study 1. The results were tallied and recorded in Table 2.

Table 2

Trial #	Direction			
	12:00	**3:00**	**6:00**	**9:00**
1	470	6	15	9
2	8	483	3	6
3	17	4	474	5
4	5	19	9	467

12. The theory that light was not the primary stimulus affecting the direction of bacterial migration is:

   F. supported by a comparison of the results of Studies 1 and 2.

   G. supported by a comparison of the results of Trials 1 and 2 of Study 1.

   H. supported by a comparison of the results of Trials 3 and 4 of Study 1.

   J. not supported by any of the results noted in the passage.

13. If the south pole of the permanent magnet used in Study 2 had been placed near the microscope slide, what would the most likely result have been?

   A. The figures for each trial would have remained approximately the same, since the strength of the magnetic field would be unchanged.

   B. The bacteria would have become disoriented, with approximately equal numbers moving in each direction.

   C. The major direction of travel would have shifted by 180° because of the reversed direction of the magnetic field.

   D. The bacteria would still have tended to migrate toward Earth's magnetic north, but would have taken longer to orient themselves.

14. It has been suggested that magnetic sensitivity helps magnetotactic bacteria orient themselves downward. Such an orientation would be most advantageous from an evolutionary standpoint if:

   F. organisms that consume magnetotactic bacteria were mostly bottom-dwellers.

   G. the bacteria could only reproduce by migrating upwards to the water's surface.

   H. bacteria that stayed in the top layers of water tended to be dispersed by currents.

   J. the nutrients necessary for the bacteria's survival were more abundant in bottom sediments.

GO ON TO THE NEXT PAGE

15. Researchers could gain the most useful new information about the relationship between magnetic field strength and bacterial migration by repeating Study 2 with:

    A. incremental position changes of less than 90°.

    B. a magnet that rotated slowly around the slide in a counterclockwise direction.

    C. more and less powerful magnets.

    D. larger and smaller samples of bacteria.

16. Which of the following statements is supported by the results of Study 1?

    F. The majority of magnetotactic bacteria migrate toward the Earth's magnetic north pole.

    G. The majority of magnetotactic bacteria migrate toward the north pole of the nearest magnet.

    H. The majority of magnetotactic bacteria migrate toward the 12:00 position.

    J. The effect of the Earth's magnetic field on magnetotactic bacteria is counteracted by electric fields.

17. What is the control in Study 1?

    A. Trial 1
    B. Trial 2
    C. Trial 3
    D. Trial 5

18. If each diagram below represents a microscopic field, which diagram best reflects the results of Trial 2 from Study 2?

    F.

    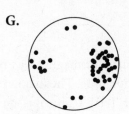

    G.

    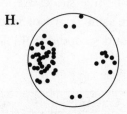

    H.

    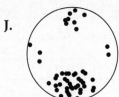

    J.

PASSAGE IV

A process has been developed by which plastic bottles can be recycled into a clear, colorless material. This material, called *nu-PVC*, can be used to form park benches and other similar structures. A series of experiments was performed to determine the weathering abilities of *nu-PVC*.

EXPERIMENT 1

Fifteen boards of *nu-PVC*, each 150 × 25 × 8 cm in size, were sprayed with distilled water for 10 hours a day for 32 weeks. All 15 boards remained within 0.1 cm of their original dimensions. The surfaces of the boards displayed no signs of cracking, bubbling, or other degradation.

EXPERIMENT 2

Fifteen sheets of *nu-PVC*, each 2 m × 2 m × 5 cm, were hung in a chamber in which the humidity and temperature were held constant. The sheets were irradiated with ultraviolet light for 12 hours a day for 32 weeks. At the end of the experiment, the sheets' flexibility had decreased to an average of 17.5%. The surface of the sheets showed no signs of degradation, but they had all become milky white in color. Results from this experiment are shown in Figure 1.

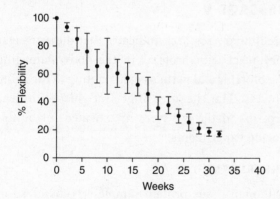

Figure 1

EXPERIMENT 3

Fifteen boards, as in Experiment 1, were each found to be capable of supporting an average of 963 pounds for 15 days without breaking or bending. These same boards were then kept for 32 weeks at temperatures ranging from 5°C to –15°C. At the end of the experiment, the boards were able to support an average of only 400 pounds without bending or breaking. This data is shown in Figure 2.

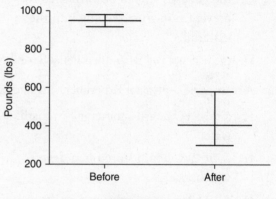

Figure 2

19. According to Figure 1, what would be the estimated percent flexibility if a sheet of *nu-PVC* was measured at 40 weeks?

    A. 3%

    B. 12%

    C. 20%

    D. 22.5%

20. Based on the results of Experiment 1, which of the following conclusions concerning the effect of rain on *nu-PVC* is valid?

    F. The material absorbs water over time, causing it to permanently swell.

    G. The material will be useful only in areas where there is no acid rain.

    H. The material's surface does not appear to require a protective coating to avoid water damage.

    J. The material loses flexibility after prolonged exposure to precipitation.

GO ON TO THE NEXT PAGE

21. Park benches in an often snow-covered region of Minnesota are to be replaced with new ones, made of *nu-PVC*. The experimental results indicate that:

   A. the benches will suffer little degradation due to weathering.

   B. the benches would have to be stored indoors during the winter to retain their initial strength.

   C. the benches should be varnished to prevent rain from seeping into the material.

   D. the benches will retain their initial flexibility.

22. A reasonable control for Experiment 1 would be:

   F. a *nu-PVC* board submerged in distilled water.

   G. a *nu-PVC* board stored in a dry warehouse.

   H. a wooden board subjected to the same conditions.

   J. a wooden board stored in a dry warehouse.

23. The purpose of the ultraviolet radiation used in Experiment 2 was to:

   A. simulate the effects of sunlight.

   B. avoid damage to the material's finish.

   C. turn the boards a uniform color.

   D. test the strength of the *nu-PVC*.

24. From the information given, it can be inferred that *nu-PVC*'s advantage over standard building materials, such as wood, is that it:

   F. is heavier and denser than other materials.

   G. can be developed in different colors and textures.

   H. is less subject to structural cracking and failure.

   J. is made from recycled plastic wastes.

25. Which of the following experiments would be the most likely to provide useful information concerning the weathering of *nu-PVC*?

   I. Repeating Experiment 1, increasing the length of the experiment from 32 weeks to 64 weeks

   II. Investigating the effects of sea water and salt-rich air on the material

   III. Repeating Experiment 3, decreasing the minimum temperature from $-15°C$ to $-40°C$

   A. II only

   B. III only

   C. I and II only

   D. I, II, and III

26. According to Figure 2, what is the approximate range of weight that the boards could hold after being kept in the cold?

   F. 100 lbs.

   G. 300 lbs.

   H. 563 lbs.

   J. 650 lbs.

PASSAGE V

Preliminary research indicates that dietary sugar may react with proteins in the body, damaging the proteins and perhaps contributing to the aging process. The chemical effects of glucose on lens proteins in the eye were investigated in the following experiments.

EXPERIMENT 1

A human tissue protein sample was dissolved in a glucose and water solution, resulting in a clear, yellow solution. After 30 minutes, the solution became opaque. Spectrographic analysis (Figure 1) revealed that an *Amadori* product had formed on the protein. It was determined that the *Amadori*

GO ON TO THE NEXT PAGE

products on one protein had combined with free amino groups on nearby proteins, forming brown pigmented cross-links between the two proteins. The cross-links are termed advanced glycosylation end products (AGE).

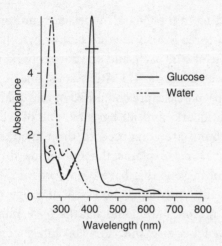

Figure 1

EXPERIMENT 2

Forty-six samples of human lens proteins taken from subjects ranging in age from 12–80 years were studied under an electron microscope. The lens proteins in the samples from older subjects occurred much more often in aggregates formed by cross-linked bonds than did the lens proteins in the samples from younger subjects. Fluorescent characteristics revealed the cross-links to be of two types: disulfide bonds and an indeterminate formation with brownish pigmentation.

EXPERIMENT 3

Two solutions containing lens proteins from cow lenses were prepared, one with glucose and one without. Only the glucose solution turned opaque. Analysis revealed that the lens proteins in the glucose solution had formed pigmented cross-links with the brownish color and fluorescence characteristics of those observed in Experiment 1.

27. According to Figure 1, what is the wavelength of the maximum absorbance of *Amadori* products?

   A. 250 nm

   B. 300 nm

   C. 350 nm

   D. 400 nm

28. It was assumed in the design of Experiment 3 that cow lens proteins:

   F. have a brownish pigment.

   G. react with sulfides.

   H. remain insoluble in water.

   J. react similarly to human lens proteins.

29. Based on the results in Experiment 1 only, it can be concluded that:

   A. proteins can form disulfide cross-links.

   B. glucose dissolved in water forms AGE.

   C. glucose can react with proteins to form cross-links.

   D. Amadori products are a result of glucose metabolism.

30. As people age, the lenses in their eyes sometimes turn brown and cloudy (known as senile cataracts). Based on this information and the results of Experiment 2, which of the following hypotheses is the most likely to be valid?

   F. As people age, the amount of sulfur contained in lens proteins increases.

   G. Senile cataracts are caused by cross-linked bonds between lens proteins.

   H. Lens proteins turn brown with age.

   J. Older lens proteins are more fluorescent than younger lens proteins.

GO ON TO THE NEXT PAGE

31. Which of the following hypotheses about the brown pigmented cross-links observed in Experiment 2 is best supported by the results of the three experiments?

    A. Their brownish color is caused by disulfide bonds.

    B. They are a natural formation which can be found at birth.

    C. They are caused by glucose in the diet reacting with lens proteins.

    D. They form when proteins are dissolved in water.

32. Based on the experimental results, lens proteins from a 32-year-old man would most likely have:

    F. more cross-links than lens proteins from a 32-year-old woman.

    G. more cross-links than lens proteins from an 18-year-old cow.

    H. more cross-links than lens proteins from an 18-year-old man.

    J. fewer cross-links than lens proteins from an 80-year-old man.

33. People with uncontrolled diabetes have excess levels of blood glucose. Based on this information and the results of the experiments, a likely symptom of advanced diabetes would be:

    A. senile cataracts, due to an increase of free amino groups in the urine.

    B. senile cataracts, due to glucose interacting with disulfide cross-links on lens proteins.

    C. senile cataracts, due to AGE cross-links of lens proteins.

    D. kidney failure, due to high levels of free amino groups in the urine.

PASSAGE VI

Two scientists discuss their views about the Quark Model.

SCIENTIST 1

According to the Quark Model, each proton consists of three quarks: two up quarks, which carry a charge of +2/3 each, and one down quark, which carries a charge of −1/3. All mesons, one of which is the p+ particle, are composed of one quark and one antiquark, and all baryons, one of which is the proton, are composed of three quarks. The Quark Model explains the numerous different types of mesons that have been observed. It also successfully predicted the essential properties of the Y meson. Individual quarks have not been observed because they are absolutely confined within baryons and mesons. However, the results of deep inelastic scattering experiments indicate that the proton has a substructure. In these experiments, high-energy electron beams were fired into protons. While most of the electrons incident on the proton passed right through, a few bounced back. The number of electrons scattered through large angles indicated that there are three distinct lumps within the proton.

SCIENTIST 2

The Quark Model is seriously flawed. Conventional scattering experiments should be able to split the proton into its constituent quarks, if they existed. Once the quarks were free, it would be easy to distinguish quarks from other particles using something as simple as the Millikan oil drop experiment because they would be the only particles that carry fractional charge. Furthermore, the lightest quark would be stable because there is no lighter particle for it to decay into. Quarks would be so easy to produce, identify, and store that they would have been detected if they truly existed. In addition, the Quark Model violates the Pauli exclusion principle which originally was believed

GO ON TO THE NEXT PAGE ⟶

to hold for electrons but was found to hold for all particles of half-integer spin. The Pauli exclusion principle states that no two particles of half-integer spin can occupy the same state. The Δ^{++} baryon which supposedly has three up quarks violates the Pauli exclusion principle because two of those quarks would be in the same state. Therefore, the Quark Model must be replaced.

34. Which of the following would most clearly strengthen Scientist 1's hypothesis?

    F. Detection of the Δ^{++} baryon

    G. Detection of a particle with fractional charge

    H. Detection of mesons

    J. Detection of baryons

35. Which of the following are reasons why Scientist 2 claims quarks should have been detected, if they existed?

    I. They have a unique charge.

    II. They are confined within mesons and baryons.

    III. They are supposedly fundamental particles, and so could not decay into any other particle.

    A. I only

    B. II only

    C. I and III only

    D. I, II, and III

36. Which of the following could Scientist 1 use to counter Scientist 2's point about the Pauli exclusion principle?

    F. Evidence that quarks do not have half-integer spin

    G. Evidence that the Δ^{++} baryon exists

    H. Evidence that quarks have fractional charge

    J. Evidence that quarks have the same spin as electrons

37. If Scientist 1's hypothesis is correct, the Δ^{++} baryon should have a charge of:

    A. −1

    B. 0

    C. 1

    D. 2

38. Scientist 2 says the Quark Model is flawed because:

    F. the existence of individual baryons cannot be experimentally verified.

    G. the existence of individual quarks cannot be experimentally verified.

    H. particles cannot have fractional charge.

    J. it doesn't include electrons as elementary particles.

39. Scientist 1 says that some high-energy electrons that were aimed into the proton in the deep inelastic scattering experiments bounced back because they:

    A. hit quarks.

    B. hit other electrons.

    C. were repelled by the positive charge on the proton.

    D. hit baryons.

40. The fact that deep inelastic scattering experiments revealed a proton substructure of three lumps supports the Quark Model because:

    F. protons are mesons, and mesons supposedly consist of three quarks.

    G. protons are mesons, and mesons supposedly consist of a quark and an antiquark.

    H. protons are baryons, and baryons supposedly consist of three quarks.

    J. protons are baryons, and baryons supposedly consist of one quark and one antiquark.

IF YOU FINISH BEFORE TIME IS CALLED, YOU MAY CHECK YOUR WORK ON THIS SECTION ONLY. DO NOT TURN TO ANY OTHER SECTION IN THE TEST.   **STOP**

Practice Test Three
ANSWER KEY

MATHEMATICS TEST

1. B	9. E	17. E	25. C	33. C	41. D	49. C	57. C
2. J	10. K	18. J	26. J	34. F	42. G	50. K	58. K
3. C	11. C	19. D	27. D	35. C	43. D	51. B	59. C
4. G	12. H	20. F	28. J	36. G	44. H	52. H	60. K
5. E	13. B	21. C	29. B	37. C	45. A	53. C	
6. J	14. K	22. H	30. F	38. H	46. J	54. G	
7. B	15. A	23. D	31. D	39. B	47. D	55. D	
8. G	16. J	24. G	32. G	40. F	48. J	56. J	

SCIENCE TEST

1. C	6. J	11. A	16. F	21. B	26. G	31. C	36. F
2. J	7. C	12. G	17. A	22. G	27. D	32. J	37. D
3. C	8. G	13. C	18. G	23. A	28. J	33. C	38. G
4. G	9. D	14. J	19. B	24. J	29. C	34. G	39. A
5. D	10. G	15. C	20. H	25. A	30. G	35. C	40. H

ANSWERS AND EXPLANATIONS

MATHEMATICS TEST

1. B
Category: Proportions and Probability
Difficulty: Low
Getting to the Answer: Plug the terms into the average formula and solve:

$$\text{Average} = \frac{\text{Sum of terms}}{\text{Number of terms}}$$

$$= \frac{230 + 155 + 320 + 400 + 325}{5}$$

$$= \frac{1,430}{5}$$

$$= 286$$

That's (B).

2. J
Category: Operations
Difficulty: Medium
Getting to the Answer: Begin by converting Sarah's 12 feet of wood into inches. There are 12 inches in a foot, so Sarah has 12×12 inches = 144 inches. Cutting off three 28-inch pieces removes 3×28 = 84 inches, which leaves her with $144 - 84 = 60$ inches. That's (J).

3. C
Category: Variable Manipulation
Difficulty: Low
Getting to the Answer: To evaluate x, isolate it on one side of the equation, then solve.

$$4x + 18 = 38$$

$$4x = 20$$

$$x = 5$$

Choice (C) is correct.

4. G
Category: Proportions and Probability
Difficulty: Medium
Getting to the Answer: Because John weighs *more* than Ellen, begin by eliminating J and K, as doing so will reduce the chance of a miscalculation error. According to the problem, John's 165 lb represents 1.5 times Ellen's weight. Therefore, Ellen's weight must be $\frac{165}{1.5} = 110$ lb. Choice (G) is correct.

5. E
Category: Proportions and Probability
Difficulty: Low
Getting to the Answer: To find the average of four numbers, plug them into the average formula and solve:

$$\text{Average} = \frac{\text{Sum of terms}}{\text{Number of terms}}$$

$$= \frac{237 + 482 + 375 + 210}{4}$$

$$= \frac{1,304}{4}$$

$$= 326$$

Choice (E) is correct.

6. J
Category: Operations
Difficulty: Medium
Getting to the Answer: Cube both sides to solve for x.

$$\sqrt[3]{x} = 4$$

$$x = 64$$

7. B
Category: Variable Manipulation
Difficulty: Low
Getting to the Answer: Isolate the variable, then solve for x:

$$x^2 + 14 = 63$$

$$x^2 = 49$$

$$x = \pm 7$$

This is (B).

8. G

Category: Operations

Difficulty: Medium

Getting to the Answer: You *could* use your calculator to solve this problem, but there's a much easier way. Begin by eliminating H, as it is not a perfect square. J can also be eliminated for the same reason. To simplify the radical, factor out a perfect square from 54. The largest factor of 54 that's also a perfect square is 9, so $\sqrt{54} = \sqrt{9 \times 6} = 3\sqrt{6}$.

9. E

Category: Operations

Difficulty: Medium

Getting to the Answer: You *could* punch the expression into your calculator, but it may actually be quicker to estimate. $\sqrt{90} \approx \sqrt{81} = 9$ and $\sqrt{32} \approx \sqrt{36} = 6$, so $\sqrt{90} \times \sqrt{32} \approx 9 \times 6 = 54$. With the calculator, the actual value is 53.6656. That's closest to (E).

10. K

Category: Operations

Difficulty: Medium

Getting to the Answer: When the choices are spaced far apart, estimation is generally the quickest way to the correct answer. To estimate, round 5.2 to 5 and 6.8 to 7. Because $5^3 + 7^2 = 125 + 49 = 174$, the correct answer will be very close to 174. That would be (K).

11. C

Category: Operations

Difficulty: Medium

Getting to the Answer: $x^3 = 512$, so $x = \sqrt[3]{512} = 8$. Be careful not to stop too soon. The problem asks for x^2, not x. $8^2 = 64$, which is (C).

12. H

Category: Operations

Difficulty: Medium

Getting to the Answer: To solve this problem, you'll need to follow the order of operations (PEMDAS).

First, evaluate the parentheses: $3^3 \div 9 + (6^2 - 12) \div 4 = 3^3 \div 9 + (36 - 12) \div 4 = 3^3 \div 9 + 24 \div 4$.

Next, simplify the exponent: $3^3 \div 9 + 24 \div 4 = 27 \div 9 + 24 \div 4$.

Then, take care of any multiplication and/or division, from left to right: $27 \div 9 + 24 \div 4 = 3 + 6$.

Finally, take care of any addition and/or subtraction, from left to right: $3 + 6 = 9$.

So (H) is correct.

13. B

Category: Operations

Difficulty: Low

Getting to the Answer: Each banana costs \$.24, so the price of x bananas is \$.24x. Similarly, each orange costs \$.38, so the price of y oranges is \$.38y. Therefore, the total price of x bananas and y oranges is \$.24x + \$.38y. That's (B).

14. K

Category: Variable Manipulation

Difficulty: Low

Getting to the Answer: Isolate the variable, then solve for x:

$$4x + 13 = 16$$

$$4x = 3$$

$$x = \frac{3}{4}$$

Choice (K) is correct, when you convert $\frac{3}{4}$ to the decimal 0.75.

15. A

Category: Proportions and Probability

Difficulty: Low

Getting to the Answer: The quickest way to solve this problem is to estimate. While you may or may not know 6% of 1,250 off the top of your head, 10% of 1,250 is 125. Because 6% < 10%, the correct answer must be less than 125. Only (A) works.

To solve this the more traditional way, multiply 1,250 by the decimal form of 6%: $1{,}250 \times .06 = 75$.

16. J

Category: Proportions and Probability
Difficulty: High
Getting to the Answer: When an average problem involves variables, it often helps to think in terms of sum instead. For Sarah's exam scores to average at least a 90, they must sum to at least $90 \times 4 = 360$. She already has an 89, a 93, and an 84, so she needs at least $360 - (89 + 93 + 84) = 360 - 266 = 94$ points on her last test. Choice (J) is correct.

17. E

Category: Variable Manipulation
Difficulty: Medium
Getting to the Answer: Treat inequalities just as equations. The only exception is that if you multiply or divide by a negative number, you must flip the inequality sign. With inequalities, you are solving for a range of values.

$$3x - 11 \geq 22$$
$$3x \geq 33$$
$$x \geq 11$$

This matches (E).

18. J

Category: Proportions and Probability
Difficulty: Low

Getting to the Answer: The basketball team scored 364 points in 13 games, so they scored an average of $\frac{364}{13} = 28$ points per game. Choice (J) is correct.

19. D

Category: Variable Manipulation
Difficulty: Low
Getting to the Answer: Isolate the variable, then solve for x:

$$6x + 4 = 11x - 21$$
$$4 = 5x - 21$$
$$25 = 5x$$
$$5 = x$$

That's (D).

20. F

Category: Proportions and Probability
Difficulty: Medium
Getting to the Answer: Backsolving works well for word problems with numbers in the answer choices.

First, find the number of movie theme songs:

$$10\% \text{ of } 300 = 0.10 \times 300 = 30$$

Then subtract the number of movie theme songs and marches from the total number of pieces of music:

$$300 - (30 + 80) = 300 - 110 = 190$$

If you solve this answer by Backsolving, first remember that there are always 300 pieces of music; therefore, there will always be 30 movie theme songs and 80 marches. Start with H:

$$210 + 30 + 80 = 320 \neq 300.$$

Because the value is too big, eliminate H, J, and K, and then try G.

$$198 + 30 + 80 = 308 \neq 300.$$

Choice G is incorrect, so (F) must be the correct answer:

$$190 + 30 + 80 = 300$$

21. C

Category: Proportions and Probability
Difficulty: High
Getting to the Answer: Set up an expression that can be solved to find the total number of class sessions. Letting x = the total number of sessions:

$$16 = \frac{19 + 17 + 15(x - 2)}{x}$$

$$16x = 36 + 15x - 30$$

$$16x - 15x = 36 - 30$$

$$x = 6 \text{ sessions}$$

22. H

Category: Proportions and Probability
Difficulty: Medium
Getting to the Answer: To find the percent shaded, divide the number of shaded triangles by the total number of triangles. There are 24 small triangles in all, and 8 of them are shaded.

$$\frac{8}{24} = \frac{1}{3} = 33\frac{1}{3}\%$$

Choice (H) is correct.

23. D

Category: Proportions and Probability
Difficulty: Medium

Getting to the Answer: The piece of paper is $8\frac{1}{2}$ inches wide. To find the number of $\frac{5}{8}$ inch wide strips of paper you can cut, divide:

$$8\frac{1}{2} \div \frac{5}{8} = \frac{17}{2} \div \frac{5}{8} = \frac{17}{2} \times \frac{8}{5} = \frac{136}{10} = \frac{68}{5} = 13.6$$

Thus, you can make 13 strips of paper that are $\frac{5}{8}$ inch wide and 11 inches long, and you will have a small, thin strip of paper left over.

24. G

Category: Proportions and Probability
Difficulty: Medium
Getting to the Answer: The ratio of buntings to total birds and the ratio of larks to total birds are given as 35:176 and 5:11, respectively. Assume that the total number of birds that have been counted in the meadow is 176. This automatically means that 35 buntings have been counted. The problem also gives the information that 5 out of 11 of the birds are larks. Set up a proportion:

$$\frac{5}{11} = \frac{x}{176}, \text{ where } x \text{ represents the number of}$$

larks on the field

$$x = 176 \times \frac{5}{11} = 80$$

Then find the number of sparrows, which is the number of birds remaining:

$$176 - 35 - 80 = 61$$

Because there is the greatest number of larks in the meadow (80 larks), there would be the greatest probability that a lark would be chosen at random (45% larks > 35% sparrows > 20% buntings).

25. C

Category: Variable Manipulation
Difficulty: Medium
Getting to the Answer: This problem seems long, but it actually isn't that complicated. The order of operations says that all of the multiplication should be taken care of first. Let's begin with the first two terms:

$$(x + 4)(x - 4)$$

(If you noticed the difference of squares here, that will save you some time. If not, use FOIL.)

First : $x \times x = x^2$ Outer : $x \times -4 = -4x$
Inner : $4 \times x = 4x$ Last : $4 \times -4 = -16$

Combine like terms:

$$x^2 + (-4x) + 4x + (-16) = x^2 - 16$$

Now for the other two terms:

$$(2x + 2)(x - 2)$$

First : $2x \times x = 2x^2$ Outer : $2x \times -2 = -4x$
Inner : $2 \times x = 2x$ Last : $2 \times -2 = -4$

Combine like terms:

$$2x^2 + (-4x) + 2x + (-4) = 2x^2 - 2x - 4$$

Finally, add the two polynomials:

$$(x^2 - 16) + (2x^2 - 2x - 4) = 3x^2 - 2x - 20$$

Choice (C) is correct.

26. J

Category: Variable Manipulation
Difficulty: Medium
Getting to the Answer: Use the acronym FOIL (First, Outside, Inside, Last) to multiply the binomials:

$$(4x - 1)(x + 5) =$$

$$4x^2 + 20x - x - 5 =$$

$$4x^2 + 19x - 5$$

27. D

Category: Trigonometry
Difficulty: High
Getting to the Answer: You are given the cosine of $\angle BAC$ and the hypotenuse of the triangle, so begin by using these to find the adjacent side:

$$\text{Cos } A = \frac{\text{Adjacent}}{\text{Hypotenuse}}$$

$$0.6 = \frac{\text{Adjacent}}{15}$$

$$\text{Adjacent} = 9$$

So the adjacent side, $\overline{AB}$, is 9, and triangle ABC is a right triangle with a leg of 9 and a hypotenuse of 15. ABC must therefore be a 3-4-5 right triangle, and $\overline{BC}$ must be 12. Choice (D) is correct.

28. J

Category: Proportions and Probability
Difficulty: Low
Getting to the Answer: With variables in the question stem and the answer choices, this problem is perfect for Picking Numbers. Pick 2 for x and 3 for y. Now the problem reads: "If a car drives 80 miles per hour for two hours and 60 miles per hour for three hours, what is the car's average speed, in miles, for the total distance traveled?"

In this case, the car would have driven $80 \times 2 = 160$ miles and $60 \times 3 = 180$ miles, for a total of $160 + 180 = 340$ miles in five hours. The average speed is therefore $\frac{340}{5} = 68$ miles per hour. Plug 2 in for x and 3 in for y for each of the choices and see which comes out to 68:

F: $\frac{480}{2 \times 3} = \frac{480}{6} = 80.$ Eliminate.

G: $\frac{80}{2} = \frac{60}{3} = 40 + 20 = 60.$ Eliminate.

H: $\frac{80}{2} \times \frac{60}{3} = 40 \times 20 = 800.$ Eliminate.

(J): $\frac{80(2) + 60(3)}{2 + 3} \times \frac{160 + 180}{5} = \frac{340}{5} = 68.$

K: This expression is the reciprocal of the one in (J), which gives $\frac{5}{340}$.   Eliminate.

Only (J) works, so it must be correct.

29. B

Category: Patterns, Logic, and Data
Difficulty: High
Getting to the Answer: Set up your sequence using blanks in place of the numbers you don't know. If the common difference is –7, then the second term must be $8 - 7 = 1$, and the third term must be

$1 - 7 = -6$: 8, 1, –6, 36. To find the common ratio, find the rate of difference between the second and third (or third and fourth) term: $1 \times -6 = -6 - 6 \times -6 = 36$. The common ratio is –6.

8, ___, ___, 36

If the common difference is –7, then the second term must be $8 - 7 = 1$, and the third term must be $1 - 7 = -6$:

8, 1, –6, 36

To find the common ratio, find the rate of difference between the second and third (or third and fourth) term:

$1 \times -6 = -6$

$-6 \times -6 = 36$

The common ratio is –6, (B).

30. F

Category: Operations
Difficulty: Medium
Getting to the Answer: When an exponent equation looks difficult on Test Day, try to rewrite the problem so that either the bases or the exponents themselves are the same. In this problem, the two bases seem different at first glance but, because 27 is actually 3^3, you can rewrite the equation as:

$$3^{3x+3} = 3^{3\left(\frac{2}{3}x - \frac{1}{3}\right)}$$

This simplifies to $3^{3x+3} = 3^{2x-1}$. Now that the bases are equal, set the exponents equal to each other and solve for x:

$3x + 3 = 2x - 1$

$x + 3 = -1$

$x = -4$

Choice (F) is correct.

31. D

Category: Operations
Difficulty: Medium
Getting to the Answer: The temperature of the container of liquid nitrogen is lower than the temperature of the room, so it must rise to match the room's temperature. Eliminate A and B, which indicate that a drop in temperature is needed. To find the positive difference, subtract:

$72°F - (-330°F) =$

$72°F + 330°F =$

$+402°F$

32. G

Category: Variable Manipulation
Difficulty: Medium
Getting to the Answer: This is an area problem with a twist—we're cutting a piece out of the rectangle. To find the area of the remaining space, you will need to subtract the area of the sandpit from the area of the original playground. Recall that the area of a rectangle is length × width. The dimensions of the original playground are $x + 7$ and $x + 3$, so its area is $(x + 7)(x + 3) = x^2 + 10x + 21$. The sandpit is a square with side x, so its area is x^2. Remove the pit from our playground, and the remaining area is $x^2 + 10x + 21 - x^2 = 10x + 21$. Choice (G) is correct.

33. C

Category: Number Properties
Difficulty: Medium
Getting to the Answer: When a problem tests a number property, the easiest way to solve it is to Pick Numbers. Because u is an integer, pick some integers for u. If $u = 2$, then $(u - 3)^2 + 5 = (2 - 3)^2 + 5 = (-1)^2 + 5 = 1 + 5 = 6$. This eliminates B, D, and E. If $u = 3$, then $(u - 3)^2 + 5 = (3 - 3)^2 + 5 = 5$. This eliminates A, leaving (C) as the correct answer.

34. F

Category: Coordinate Geometry
Difficulty: High
Getting to the Answer: Use the midpoint formula to solve:

$$\left(\frac{x_1 + x_2}{2}, \frac{y_1 + y_2}{2} \right)$$

$$1 + \frac{m}{2} = -3 \quad 9 + \frac{n}{2} = -2$$

$$m = -3, n = -13$$

$$(m, n) = (-3, -13)$$

35. C

Category: Variable Manipulation
Difficulty: Medium
Getting to the Answer: With nested functions, work from the inside out. To solve this problem, substitute the entire function of $g(x)$ for x in the function $f(x)$, then solve:

$$f(g(x)) = \frac{1}{3}(3x^2 + 6x + 12) + 13$$

$$f(g(x)) = x^2 + 2x + 4 + 13$$

$$f(g(x)) = x^2 + 2x + 17$$

Choice (C) is correct.

36. G

Category: Coordinate Geometry
Difficulty: Low
Getting to the Answer: To find the length of a line segment on the coordinate plane, you would normally need to use the distance formula. This requires the coordinates of the segment's two endpoints. Because A (1,5) and C (1,1) have the same x-coordinate, a much faster way is to simply subtract the y-coordinate of C from the y-coordinate of A. The length of segment AC is $5 - 1 = 4$. Choice (G) is correct.

37. C

Category: Operations
Difficulty: High
Getting to the Answer: When given a function and a value of x, plug in the number value for x in the equation and simplify. Make sure to follow the order of operations.

$$f(x) = 16x^2 - 20x$$

$$f(3) = 16(3)^2 - 20(3)$$

$$f(3) = 16(9) - 60$$

$$f(3) = 144 - 60 = 84.$$

Choice (C) is the answer.

38. H

Category: Coordinate Geometry
Difficulty: Medium
Getting to the Answer: Perpendicular lines have negative reciprocal slopes. Because the line in the problem has a slope of $\frac{2}{3}$, the line you are looking for must have a slope of $-\frac{3}{2}$. The problem also says that this line contains the point (4,−3). Plugging all of this information into the equation of a line, $y = mx + b$, will allow us to find the final missing piece—the y-intercept:

$$y = mx + b$$

$$-3 = -\frac{3}{2}(4) + b$$

$$-3 = -6 + b$$

$$3 = b$$

With a slope of $-\frac{3}{2}$ and a y-intercept of 3, the line is $y = -\frac{3}{2}x + 3$. The correct answer is (H).

39. B

Category: Coordinate Geometry
Difficulty: Medium
Getting to the Answer: Use the slope formula to find the slope m of the line:

$$m = \frac{y_2 - y_1}{x_2 - x_1}$$
$$= \frac{-6 - 0}{0 - (-10)}$$
$$= \frac{-6}{10}$$
$$= -\frac{3}{5}$$

40. F

Category: Coordinate Geometry
Difficulty: Low
Getting to the Answer: Because the rectangle lies on a coordinate plane, Point B will have the same x-coordinate as the point directly above it (–2), and the same y-coordinate as the point to its right (–3). Therefore, the coordinates of point B are (–2,–3), so (F) is the correct answer.

41. D

Category: Coordinate Geometry
Difficulty: Medium
Getting to the Answer: To find the distance between two points, plug them into the distance formula and evaluate:

$$\text{Distance} = \sqrt{(x_2 - x_1)^2 + (y_2 - y_1)^2}$$
$$= \sqrt{(-2 - 3)^2 + (6 - (-6))^2}$$
$$= \sqrt{(-5)^2 + 12^2}$$
$$= \sqrt{25 + 144}$$
$$= \sqrt{169}$$
$$= 13$$

Choice (D) is correct.

42. G

Category: Coordinate Geometry
Difficulty: Low
Getting to the Answer: Plug given points [(–1,1) and (1,3)] into the midpoint formula and solve:

$$\text{Midpoint} = \left(\frac{x_1 + x_2}{2}, \frac{y_1 + y_2}{2}\right)$$
$$= \left(\frac{1 + (-1)}{2}, \frac{3 + 1}{2}\right)$$
$$= \left(\frac{0}{2}, \frac{4}{2}\right)$$
$$= (0, 2)$$

Choice (G) is correct.

43. D

Category: Plane Geometry
Difficulty: High
Getting to the Answer:

$$\text{Volume of a Sphere} = \frac{4}{3}\pi r^3$$

Because you are told the diameter is 6, you know the radius of the sphere is 3. Watch out for this terminology on the ACT. If you knew the correct formula but use the diameter of the sphere instead of the radius, you would have selected E. If you don't recall the formula for the volume of a sphere, you should be able to eliminate some of the options and make an educated guess. Choice B is the area of a cross section of the sphere—too small—and A is even smaller, so you can eliminate both of these choices. Choice E seems pretty big (unless you fell for the trap), so you're down to C and (D). Hopefully you'll remember that the volume of a sphere is a little bigger than the radius cubed times π, C, and will select (D). But even if you guess between the two answers, you have a 50 percent chance of guessing correctly.

Choice (D) is correct.

44. H

Category: Plane Geometry

Difficulty: Medium

Getting to the Answer: The perimeter of a rectangle is twice its length plus twice its width, or Perimeter = $2l + 2w$. To find the area, you must first determine the value of w, so plug in the values for perimeter and length to solve:

$$\text{Perimeter} = 2l + 2w$$

$$24 = 2(8) + 2w$$

$$24 = 16 + 2w$$

$$8 = 2w$$

$$4 = w$$

So the width is 4. The area of the rectangle is length × width, or $8 \times 4 = 32$. Choice (H) is correct.

45. A

Category: Plane Geometry

Difficulty: Medium

Getting to the Answer: With only one known side, you cannot find the area directly, as you will need to figure out more sides first. Given the area of triangle ABC and its base, the first step is to find height $\overline{CD}$:

$$\text{Area} = \frac{1}{2}bh$$

$$48 = \frac{1}{2}(12)h$$

$$48 = 6h$$

$$8 = h$$

So $\overline{CD} = 8$. Triangle ABC is an isosceles triangle, so $\overline{CD}$ also happens to be the perpendicular bisector of $\overline{AB}$, meaning $\overline{AD} = \overline{DB} = 6$. With legs of 6 and 8, each of the smaller right triangles must be 3-4-5 right triangles, making the hypotenuse of each—$\overline{AC}$ and $\overline{CB}$—10. (You can also use the Pythagorean theorem; if x equals the hypotenuse, then $6^2 + 8^2 = x^2$ which simplifies to $36 + 64 = 100 = x^2$. Therefore,

x equals 10.) Therefore, the perimeter of triangle ABC is $10 + 10 + 12 = 32$. Choice (A) is correct.

46. J

Category: Plane Geometry

Difficulty: Low

Getting to the Answer: It may sound a bit more complex, but this problem is only asking you for the perimeter of a rectangle with the given dimensions, so plug them into the perimeter formula and solve:

$$\text{Perimeter} = 2l + 2w$$

$$= 2(130) + 2(70)$$

$$= 260 + 140$$

$$= 400$$

Choice (J) is correct.

47. D

Category: Plane Geometry

Difficulty: Low

Getting to the Answer: When two parallel lines are cut by a transversal, half of the angles will be acute and half will be obtuse. Each acute angle will have the same measure as each other acute angle. The same is true of every obtuse angle. Furthermore, the acute angles will be supplementary to the obtuse angles. $\angle a$ is an acute angle measuring 68° while $\angle f$ is an obtuse angle, so $\angle a$ must be supplementary to $\angle f$. Therefore, $\angle f = 180° - 68° = 112°$.

Choice (D) is correct.

48. J

Category: Plane Geometry

Difficulty: Low

Getting to the Answer: Because $ABCD$ is a square, each side has the same length, which means the base and height of triangle BCD both equal 6 centimeters. Substitute these numbers in the formula for the area of a triangle:

$$\text{Area} = \frac{1}{2}bh$$
$$= \frac{1}{2}(6)6$$
$$= 18$$

The area of the triangle *BCD* is 18 cm².

49. C

Category: Plane Geometry
Difficulty: Medium

Getting to the Answer: You are told that triangle *RST* is a right triangle and that one of its angles is 30°, so *RST* must be a 30-60-90 right triangle, meaning its sides must be in the proportion $x : x\sqrt{3} : 2x$. Hypotenuse *RT* is 16, so *x* must be $\frac{16}{2} = 8$, and *RS* (the longer leg) must be $8\sqrt{3}$. (You can also use the Pythagorean theorem.) That matches (C).

50. K

Category: Plane Geometry
Difficulty: Medium
Getting to the Answer: Write the formula for the area of a circle, using r to represent the radius of the original circle in the problem. $A = \pi r^2$. This is the area of the original circle. Then write the formula for the area of the resulting circle, using 2*r* as the radius:

$$A = \pi(2r)^2 = \pi(4r^2) = 4\pi r^2$$

Now, divide the two areas (area of resulting circle/ area of original circle) to find out how many times larger the area of the resulting circle is compared to the area of the original circle.

$$\frac{\text{Area of resulting circle}}{\text{Area of original circle}} = \frac{4\pi r^2}{\pi r^2} = 4$$

51. B

Category: Plane Geometry
Difficulty: Low
Getting to the Answer: A square has four right angles and four equal sides. Its diagonal cuts the square into two identical isosceles right triangles. The square in this problem has a side length of 7, so the base and height of each isosceles right triangle is also 7. The sides of an isosceles right triangle are in the proportion $x : x : x\sqrt{2}$, so the length of the diagonal (the hypotenuse of both triangles) is $7\sqrt{2}$. (Again, you can also use the Pythagorean theorem.) Choice (B) is correct.

52. H

Category: Plane Geometry
Difficulty: Low
Getting to the Answer: A regular polygon is equilateral, so a regular hexagon is a hexagon with six equal sides. The regular hexagon in the problem has a side of 11, so its perimeter is $6 \times 11 = 66$.

Choice (H) is correct.

53. C

Category: Plane Geometry
Difficulty: Medium
Getting to the Answer: The perimeter of the rectangle is 28, and one of its sides is 2.5 times the length of the other, so call *x* the shorter side. Our rectangle now has sides of *x* and 2.5*x*. Draw a figure to help visualize this problem.

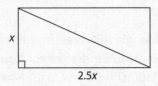

To find *x*, plug the information into the perimeter formula and solve:

Perimeter = 2*l* + 2*w*

$$28 = 2(x) + 2(2.5x)$$

$$28 = 2x + 5x$$

$$28 = 7x$$

$$4 = x$$

So *x* = 4, and the dimensions of the rectangle must be 4 × 1 = 4 and 2.5 × 4 = 10. These values are not parts of a special right triangle, so use the Pythagorean theorem to find the diagonal:

$$a^2 + b^2 = c^2$$
$$4^2 + 10^2 = c^2$$
$$16 + 100 = c^2$$
$$116 = c^2$$
$$\sqrt{116} = c$$

Because 116 isn't a perfect square but it lies between $10^2 = 100$ and $11^2 = 121$, $\sqrt{116}$ must be somewhere between 10 and 11 (it's approximately 10.77). The only choice that fits is (C).

54. G

Category: Plane Geometry
Difficulty: Medium
Getting to the Answer: This is a pair of parallel lines cut by a transversal, but this time, there's also a triangle thrown into the mix. Begin with *AB*. This is a transversal, so ∠*MAB* and ∠*ABC* are alternate interior angles and ∠*MAB* = ∠*ABC* = 55°. Because triangle *ABC* is isosceles with *AB* = *AC*, ∠*ACB* is also 55°.

Choice (G) is correct.

55. D

Category: Plane Geometry
Difficulty: Medium
Getting to the Answer: The chord is perpendicular to the line segment from the center of the circle, so that line segment must be its perpendicular bisector. This allows us to add the following to the figure:

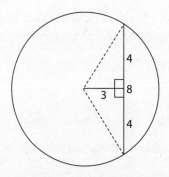

The two right triangles have legs 3 and 4, so they are both 3-4-5 right triangles with hypotenuse 5. This hypotenuse is also the radius of the circle, so plug that into the area formula to solve:

$$Area = \pi r^2$$
$$= \pi(5)^2$$
$$= 25\pi$$

The correct answer is (D).

56. J

Category: Plane Geometry
Difficulty: High
Getting to the Answer: Triangle *QRS* is an isosceles triangle, so its height is also the perpendicular bisector of *RS*. Each half of *RS* is $\frac{16}{2} = 8$ units long, so each of the smaller right triangles has a leg of 8 and a hypotenuse of 17. Using the Pythagorean theorem, a leg of 8 and hypotenuse of 17 means the other leg—the height—is 15.

$$\frac{1}{2} \times 16 \times 15 = \frac{1}{2} \times 240 = 120.$$

Choice (J) is correct.

57. C

Category: Proportions and Probability
Difficulty: Medium
Getting to the Answer: The ratio of girls to boys is 5:3, so the ratio of girls to the total number of seniors

is 5:(3 + 5), or 5:8. Call x the number of girls in the senior class. Set up the proportion and solve for x:

$$\frac{5}{8} = \frac{x}{168}$$
$$8x = 840$$
$$x = 105$$

There are 105 girls in the senior class, which is (C).

58. K

Category: Trigonometry
Difficulty: Medium

Getting to the Answer: The tangent of an angle is defined by $\text{Tan } A = \dfrac{\text{Opposite}}{\text{Adjacent}}$. The side opposite $\angle EFD$ is 12 and the side adjacent to $\angle EFD$ is 5, so $\tan \angle EFD = \dfrac{12}{5}$. Choice (K) is correct.

59. C

Category: Trigonometry
Difficulty: Medium

Getting to the Answer: Because $QS = QR$, triangle QRS must be a 45-45-90 right triangle and the hypotenuse is $5\sqrt{2}$. Remember that $\sin \angle QRS = \dfrac{\text{opposite}}{\text{hypotenuse}}$. Therefore,

$$\sin \angle QRS = \frac{5}{5\sqrt{2}} = \frac{1}{\sqrt{2}} = \frac{\sqrt{2}}{2}.$$

Choice (C) is correct.

60. K

Category: Trigonometry
Difficulty: Medium

Getting to the Answer: Using the Pythagorean theorem, a right triangle with leg 8 and hypotenuse 17 must have another leg of 15, so $JK = 15$. Because JK is opposite $\angle JLK$, $\sin \angle JLK = \dfrac{15}{17}$. Choice (K) is correct.

SCIENCE TEST

PASSAGE I

1. C

Category: Figure Interpretation
Difficulty: High

Getting to the Answer: Point c is the point on the light curve at which the eclipsing binary is the darkest. Take a look at Figure 1. From the point of view of the observer, System Q is going to be darkest when the light from the brighter star, Star 1, is being blocked by the light from the less bright Star 2. Star 2 interposes itself between Star 1 and the observer when Star 2 is in position 3, (C).

2. J

Category: Figure Interpretation
Difficulty: Medium

Getting to the Answer: According to the x-axis of the light curve, the complete cycle of changes in the system's surface brightness (from point a through points b, c, and d and back to point a again) lasts 20 days. This means that Star 2 requires 20 days to complete its orbit around Star 1 (it is this orbit, after all, that is causing the changes in the system's brightness). Choice (J) is the correct answer.

3. C

Category: Figure Interpretation
Difficulty: High

Getting to the Answer: The drops in brightness on the light curve (Figure 2) indicate when one star is eclipsing the other. The sharp drop known as the primary minimum—when the darker star eclipses the brighter star—lasts approximately two days. The secondary minimum—when the brighter star eclipses the darker star—goes from day 16 through day 20 to day 4, a total of eight days.

4. G

Category: Scientific Reasoning
Difficulty: Medium
Getting to the Answer: This question introduces two new light curve graphs. What gives System Z a deeper primary minimum than X? The reason Q had a deep primary minimum was that one star was brighter than the other; during the time that the brighter star's light was eclipsed by the darker star, the whole system became much darker. You can safely assume that this is the reason Z has a deep primary minimum as well. X's lack of a deep primary minimum must mean, then, that neither of its stars is significantly brighter than the other. As you can see from X's light curve, the drop in brightness of the system is about the same no matter which star is being eclipsed, so the two stars must be equally bright. Because the more extreme difference in the magnitudes of System Z's stars is the reason for Z's deeper primary minimum, (G) is correct.

5. D

Category: Scientific Reasoning
Difficulty: Medium
Getting to the Answer: If either star in an eclipsing binary system is in front of the other, the brightness of the system will be reduced. The only time the system reaches maximum brightness is when both stars are completely visible—when the full brightness of one star is added to the full brightness of the other, (D).

PASSAGE II

6. J

Category: Scientific Reasoning
Difficulty: Medium
Getting to the Answer: The question requires you to identify what is formed when CO_2 leaves the air in the diagramed carbon cycle. The figure indicates that CO_2 leaves the air via photosynthesis and goes to green plants and carbohydrates. In the same box, proteins, fats, and other carbon compounds are also mentioned. Therefore, Roman numerals I and III are true. The only answer choice to have both I and III is (J). II is also true because starch is an example of a carbohydrate. While this question is subtly testing background knowledge about carbohydrates, you are still able to get to the correct answer using the figure.

7. C

Category: Figure Interpretation
Difficulty: Medium
Getting to the Answer: Decay bacteria, according to the diagram, get carbon from the carbon compounds of dead plants and animals and add to the supply of carbon dioxide in air and water through fermentation and putrefaction. Because there is no other way to get carbon from dead plants and animals back to carbon dioxide, a drop in decay bacteria will reduce the amount of carbon available for forming carbon dioxide. The carbon dioxide level will not be greatly affected, though, because there are three other sources of carbon for carbon dioxide. Choice (C) is correct.

8. G

Category: Figure Interpretation
Difficulty: Medium
Getting to the Answer: Consider each of the statements one by one. Statement I is false because the diagram shows that the carbon in animals can move to "other animals" through the process of "animal digestion." An animal that only eats other animals is participating in the carbon cycle when it digests its prey. Statement II is clearly true; there are arrows labeled *respiration* going from the green plants, the animals, and the other animals stages back to the carbon dioxide stage. Statement III, however, is not true, because some carbon dioxide is released into the air by fermentation and putrefaction. Because only statement II is true, (G) is correct.

9. D

Category: Figure Interpretation

Difficulty: High

Getting to the Answer: There are four direct sources of atmospheric carbon dioxide, according to the diagram: fermentation by decay bacteria and respiration by green plants, by animals, and by "other animals or animal parasites." Choice (D) correctly cites the respiration of animal parasites. If you were tempted by B, note the direction of the arrow for photosynthesis. Photosynthesis removes atmospheric carbon dioxide, using it as a source for carbon.

10. G

Category: Figure Interpretation

Difficulty: Medium

Getting to the Answer: The arrow signifying animal respiration and the arrow signifying photosynthesis are linked in the diagram by the "carbon dioxide in air or dissolved in water" box. Animal respiration is one of the sources of carbon for carbon dioxide, and the carbon dioxide in turn provides carbon for the process of photosynthesis. In this sense, "animal respiration provides vital gases for green plants," (G).

11. A

Category: Figure Interpretation

Difficulty: Medium

Getting to the Answer: Look closely at the diagram. The only way that carbon dioxide can enter the carbon cycle is through the process of photosynthesis in which it is taken up by green plants. Note that green plants also emit carbon dioxide back into the air via the process of respiration. Therefore, if green plants were eliminated, the carbon cycle would come to a complete stop, and (A) is correct.

PASSAGE III

12. G

Category: Scientific Reasoning

Difficulty: Medium

Getting to the Answer: In order to be able to tell whether light was the stimulus affecting the direction of bacterial migration, you have to compare two trials, one with the light on and one with the light off. If there is no difference in the direction of bacterial migration in the two trials, then light does not have an effect and is not the primary stimulus. The two trials you need to compare are Trial 1 and Trial 2 of Study 1, because Trial 1 was conducted under standard lab conditions (with the light on), and in Trial 2 the microscope was shielded from all external light. Because the results of the trials differed only minimally, they support the theory that light was not the primary stimulus. Choice (G) is correct.

13. C

Category: Scientific Reasoning

Difficulty: Medium

Getting to the Answer: The data shows that bacteria are sensitive to magnetic fields and tend to migrate in the direction of magnetic north. In Study 2, this meant that the bacteria moved toward the magnet because the magnet's north pole was near the slide. If the magnet's south pole were placed near the slide, the magnetic field would be reversed and the bacteria would migrate in exactly the opposite direction, away from the magnet. That makes (C) the correct answer.

14. J

Category: Scientific Reasoning

Difficulty: Medium

Getting to the Answer: You don't need experimental data to answer this question; you just have to figure out which answer choice provides the best reason a bacteria should be able to move downward. Choices F and G are out immediately because both are good reasons a bacteria should move upward, not downward. While it may have been somewhat advantageous for bacteria not to be dispersed by currents, H, it would have been much more important for them to move downward to find food, so (J) is the best answer.

15. C
Category: Scientific Reasoning
Difficulty: High
Getting to the Answer: Here, you have to determine which new study would yield new and useful information about the relationship between magnetic field strength and bacterial migrations. You should try to determine the experimental condition that the researchers should vary before you look at the answer choices. To gain new information about magnetic field strength and bacterial migrations, the researchers should vary the magnetic field strength and observe the effect on bacterial migrations. Choice (C) is correct because it suggests using more and less powerful magnets, which would produce stronger and weaker magnetic fields than that of the magnet in Study 2.

16. F
Category: Figure Interpretation
Difficulty: Medium
Getting to the Answer: In each of the trials of Study 1, bacterial migrations were largely found to be in the direction of magnetic north. Shielding from light and electric fields, rotation of the microscope, and movement of the microscope to another lab all had no distinct effect on the direction of migration. It is fairly easy to conclude from Study 1 that "the majority of magnetotactic bacteria migrate toward the earth's magnetic north pole," (F).

17. A
Category: Scientific Reasoning
Difficulty: Low
Getting to the Answer: In any experiment, the control condition is the one used as a standard of comparison in judging the experimental effects of the other conditions. In Study 1, there would be no

way to know what the effect of, say, rotating the microscope was on bacterial migration if you didn't know how the bacteria migrated before you rotated the microscope. The control condition is the trial that is run without any experimental manipulations: in this case, Trial 1, (A).

18. G
Category: Patterns
Difficulty: High
Getting to the Answer: To answer this question, refer to Table 2. Based on the data for Trial 2 in the table, 483 bacteria accumulated at the 3:00 direction on the microscope slide, whereas only a few accumulated at the other directions on the slide. Only (G) has a high density of dots at the 3:00 PM location and a low density of dots at the other locations.

PASSAGE IV

19. B
Category: Patterns
Difficulty: Low
Getting to the Answer: A data point does not exist for week 40 on Figure 1. In order to estimate the percent flexibility of the *nu-PVC*, extend the line of the graph and make sure that it is consistent with the downward trend. Use 40 on the *x*-axis to trace back to the *y*-axis, which reveals that the closest value is indeed 12%. The answer cannot be H or J because they are greater than 17.5%, the last data point that is at 32 weeks. The downward trend of the graph tapers off at the end, between weeks 24 and 32. If the extended line did not follow the trend and was too steep, A would be incorrectly selected.

20. H
Category: Figure Interpretation
Difficulty: Medium
Getting to the Answer: According to the results of Experiment 1, *nu-PVC* does not suffer any kind of damage when exposed to water for a long period of time. It seems safe to say, then, that *nu-PVC*

would not need a protective coating to avoid water damage, (H).

21. B

Category: Scientific Reasoning
Difficulty: Medium
Getting to the Answer: The results that are relevant here are those from Experiment 3, in which *nu-PVC*'s ability to withstand cold temperatures was tested. The *nu-PVC* boards did not fare well during this experiment; they lost more than half of their capacity to support weight without bending or breaking. In a cold environment, the benches could not be kept outside during the winter without sustaining damage. Choice (B) is correct.

22. G

Category: Scientific Reasoning
Difficulty: Medium
Getting to the Answer: Remember that a control is an experimental condition in which nothing special is done to the thing being tested. The control serves as a standard of comparison for the experimental effects found in other conditions. In Experiment 1, the *nu-PVC* boards were sprayed with distilled water for a long period of time to see what effect the water would have. In order to assess the water's effects accurately, though, you have to compare the results of Experiment 1 to the results of a condition in which *nu-PVC* boards are not exposed to water, (G).

23. A

Category: Scientific Reasoning
Difficulty: Medium
Getting to the Answer: Although at first this question seems to require outside knowledge, it can be answered by the process of elimination using the information in the introduction. The introduction of the passage states that the series of experiments was performed to test the weathering abilities of *nu-PVC*. The only weather phenomenon in the answer choices is sunlight, (A).

24. J

Category: Scientific Reasoning
Difficulty: High
Getting to the Answer: You know nothing in particular about standard building materials from the passage—you don't know how well they stand up to rain or ultraviolet radiation or how well they support weight after exposure to cold weather. This means you cannot compare *nu-PVC* to other materials at all, which rules out F and H. Choice G is out because there is no information in the passage about the "different colors and textures" of *nu-PVC*. That leaves (J). Choice (J) is correct because it is an advantage of *nu-PVC* (perhaps the only one) that it is made of recycled plastic wastes. *Nu-PVC* helps to solve the garbage problem at the same time that it fills the need for building materials; wood and other standard building materials don't do that.

25. A

Category: Scientific Reasoning
Difficulty: High
Getting to the Answer: The conditions of the original experiments were clearly chosen so that *nu-PVC* would be subjected to extremes of water, radiation, and cold. There is little point in making the conditions even more extreme when a park bench would never have to survive such weather. It is unlikely, for example, that a park bench will be exposed to 32 weeks of continuous rain, much less 64 weeks, or that a bench will have to endure eight months of −40°C temperature. Thus, statements I and III are not going to provide useful information. Statement II, on the other hand, is an important experiment, because sea water and salt-rich air may well have a corrosive effect on *nu-PVC* whereas distilled water did not. Choice (A) is correct.

26. G

Category: Figure Interpretation
Difficulty: Medium
Getting to the Answer: The data for each treatment in Figure 2 is represented with three lines: One

that reflects the greatest value, one that reflects the mean, and one that reflects the lowest value. In order to answer the question, direct your attention to the After data in the graph. Draw a line from the top line of the data to the y-axis, which is approximately 600 lbs. Repeat with the bottom line of the data to find that it is approximately 300 lbs. The distance between the two is 300 lbs, which matches (G). Choice F is a trap because it is the range for the Before data.

PASSAGE V

27. D

Category: Figure Interpretation

Difficulty: Medium

Getting to the Answer: The passage states that *Amadori* products form when the protein samples are incubated with glucose but not with water. On Figure 1, locate the peak that represents the glucose treated protein (the solid line). Draw a line from the top of the glucose peak, which is the maximum absorbance of the sample, to the x-axis to find that the wavelength is closest to 400 nm. Choice A is a trap because it is the wavelength for the maximum absorbance of the sample that had been incubated in water.

28. J

Category: Scientific Reasoning

Difficulty: Medium

Getting to the Answer: The researcher who designed the experiments was interested in the effect of dietary glucose on lens proteins in the human eye, not the cow eye. There would be no reason to use cow lens proteins in Experiment 3 if cow lens proteins were expected to react any differently from human lens proteins, especially when human lens proteins were readily available for use—they were used, after all, for Experiment 1. Therefore, you know that the researcher assumed that cow lens proteins would react the same as human lens proteins, (J).

29. C

Category: Scientific Reasoning

Difficulty: Medium

Getting to the Answer: Make sure that you stick to the results of Experiment 1 only when you answer this question. All you know from Experiment 1 is that when a human tissue protein was dissolved in a glucose and water solution, the proteins formed Amadori products that combined with other proteins to make brown cross-links. You can conclude from this that the glucose reacted with the proteins to form cross-links, (C). Based on Experiment 1, though, you know nothing about disulfide cross links, A, or glucose metabolism, D. Choice B contradicts the results of Experiment 1 because it is protein, not glucose, that forms AGE.

30. G

Category: Scientific Reasoning

Difficulty: Medium

Getting to the Answer: Take another look at the results of Experiment 2. It was found that in the samples from older subjects, the lens proteins often formed cross-linked bonds, some of which were brown. The senile cataracts in the lenses of older people are also brown. The conclusion suggested by the identical colors of the cataracts and the cross-linked bonds is that the senile cataracts are made up of, or caused by, cross-linked bonds, (G).

31. C

Category: Scientific Reasoning

Difficulty: Medium

Getting to the Answer: You don't know from the results of Experiment 2 how the brown pigmented cross-links developed among the lens proteins of older humans. Experiments 1 and 3 indicate, however, that glucose reacts with lens proteins in such a way that brown pigmented cross-links form among the proteins. And remember the main purpose of the experiments: The researcher is investigating the effects of glucose on lens proteins in order to see whether dietary sugar (glucose) damages proteins. The hypothesis that dietary sugar reacted with lens proteins to cause the brown pigmented cross-links found in older subjects would seem to be supported by the results of the three experiments, (C).

32. J

Category: Scientific Reasoning

Difficulty: Medium

Getting to the Answer: The relevant results are those from Experiment 2: the lens proteins of younger subjects were found to have formed cross-linked aggregates much less frequently than the lens proteins of older subjects did. So you would expect that the lens proteins of a 32-year-old man would have fewer cross-links than the lens proteins of an 80-year-old man, (J). Choice H appears true, but the lens proteins appear "much more often" in older samples. The age difference is greater from 32 years to 80 than 18 years to 32; (J) is, thus, the most likely, and correct, answer.

33. C

Category: Scientific Reasoning

Difficulty: High

Getting to the Answer: String together the hypotheses that were the correct answers from questions 30 and 31, and you have the following overall hypothesis: dietary glucose causes brown pigmented (AGE) cross-links to form among lens proteins, and these brown cross-links in turn cause the formation of brown senile cataracts. According to this hypothesis, the excess glucose in an uncontrolled diabetic's blood should cause the formation of AGE cross-links among lens proteins and subsequently the development of senile cataracts, (C).

PASSAGE VI

34. G

Category: Scientific Reasoning

Difficulty: High

Getting to the Answer: Scientist 1 is a proponent of the Quark Model, which says that baryons (including the proton) and mesons are made up of quarks, which have fractional charge. Quarks have never been observed, however. You find out from Scientist 2 that quarks should be easy to distinguish from other particles because they would be the only ones with fractional charge. If a particle with fractional charge was detected, then, it would most likely be a quark, and this would strengthen Scientist 1's hypothesis that the Quark Model is correct. Mesons, baryons, and the Δ^{++} baryon have all been detected, but mere detection of them does not tell us anything about their substructure, so it cannot be used to support the Quark Model.

35. C

Category: Scientific Reasoning

Difficulty: Medium

Getting to the Answer: Scientist 2 says that it should be easy to split the proton into quarks, that the quarks should be easy to distinguish because of their unique charge (statement I), and that they should be stable because they can't decay into lighter particles (statement III). Statement II is wrong because it is Scientist 1's explanation of why quarks cannot be detected. Therefore, (C) is correct.

36. F

Category: Scientific Reasoning

Difficulty: Low

Getting to the Answer: Scientist 2 says that the Quark Model is wrong because it violates the Pauli exclusion principle, which states that no two particles of half-integer spin can occupy the same state. He says that in the Δ^{++} baryon, for example, the presence of two up quarks in the same state would violate the principle, so the model must be incorrect. If Scientist 1 were able to show, however, that quarks do not have half-integer spin, (F), she could argue that the Pauli exclusion principle does not apply to quarks and thus counter Scientist 2's objections. Evidence that the Δ^{++} baryon exists, G, or that quarks have fractional charge, H, isn't going to help Scientist 1 because neither has anything to do with the Pauli exclusion principle. Evidence that quarks have the same spin as electrons, J, would only support Scientist 2's position.

37. D

Category: Scientific Reasoning
Difficulty: Medium
Getting to the Answer: According to Scientist 2, the Δ^{++} baryon has three up quarks. Each up quark has a charge of +2/3 each, so the three quarks together have a total charge of 2, (D).

38. G

Category: Scientific Reasoning
Difficulty: High
Getting to the Answer: Scientist 2 thinks that the Quark Model is flawed for two reasons: 1) quarks have not been detected experimentally, and they would have been if they existed, and 2) the Quark Model violates the Pauli exclusion principle. The first reason is paraphrased in (G), the correct answer. Choice F is incorrect because the existence of individual baryons, including protons, has been verified experimentally. Scientist 2 never says that he thinks particles cannot have fractional charge, nor does he complain that the Quark Model doesn't include electrons as elementary particles, so H and J are incorrect as well.

39. A

Category: Scientific Reasoning
Difficulty: Medium
Getting to the Answer: The deep inelastic scattering experiments, according to Scientist 1, showed that the proton has a substructure. The three distinct lumps that were found to bounce high-energy electrons back and scatter them through large angles were the three quarks that make up the proton (at least in Scientist 1's view), so (A) is correct.

40. H

Category: Scientific Reasoning
Difficulty: High
Getting to the Answer: This question is a follow-up to the last one. If the three lumps were indeed quarks, then this supports the Quark Model because in the Quark Model, the proton consists of three quarks. "Protons supposedly consist of three quarks" is not one of the choices, though, so you have to look for a paraphrase of this idea. Protons are baryons and not mesons, so F and G are out. Baryons, like protons, are all supposed to consist of three quarks, so this rules out J and makes (H) the correct answer.